KT-583-436

Tim Kirk

WITHDRAWN

Physics

FOR THE IB DIPLOMA

2014 edition

OXFORD

UNIVERSITY PRESS

OXFORD
UNIVERSITY PRESS

Great Clarendon Street, Oxford, OX2 6DP, United Kingdom

Oxford University Press is a department of the University of Oxford. It furthers the University's objective of excellence in research, scholarship, and education by publishing worldwide. Oxford is a registered trade mark of Oxford University Press in the UK and in certain other countries

British Library Cataloguing in Publication Data
Data available

978-0-19-839355-9

5 7 9 10 8 6 4

Paper used in the production of this book is a natural, recyclable product made from wood grown in sustainable forests. The manufacturing process conforms to the environmental regulations of the country of origin.

Printed in Great Britain by Bell and Bain Ltd, Glasgow

Acknowledgements

This work has been developed independently from and is not endorsed by the International Baccalaureate (IB).

Cover: © James Brittain/Corbis; **p191:** Chase Preuninger; **p211:** NASA/WMAP Science Team; **p117:** vilax/Shutterstock; **p205:** NASA/WMAP

Artwork by Six Red Marbles and Oxford University Press.

We have tried to trace and contact all copyright holders before publication. If notified the publishers will be pleased to rectify any errors or omissions at the earliest opportunity.

Introduction and acknowledgements

Many people seem to think that you have to be really clever to understand Physics and this puts some people off studying it in the first place. So do you really need a brain the size of a planet in order to cope with IB Higher Level Physics? The answer, you will be pleased to hear, is 'No'. In fact, it is one of the world's best kept secrets that Physics is easy! There is very little to learn by heart and even ideas that seem really difficult when you first meet them can end up being obvious by the end of a course of study. But if this is the case why do so many people seem to think that Physics is really hard?

I think the main reason is that there are no 'safety nets' or 'short cuts' to understanding Physics principles. You won't get far if you just learn laws by memorising them and try to plug numbers into equations in the hope of getting the right answer. To really make progress you need to be familiar with a concept and be completely happy that you understand it. This will mean that you are able to apply your understanding in unfamiliar situations. The hardest thing, however, is often not the learning or the understanding of new ideas but the getting rid of wrong and confused 'every day explanations'.

This book should prove useful to anyone following a pre-university Physics course but its structure sticks very closely to the recently revised International Baccalaureate syllabus. It aims to provide an explanation (albeit very brief) of all of the core ideas that are needed throughout the whole IB Physics course. To this end each of the sections is clearly marked as either being appropriate for everybody or only being needed by those studying at Higher level. The same is true of the questions that can be found at the end of the chapters.

I would like to take the opportunity to thank the many people that have helped and encouraged me during the writing of this book. In particular I need to mention David Jones and Paul Ruth who provided many useful and detailed suggestions for improvement – unfortunately there was not enough space to include everything. The biggest thanks, however, need to go to Betsan for her support, patience and encouragement throughout the whole project.

Tim Kirk
October 2002

Third edition

Since the IB Study Guide's first publication in 2002, there have been two significant IB Diploma syllabus changes. The aim, to try and explain all the core ideas essential for the IB Physics course in as concise a way as possible, has remained the same. I continue to be grateful to all the teachers and students who have taken time to comment and I would welcome further feedback. In addition to the team at OUP, I would particularly like to thank my exceptional colleagues and all the outstanding students at my current school, St. Dunstan's College, London. It goes without saying that this third edition could not have been achieved without Betsan's continued support and encouragement.

This book is dedicated to the memory of my father, Francis Kirk.

Tim Kirk
August 2014

Contents

(Italics denote topics which are exclusively Higher Level)

The realm of physics – range of magnitudes of quantities in our universe

ORDERS OF MAGNITUDE – INCLUDING THEIR RATIOS

Physics seeks to explain nothing less than the Universe itself. In attempting to do this, the range of the magnitudes of various quantities will be huge.

If the numbers involved are going to mean anything, it is important to get some feel for their relative sizes. To avoid 'getting lost' among the numbers it is helpful to state them to the nearest **order of magnitude** or power of ten. The numbers are just rounded up or down as appropriate.

Comparisons can then be easily made because working out the ratio between two powers of ten is just a matter of adding or subtracting whole numbers. The diameter of an atom, 10^{-10} m, does not sound that much larger than the diameter of a proton in its nucleus, 10^{-15} m, but the ratio between them is 10^5 or 100,000 times bigger. This is the same ratio as between the size of a railway station (order of magnitude 10^2 m) and the diameter of the Earth (order of magnitude 10^7 m).

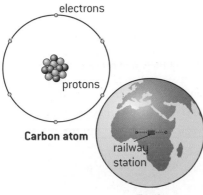

Carbon atom

Earth

For example, you would probably feel very pleased with yourself if you designed a new, environmentally friendly source of energy that could produce 2.03×10^3 J from 0.72 kg of natural produce. But the meaning of these numbers is not clear – is this a lot or is it a little? In terms of orders of magnitudes, this new source produces 10^3 joules per kilogram of produce. This does not compare terribly well with the 10^5 joules provided by a slice of bread or the 10^8 joules released per kilogram of petrol.

You do NOT need to memorize all of the values shown in the tables, but you should try and develop a familiarity with them.

RANGE OF MASSES

Mass / kg

10^{52}	total mass of observable
10^{48}	Universe
10^{44}	
10^{40}	mass of local galaxy
10^{36}	(Milky Way)
10^{32}	mass of Sun
10^{28}	
10^{24}	mass of Earth
10^{20}	total mass of oceans
10^{16}	total mass of atmosphere
10^{12}	
10^{8}	laden oil supertanker
10^{4}	elephant
10^{0}	human
	mouse
10^{-4}	grain of sand
10^{-8}	blood corpuscle
10^{-12}	bacterium
10^{-16}	
10^{-20}	
10^{-24}	haemoglobin molecule
10^{-20}	proton
	electron
10^{-32}	

RANGE OF LENGTHS

Size / m

	radius of observable Universe
10^{26}	
10^{24}	
10^{22}	radius of local galaxy (Milky Way)
10^{20}	
10^{18}	distance to nearest star
10^{16}	
10^{14}	
10^{12}	distance from Earth to Sun
10^{10}	distance from Earth to Moon
10^{8}	radius of the Earth
10^{6}	deepest part of the
10^{4}	ocean / highest mountain
10^{2}	tallest building
10^{0}	
10^{-2}	length of fingernail
10^{-4}	thickness of piece of paper
	human blood corpuscle
10^{-6}	
10^{-8}	wavelength of light
10^{-10}	diameter of hydrogen atom
10^{-12}	wavelength of gamma ray
10^{-14}	diameter of proton
10^{-16}	

RANGE OF TIMES

Time / s

10^{20}	age of the Universe
10^{18}	
10^{16}	age of the Earth
10^{14}	age of species – *Homo sapiens*
10^{12}	
10^{10}	
10^{8}	typical human lifespan
10^{6}	1 year
10^{4}	1 day
10^{2}	
10^{0}	heartbeat
10^{-2}	
10^{-4}	period of high-frequency sound
10^{-6}	
10^{-8}	passage of light across a room
10^{-10}	
10^{-12}	
10^{-14}	vibration of an ion in a solid
10^{-16}	period of visible light
10^{-18}	
10^{-20}	passage of light across an atom
10^{-22}	
10^{-24}	passage of light across a nucleus

RANGE OF ENERGIES

Energy / J

10^{44}	energy released in a supernova
10^{34}	
10^{30}	energy radiated by Sun in 1 second
10^{26}	
10^{22}	energy released in an earthquake
10^{18}	energy released by annihilation of 1 kg of matter
10^{14}	
10^{10}	energy in a lightning discharge
10^{6}	energy needed to charge a car battery
10^{2}	kinetic energy of a tennis ball during game
10^{-2}	energy in the beat of a fly's wing
10^{-6}	
10^{-10}	
10^{-14}	
10^{-18}	energy needed to remove electron from the surface of a metal
10^{-22}	
10^{-26}	

The SI system of fundamental and derived units

FUNDAMENTAL UNITS

Any measurement and every quantity can be thought of as being made up of two important parts:

1. the number and
2. the units.

Without **both** parts, the measurement does not make sense. For example a person's age might be quoted as 'seventeen' but without the 'years' the situation is not clear. Are they 17 minutes, 17 months or 17 years old? In this case you would know if you saw them, but a statement like

length = 4.2

actually says nothing. Having said this, it is really surprising to see the number of candidates who forget to include the units in their answers to examination questions.

In order for the units to be understood, they need to be defined. There are many possible systems of measurement that have been developed. In science we use the International System of units (SI). In SI, the **fundamental** or **base** units are as follows

Quantity	SI unit	SI symbol
Mass	kilogram	kg
Length	metre	m
Time	second	s
Electric current	ampere	A
Amount of substance	mole	mol
Temperature	kelvin	K
(Luminous intensity	candela	cd)

You do not need to know the precise definitions of any of these units in order to use them properly.

DERIVED UNITS

Having fixed the fundamental units, all other measurements can be expressed as different combinations of the fundamental units. In other words, all the other units are **derived units**. For example, the fundamental list of units does not contain a unit for the measurement of speed. The definition of speed can be used to work out the derived unit.

$$\text{Since speed} = \frac{\text{distance}}{\text{time}}$$

$$\text{Units of speed} = \frac{\text{units of distance}}{\text{units of time}}$$

$$= \frac{\text{metres}}{\text{seconds}} \quad \text{(pronounced 'metres per second')}$$

$$= \frac{m}{s}$$

$$= m\ s^{-1}$$

Of the many ways of writing this unit, the last way ($m\ s^{-1}$) is the best.

Sometimes particular combinations of fundamental units are so common that they are given a new derived name. For example, the unit of force is a derived unit – it turns out to be $kg\ m\ s^{-2}$. This unit is given a new name the newton (N) so that $1N = 1\ kg\ m\ s^{-2}$.

The great thing about SI is that, so long as the numbers that are substituted into an equation are in SI units, then the answer will also come out in SI units. You can always 'play safe' by converting all the numbers into proper SI units. Sometimes, however, this would be a waste of time.

There are some situations where the use of SI becomes awkward. In astronomy, for example, the distances involved are so large that the SI unit (the metre) always involves large orders of magnitudes. In these cases, the use of a different (but non SI) unit is very common. Astronomers can use the astronomical unit (AU), the light-year (ly) or the parsec (pc) as appropriate. Whatever the unit, the conversion to SI units is simple arithmetic.

$$1\ AU = 1.5 \times 10^{11}\ m$$
$$1\ ly = 9.5 \times 10^{15}\ m$$
$$1\ pc = 3.1 \times 10^{16}\ m$$

There are also some units (for example the hour) which are so common that they are often used even though they do not form part of SI. Once again, before these numbers are substituted into equations they need to be converted. Some common unit conversions are given on page 3 of the IB data booklet.

The table below lists the SI derived units that you will meet.

SI derived unit	SI base unit	Alternative SI unit
newton (N)	$kg\ m\ s^{-2}$	–
pascal (Pa)	$kg\ m^{-1}\ s^{-2}$	$N\ m^{-2}$
hertz (Hz)	s^{-1}	–
joule (J)	$kg\ m^2\ s^{-2}$	$N\ m$
watt (W)	$kg\ m^2\ s^{-3}$	$J\ s^{-1}$
coulomb (C)	$A\ s$	–
volt (V)	$kg\ m^2\ s^{-3}\ A^{-1}$	WA^{-1}
ohm (Ω)	$kg\ m^2\ s^{-3}\ A^{-2}$	VA^{-1}
weber (Wb)	$kg\ m^2\ s^{-2}\ A^{-1}$	$V\ s$
tesla (T)	$kg\ s^{-2}\ A^{-1}$	$Wb\ m^{-2}$
becquerel (Bq)	s^{-1}	–

PREFIXES

To avoid the repeated use of scientific notation, an alternative is to use one of the list of agreed prefixes given on page 2 in the IB data booklet. These can be very useful but they can also lead to errors in calculations. It is very easy to forget to include the conversion factor.

For example, 1 kW = 1000 W. 1 mW = 10^{-3} W (in other words, $\frac{1W}{1000}$)

Estimation

ORDERS OF MAGNITUDE

It is important to develop a 'feeling' for some of the numbers that you use. When using a calculator, it is very easy to make a simple mistake (eg by entering the data incorrectly). A good way of checking the answer is to first make an estimate before resorting to the calculator. The multiple-choice paper (paper 1) does not allow the use of calculators.

Approximate values for each of the fundamental SI units are given below.

1 kg	A packet of sugar, 1 litre of water. A person would be about 50 kg or more
1 m	Distance between one's hands with arms outstretched
1 s	Duration of a heart beat (when resting – it can easily double with exercise)
1 amp	Current flowing from the mains electricity when a computer is connected. The maximum current to a domestic device would be about 10 A or so
1 kelvin	1K is a very low temperature. Water freezes at 273 K and boils at 373 K. Room temperature is about 300 K
1 mol	12 g of carbon–12. About the number of atoms of carbon in the 'lead' of a pencil

The same process can happen with some of the derived units.

1 m s^{-1}	Walking speed. A car moving at 30 m s^{-1} would be fast
1 m s^{-2}	Quite a slow acceleration. The acceleration of gravity is 10 m s^{-2}
1 N	A small force – about the weight of an apple
1 V	Batteries generally range from a few volts up to 20 or so, the mains is several hundred volts
1 Pa	A very small pressure. Atmospheric pressure is about 10^5 Pa
1 J	A very small amount of energy – the work done lifting an apple off the ground

POSSIBLE REASONABLE ASSUMPTIONS

Everyday situations are very complex. In physics we often simplify a problem by making simple assumptions. Even if we know these assumptions are not absolutely true they allow us to gain an understanding of what is going on. At the end of the calculation it is often possible to go back and work out what would happen if our assumption turned out not to be true.

The table below lists some common assumptions. Be careful not to assume too much! Additionally we often have to assume that some quantity is constant even if we know that in reality it is varying slightly all the time.

Assumption	Example
Object treated as point particle	Mechanics: Linear motion and translational equilibrium
Friction is negligible	Many mechanics situations – but you need to be very careful
No thermal energy ("heat") loss	Almost all thermal situations
Mass of connecting string, etc. is negligible	Many mechanics situations
Resistance of ammeter is zero	Circuits
Resistance of voltmeter is infinite	Circuits
Internal resistance of battery is zero	Circuits
Material obeys Ohm's law	Circuits
Machine 100% efficient	Many situations
Gas is ideal	Thermodynamics
Collision is elastic	Only gas molecules have perfectly elastic collisions
Object radiates as a perfect black body	Thermal equilibrium, e.g. planets

SCIENTIFIC NOTATION

Numbers that are too big or too small for decimals are often written in **scientific notation**:

$$a \times 10^b$$

where a is a number between 1 and 10 and b is an integer.

e.g. $153.2 = 1.532 \times 10^2$; $0.00872 = 8.72 \times 10^{-3}$

SIGNIFICANT FIGURES

Any experimental measurement should be quoted with its uncertainty. This indicates the possible range of values for the quantity being measured. At the same time, the number of **significant figures** used will act as a guide to the amount of uncertainty. For example, a measurement of mass which is quoted as 23.456 g implies an uncertainty of ± 0.001 g (it has five significant figures), whereas one of 23.5 g implies an uncertainty of ± 0.1 g (it has three significant figures).

A simple rule for calculations (multiplication or division) is to quote the answer to the same number of significant digits as the LEAST precise value that is used.

For a more complete analysis of how to deal with uncertainties in calculated results, see page 5.

Uncertainties and error in experimental measurement

ERRORS – RANDOM AND SYSTEMATIC (PRECISION AND ACCURACY)

An experimental error just means that there is a difference between the recorded value and the 'perfect' or 'correct' value. Errors can be categorized as **random** or **systematic**.

Repeating readings does not reduce systematic errors.

Sources of random errors include

- The readability of the instrument.
- The observer being less than perfect.
- The effects of a change in the surroundings.

Sources of systematic errors include

- An instrument with **zero error**. To correct for zero error the value should be subtracted from every reading.
- An instrument being wrongly **calibrated**.
- The observer being less than perfect in the same way every measurement.

An **accurate** experiment is one that has a small systematic error, whereas a **precise** experiment is one that has a small random error.

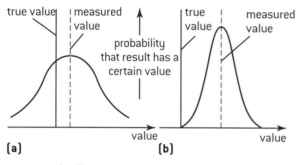

Two examples illustrating the nature of experimental results:
(a) an accurate experiment of low precision
(b) a less accurate but more precise experiment.

GRAPHICAL REPRESENTATION OF UNCERTAINTY

In many situations the best method of presenting and analysing data is to use a graph. If this is the case, a neat way of representing the uncertainties is to use **error bars**. The graphs below explains their use.

Since the error bar represents the uncertainty range, the 'best-fit' line of the graph should pass through ALL of the rectangles created by the error bars.

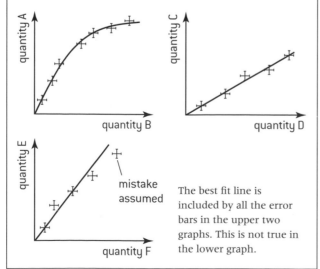

The best fit line is included by all the error bars in the upper two graphs. This is not true in the lower graph.

Systematic and random errors can often be recognized from a graph of the results.

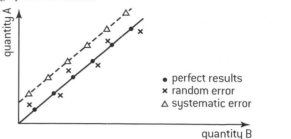

Perfect results, random and systematic errors of two proportional quantities.

ESTIMATING THE UNCERTAINTY RANGE

An **uncertainty range** applies to any experimental value. The idea is that, instead of just giving one value that implies perfection, we give the likely range for the measurement.

1. **Estimating from first principles**

All measurement involves a readability error. If we use a measuring cylinder to find the volume of a liquid, we might think that the best estimate is 73 cm³, but we know that it is not exactly this value (73.000 000 000 00 cm³).

Uncertainty range is $\pm$ 5 cm³. We say volume = 73 $\pm$ 5 cm³.

Normally the uncertainty range due to readability is estimated as below.

Device	Example	Uncertainty
Analogue scale	Rulers, meters with moving pointers	$\pm$ (half the smallest scale division)
Digital scale	Top-pan balances, digital meters	$\pm$ (the smallest scale division)

2. **Estimating uncertainty range from several repeated measurements**

If the time taken for a trolley to go down a slope is measured five times, the readings in seconds might be 2.01, 1.82, 1.97, 2.16 and 1.94. The average of these five readings is 1.98 s. The deviation of the largest and smallest readings can be calculated (2.16 − 1.98 = 0.18; 1.98 − 1.82 = 0.16). The largest value is taken as the uncertainty range. In this example the time is 1.98 s $\pm$ 0.18 s. It would also be appropriate to quote this as 2.0 $\pm$ 0.2 s.

SIGNIFICANT FIGURES IN UNCERTAINTIES

In order to be cautious when quoting uncertainties, final values from calculations are often rounded up to one significant figure, e.g. a calculation that finds the value of a force to be 4.264 N with an uncertainty of $\pm$ 0.362 N is quoted as 4.3 $\pm$ 0.4 N. This can be unnecessarily pessimistic and it is also acceptable to express uncertainties to two significant figures. For example, the charge on an electron is $1.602176565 \times 10^{-19}$ C $\pm 0.000000035 \times 10^{-19}$ C. In data booklets this is sometimes expressed as $1.602176565(35) \times 10^{-19}$ C.

Uncertainties in calculated results

MATHEMATICAL REPRESENTATION OF UNCERTAINTIES

For example if the mass of a block was measured as 10 ± 1 g and the volume was measured as 5.0 ± 0.2 cm³, then the full calculations for the density would be as follows.

Best value for density $= \frac{\text{mass}}{\text{volume}} = \frac{10}{5} = 2.0$ g cm⁻³

The largest possible value of density $= \frac{11}{4.8} = 2.292$ g cm⁻³

The smallest possible value of density $= \frac{9}{5.2} = 1.731$ g cm⁻³

Rounding these values gives density $= 2.0 \pm 0.3$ g cm⁻³

We can express this uncertainty in one of three ways – using **absolute**, **fractional** or **percentage uncertainties**.

If a quantity p is measured then the absolute uncertainty would be expressed as $\pm \Delta p$.

Then the fractional uncertainty is
$$\frac{\pm \Delta p}{p},$$
which makes the percentage uncertainty
$$\frac{\pm \Delta p}{p} \times 100\%.$$

In the example above, the fractional uncertainty of the density is ± 0.15 or $\pm 15\%$.

Thus equivalent ways of expressing this error are

$\qquad$ density $= 2.0 \pm 0.3$ g cm⁻³

$\qquad$ OR density $= 2.0$ g cm⁻³ $\pm 15\%$

Working out the uncertainty range is very time consuming. There are some mathematical 'short-cuts' that can be used. These are introduced in the boxes below.

MULTIPLICATION, DIVISION OR POWERS

Whenever two or more quantities are multiplied or divided and they each have uncertainties, the overall uncertainty is approximately equal to the **addition** of the **percentage** (fractional) uncertainties.

Using the same numbers from above,

$\quad \Delta m = \pm 1$ g

$\quad \dfrac{\Delta m}{m} = \pm \left(\dfrac{1 \text{ g}}{10 \text{ g}} \right) = \pm 0.1 = \pm 10\%$

$\quad \Delta V = \pm 0.2$ cm³

$\quad \dfrac{\Delta V}{V} = \pm \left(\dfrac{0.2 \text{ cm}^3}{5 \text{ cm}^3} \right) = \pm 0.04 = \pm 4\%$

The total % uncertainty in the result $= \pm (10 + 4)\%$
$$= \pm 14\%$$

14% of 2.0 g cm⁻³ $= 0.28$ g cm⁻³ ≈ 0.3 g cm⁻³

So density $= 2.0 \pm 0.3$ g cm⁻³ as before.

In symbols, if $y = \dfrac{ab}{c}$

$\quad$ Then $\dfrac{\Delta y}{y} = \dfrac{\Delta a}{a} + \dfrac{\Delta b}{b} + \dfrac{\Delta c}{c}$ [note this is ALWAYS added]

Power relationships are just a special case of this law.

$\quad$ If $y = a^n$

$\quad$ Then $\dfrac{\Delta y}{y} = \left| n \dfrac{\Delta a}{a} \right|$ (always positive)

For example if a cube is measured to be 4.0 ± 0.1 cm in length along each side, then

$\quad$ % Uncertainty in length $= \pm \dfrac{0.1}{4.0} = \pm 2.5\%$

$\quad$ Volume $= (\text{length})^3 = (4.0)^3 = 64$ cm³

$\quad$ % Uncertainty in [volume] $= $ % uncertainty in [(length)³]
$$= 3 \times (\% \text{ uncertainty in [length]})$$
$$= 3 \times (\pm 2.5\%)$$
$$= \pm 7.5\%$$

$\quad$ Absolute uncertainty $= 7.5\%$ of 64 cm³
$$= 4.8 \text{ cm}^3 \approx 5 \text{ cm}^3$$

$\quad$ Thus volume of cube $= 64 \pm 5$ cm³

OTHER MATHEMATICAL OPERATIONS

If the calculation involves mathematical operations other than multiplication, division or raising to a power, then one has to find the highest and lowest possible values.

Addition or subtraction

Whenever two or more quantities are added or subtracted and they each have uncertainties, the overall uncertainty is equal to the **addition** of the **absolute** uncertainties.

In symbols

If $y = a \pm b$

$\Delta y = \Delta a + \Delta b$ (note ALWAYS added)

uncertainty of thickness in a pipe wall

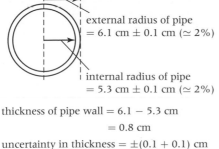

external radius of pipe $= 6.1$ cm ± 0.1 cm ($\simeq 2\%$)

internal radius of pipe $= 5.3$ cm ± 0.1 cm ($\simeq 2\%$)

thickness of pipe wall $= 6.1 - 5.3$ cm
$$= 0.8 \text{ cm}$$

uncertainty in thickness $= \pm (0.1 + 0.1)$ cm
$$= 0.2 \text{ cm}$$
$$= \pm 25\%$$

Other functions

There are no 'short-cuts' possible. Find the highest and lowest values.

e.g. uncertainty of $\sin \theta$ if $\theta = 60° \pm 5°$

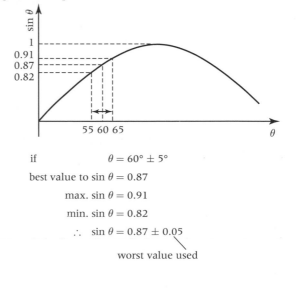

$\qquad$ if $\qquad\qquad \theta = 60° \pm 5°$

$\qquad$ best value to $\sin \theta = 0.87$

$\qquad\qquad$ max. $\sin \theta = 0.91$

$\qquad\qquad$ min. $\sin \theta = 0.82$

$\qquad \therefore \quad \sin \theta = 0.87 \pm 0.05$

$\qquad\qquad\qquad\qquad$ worst value used

Uncertainties in graphs

ERROR BARS

Plotting a graph allows one to visualize all the readings at one time. Ideally all of the points should be plotted with their error bars. In principle, the size of the error bar could well be different for every single point and so they should be individually worked out.

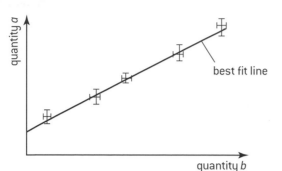

A full analysis in order to determine the uncertainties in the gradient of a best straight-line graph should **always make use of the error bars for all of the data points.**

In practice, it would often take too much time to add all the correct error bars, so some (or all) of the following short-cuts could be considered.

- Rather than working out error bars for each point – use the worst value and assume that all of the other error bars are the same.

- Only plot the error bar for the 'worst' point, i.e. the point that is furthest from the line of best fit. If the line of best fit is within the limits of this error bar, then it will probably be within the limits of all the error bars.

- Only plot the error bars for the first and the last points. These are often the most important points when considering the uncertainty ranges calculated for the gradient or the intercept (see right).

- Only include the error bars for the axis that has the worst uncertainty.

UNCERTAINTY IN SLOPES

If the gradient of the graph has been used to calculate a quantity, then the uncertainties of the points will give rise to an uncertainty in the gradient. Using the steepest and the shallowest lines possible (i.e. the lines that are still consistent with the error bars) the uncertainty range for the gradient is obtained. This process is represented below.

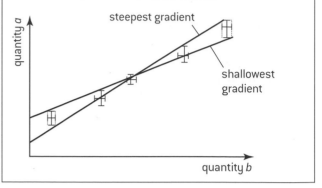

UNCERTAINTY IN INTERCEPTS

If the intercept of the graph has been used to calculate a quantity, then the uncertainties of the points will give rise to an uncertainty in the intercept. Using the steepest and the shallowest lines possible (i.e. the lines that are still consistent with the error bars) we can obtain the uncertainty in the result. This process is represented below.

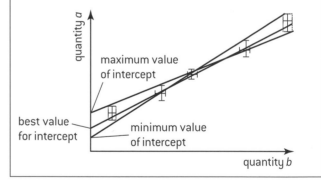

Vectors and scalars

DIFFERENCE BETWEEN VECTORS AND SCALARS

If you measure any quantity, it must have a number AND a unit. Together they express the **magnitude** of the quantity. Some quantities also have a direction associated with them. A quantity that has magnitude and direction is called a **vector** quantity whereas one that has only magnitude is called a **scalar** quantity. For example, all forces are vectors.

The table lists some common quantities. The first two quantities in the table are linked to one another by their definitions (see page 9). All the others are in no particular order.

Vectors	Scalars
Displacement ⟷	Distance
Velocity ⟷	Speed
Acceleration	Mass
Force	Energy (all forms)
Momentum	Temperature
Electric field strength	Potential or potential difference
Magnetic field strength	Density
Gravitational field strength	Area

Although the vectors used in many of the given examples are forces, the techniques can be applied to all vectors.

COMPONENTS OF VECTORS

It is also possible to 'split' one vector into two (or more) vectors. This process is called **resolving** and the vectors that we get are called the **components** of the original vector. This can be a very useful way of analysing a situation if we choose to resolve all the vectors into two directions that are at right angles to one another.

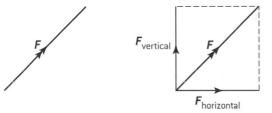

Splitting a vector into components

These 'mutually perpendicular' directions are totally independent of each other and can be analysed separately. If appropriate, both directions can then be combined at the end to work out the final resultant vector.

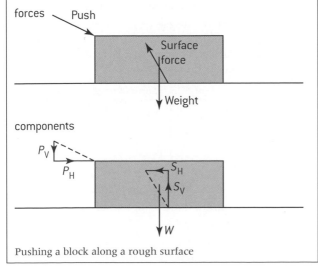

Pushing a block along a rough surface

REPRESENTING VECTORS

In most books a bold letter is used to represent a vector whereas a normal letter represents a scalar. For example **F** would be used to represent a force in magnitude AND direction. The list below shows some other recognized methods.

$$\overrightarrow{F}, \overline{F} \text{ or } \underline{F}$$

Vectors are best shown in diagrams using arrows:

- the relative magnitudes of the vectors involved are shown by the relative length of the arrows
- the direction of the vectors is shown by the direction of the arrows.

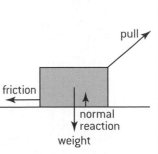

ADDITION / SUBTRACTION OF VECTORS

If we have a 3 N and a 4 N force, the overall force (resultant force) can be anything between 1 N and 7 N depending on the directions involved.

The way to take the directions into account is to do a scale diagram and use the parallelogram law of vectors. This process is the same as adding vectors in turn – the 'tail' of one vector is drawn starting from the head of the previous vector.

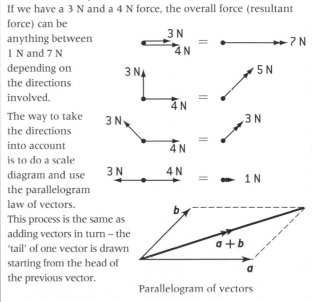

Parallelogram of vectors

TRIGONOMETRY

Vector problems can always be solved using scale diagrams, but this can be very time consuming. The mathematics of trigonometry often makes it much easier to use the mathematical functions of sine or cosine. This is particularly appropriate when resolving. The diagram below shows how to calculate the values of either of these components.

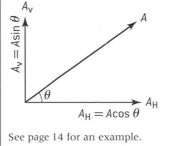

See page 14 for an example.

IB Questions – measurement and uncertainties

1. An object is rolled from rest down an inclined plane. The distance travelled by the object was measured at seven different times. A graph was then constructed of the distance travelled against the (time taken)² as shown below.

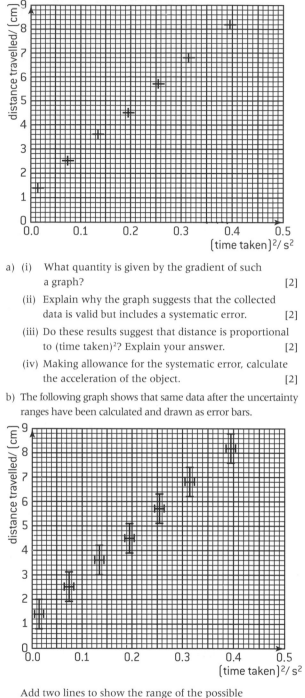

a) (i) What quantity is given by the gradient of such a graph? [2]

(ii) Explain why the graph suggests that the collected data is valid but includes a systematic error. [2]

(iii) Do these results suggest that distance is proportional to (time taken)²? Explain your answer. [2]

(iv) Making allowance for the systematic error, calculate the acceleration of the object. [2]

b) The following graph shows that same data after the uncertainty ranges have been calculated and drawn as error bars.

Add two lines to show the range of the possible acceptable values for the gradient of the graph. [2]

2. The lengths of the sides of a rectangular plate are measured, and the diagram shows the measured values with their uncertainties.

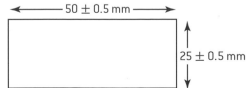

Which one of the following would be the best estimate of the percentage uncertainty in the calculated area of the plate?

A. ± 0.02% C. ± 3%

B. ± 1% D. ± 5%

3. A stone is dropped down a well and hits the water 2.0 s after it is released. Using the equation $d = \frac{1}{2}g\,t^2$ and taking $g = 9.81$ m s⁻², a calculator yields a value for the depth d of the well as 19.62 m. If the time is measured to ±0.1 s then the best estimate of the absolute error in d is

A. ±0.1 m C. ±1.0 m

B. ±0.2 m D. ±2.0 m

4. In order to determine the density of a certain type of wood, the following measurements were made on a **cube** of the wood.

Mass $= 493$ g

Length of **each** side $= 9.3$ cm

The percentage uncertainty in the measurement of mass is ±0.5% and the percentage uncertainty in the measurement of length is ±1.0%.

The best estimate for the uncertainty in the density is

A. ±0.5% C. ±3.0%

B. ±1.5% D. ±3.5%

5. Astronauts wish to determine the gravitational acceleration on Planet X by dropping stones from an overhanging cliff. Using a steel tape measure they measure the height of the cliff as $s = 7.64$ m ± 0.01 m. They then drop three similar stones from the cliff, timing each fall using a hand-held electronic stopwatch which displays readings to one-hundredth of a second. The recorded times for three drops are 2.46 s, 2.31 s and 2.40 s.

a) Explain why the time readings vary by more than a tenth of a second, although the stopwatch gives readings to one hundredth of a second. [1]

b) Obtain the average time t to fall, and write it in the form (value ± uncertainty), to the appropriate number of significant digits. [1]

c) The astronauts then determine the gravitational acceleration a_g on the planet using the formula $a_g = \frac{2s}{t^2}$. Calculate a_g from the values of s and t, and determine the uncertainty in the calculated value. Express the result in the form

a_g = (value ± uncertainty),

to the appropriate number of significant digits. [3]

HL

6. This question is about finding the relationship between the forces between magnets and their separations.

In an experiment, two magnets were placed with their North-seeking poles facing one another. The force of repulsion, f, and the separation of the magnets, d, were measured and the results are shown in the table below.

Separation d/m	Force of repulsion f/N
0.04	4.00
0.05	1.98
0.07	0.74
0.09	0.32

a) Plot a graph of log (force) against log (distance). [3]

b) The law relating the force to the separation is of the form

$f = kd^n$

(i) Use the graph to find the value of n. [2]

(ii) Calculate a value for k, giving its units. [3]

Motion

DEFINITIONS

These technical terms should not be confused with their 'everyday' use. In particular one should note that

- Vector quantities always have a direction associated with them.
- Generally, velocity and speed are NOT the same thing. This is particularly important if the object is not going in a straight line.
- The units of acceleration come from its definition. $(m\ s^{-1}) \div s = m\ s^{-2}$.
- The definition of acceleration is precise. It is related to the change in **velocity** (not the same thing as the change in speed). Whenever the motion of an object changes, it is called acceleration. For this reason acceleration does not necessarily mean constantly increasing speed – it is possible to accelerate while at constant speed if the direction is changed.
- A deceleration means slowing down, i.e. negative acceleration if velocity is positive.

	Symbol	Definition	Example	SI Unit	Vector or scalar?
Displacement	s	The distance moved in a particular direction.	The displacement from London to Rome is 1.43×10^6 m southeast.	m	Vector
Velocity	v or u	The rate of change of displacement. $velocity = \frac{change\ of\ displacement}{time\ taken}$	The average velocity during a flight from London to Rome is 160 m s⁻¹ southeast.	m s⁻¹	Vector
Speed	v or u	The rate of change of distance. $speed = \frac{distance\ gone}{time\ taken}$	The average speed during a flight from London to Rome is 160 m s⁻¹	m s⁻¹	Scalar
Acceleration	a	The rate of change of velocity. $acceleration = \frac{change\ of\ velocity}{time\ taken}$	The average acceleration of a plane on the runway during take-off is 3.5 m s⁻² in a forwards direction. This means that on average, its velocity changes every second by 3.5 m s⁻¹	m s⁻²	Vector

INSTANTANEOUS VS AVERAGE

It should be noticed that the average value (over a period of time) is very different to the instantaneous value (at one particular time).

In the example below, the positions of a sprinter are shown at different times after the start of a race.

The average speed over the whole race is easy to work out. It is the total distance (100 m) divided by the total time (11.3 s) giving 8.8 m s⁻¹.

But during the race, her instantaneous speed was changing all the time. At the end of the first 2.0 seconds, she had travelled 10.04 m. This means that her average speed over the first 2.0 seconds was 5.02 m s⁻¹. During these first two seconds, her instantaneous speed was increasing – she was accelerating. If she started at rest (speed = 0.00 m s⁻¹) and her **average** speed (over the whole two seconds) was 5.02 m s⁻¹ then her instantaneous speed at 2 seconds must be more than this. In fact the instantaneous speed for this sprinter was 9.23 m s⁻¹, but it would not be possible to work this out from the information given.

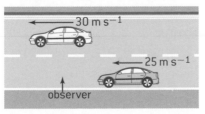

start					finish
$d = 0.00$ m	$d = 10.04$ m	$d = 28.21$ m	$d = 47.89$ m	$d = 69.12$ m	$d = 100.00$ m
$t = 0.0$ s	$t = 2.0$ s	$t = 4.0$ s	$t = 6.0$ s	$t = 8.0$ s	$t = 11.3$ s

FRAMES OF REFERENCE

If two things are moving in the same straight line but are travelling at different speeds, then we can work out their **relative velocities** by simple addition or subtraction as appropriate. For example, imagine two cars travelling along a straight road at different speeds.

If one car (travelling at 30 m s⁻¹) overtakes the other car (travelling at 25 m s⁻¹), then according to the driver of the slow car, the relative velocity of the fast car is +5 m s⁻¹.

In technical terms what we are doing is moving from one **frame of reference** into another. The velocities of 25 m s⁻¹ and 30 m s⁻¹ were measured according to a stationary observer on the side of the road. We moved from this frame of reference into the driver's frame of reference.

one car overtaking another, as seen by an observer on the side of the road.

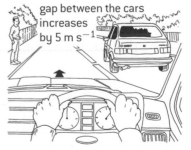

one car overtaking another, as seen by the driver of the slow car.

Graphical representation of motion

THE USE OF GRAPHS

Graphs are very useful for representing the changes that happen when an object is in motion. There are three possible graphs that can provide useful information

- displacement–time or distance–time graphs
- velocity–time or speed–time graphs
- acceleration–time graphs.

There are two common methods of determining particular physical quantities from these graphs. The particular physical quantity determined depends on what is being plotted on the graph.

1. Finding the gradient of the line.

To be a little more precise, one could find either the gradient of

- a straight-line section of the graph (this finds an average value), or
- the tangent to the graph at one point (this finds an instantaneous value).

2. Finding the area under the line.

To make things simple at the beginning, the graphs are normally introduced by considering objects that are just moving in one particular direction. If this is the case then there is not much difference between the scalar versions (distance or speed) and the vector versions (displacement or velocity) as the directions are clear from the situation. More complicated graphs can look at the component of a velocity in a particular direction.

If the object moves forward then backward (or up then down), we distinguish the two velocities by choosing which direction to call positive. It does not matter which direction we choose, but it should be clearly labelled on the graph.

Many examination candidates get the three types of graph muddled up. For example a speed–time graph might be interpreted as a distance–time graph or even an acceleration–time graph. Always look at the axes of a graph very carefully.

DISPLACEMENT–TIME GRAPHS

- The gradient of a displacement–time graph is the velocity
- The area under a displacement–time graph does not represent anything useful

Examples

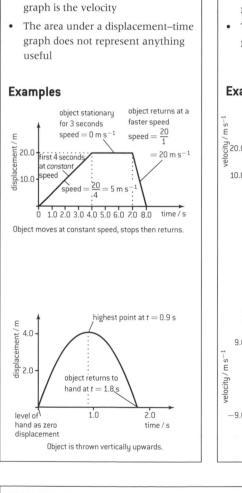

Object moves at constant speed, stops then returns.

Object is thrown vertically upwards.

VELOCITY–TIME GRAPHS

- The gradient of a velocity–time graph is the acceleration
- The area under a velocity–time graph is the displacement

Examples

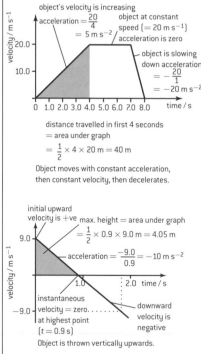

distance travelled in first 4 seconds
= area under graph
$= \frac{1}{2} \times 4 \times 20 = 40$ m

Object moves with constant acceleration, then constant velocity, then decelerates.

Object is thrown vertically upwards.

ACCELERATION–TIME GRAPHS

- The gradient of an acceleration–time graph is not often useful (it is actually the rate of change of acceleration)
- The area under an acceleration–time graph is the change in velocity

Examples

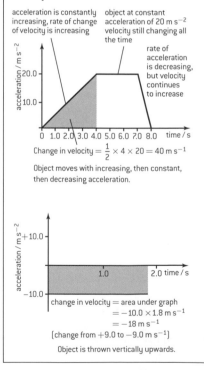

Change in velocity $= \frac{1}{2} \times 4 \times 20 = 40$ m s^{-1}

Object moves with increasing, then constant, then decreasing acceleration.

change in velocity = area under graph
$= -10.0 \times 1.8$ m s^{-1}
$= -18$ m s^{-1}
[change from $+9.0$ to -9.0 m s^{-1}]

Object is thrown vertically upwards.

EXAMPLE OF EQUATION OF UNIFORM MOTION

A car accelerates uniformly from rest. After 8 s it has travelled 120 m. Calculate: (i) its average acceleration   (ii) its instantaneous speed after 8 s

(i) $s = ut + \frac{1}{2}at^2$

$\therefore 120 = 0 \times 8 + \frac{1}{2}a \times 8^2 = 32\,a$

$\underline{a = 3.75 \text{ m s}^{-2}}$

(ii) $v^2 = u^2 + 2as$
$= 0 + 2 \times 3.75 \times 120$
$= 900$
$\therefore \underline{v = 30 \text{ m s}^{-1}}$

Uniformly accelerated motion

PRACTICAL CALCULATIONS

In order to determine how the velocity (or the acceleration) of an object varies in real situations, it is often necessary to record its motion. Possible laboratory methods include.

Light gates

A light gate is a device that senses when an object cuts through a beam of light. The time for which the beam is broken is recorded. If the length of the object that breaks the beam is known, the average speed of the object through the gate can be calculated.

Alternatively, two light gates and a timer can be used to calculate the average velocity between the two gates. Several light gates and a computer can be joined together to make direct calculations of velocity or acceleration.

Strobe photography

A strobe light gives out very brief flashes of light at fixed time intervals. If a camera is pointed at an object and the only source of light is the strobe light, then the developed picture will have captured an object's motion.

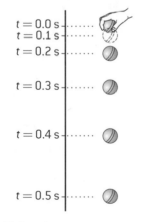

Ticker timer

A ticker timer can be arranged to make dots on a strip of paper at regular intervals of time (typically every fiftieth of a second). If the piece of paper is attached to an object, and the object is allowed to fall, the dots on the strip will have recorded the distance moved by the object in a known time.

EQUATIONS OF UNIFORM MOTION

These equations can only be used when the acceleration is constant – don't forget to check if this is the case!

The list of variables to be considered (and their symbols) is as follows

 u initial velocity

 v final velocity

 a acceleration (const)

 t time taken

 s distance travelled

The following equations link these different quantities.

$$v = u + at$$

$$s = \left(\frac{u + v}{2}\right) t$$

$$v^2 = u^2 + 2as$$

$$s = ut + \frac{1}{2} at^2$$

$$s = vt - \frac{1}{2} at^2$$

The first equation is derived from the definition of acceleration. In terms of these symbols, this definition would be

$$a = \frac{(v - u)}{t}$$

This can be rearranged to give the first equation.

$$v = u + at \qquad (1)$$

The second equation comes from the definition of average velocity.

$$\text{average velocity} = \frac{s}{t}$$

Since the velocity is changing uniformly we know that this average velocity must be given by

$$\text{average velocity} = \frac{(v + u)}{2}$$

or $\quad \dfrac{s}{t} = \dfrac{(u + v)}{2}$

This can be rearranged to give

$$s = \frac{(u + v)t}{2} \qquad (2)$$

The other equations of motion can be derived by using these two equations and substituting for one of the variables (see previous page for an example of their use).

FALLING OBJECTS

A very important example of uniformly accelerated motion is the vertical motion of an object in a uniform **gravitational field**. If we ignore the effects of air resistance, this is known as being in **free-fall**.

Taking down as positive, the graphs of the motion of any object in free-fall are

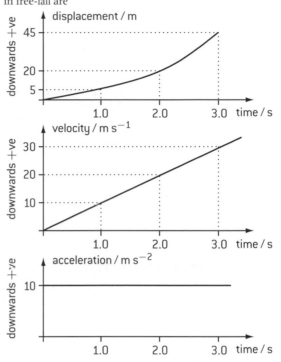

In the absence of air resistance, all falling objects have the SAME acceleration of free-fall, INDEPENDENT of their mass.

Air resistance will (eventually) affect the motion of all objects. Typically, the graphs of a falling object affected by air resistance become the shapes shown below.

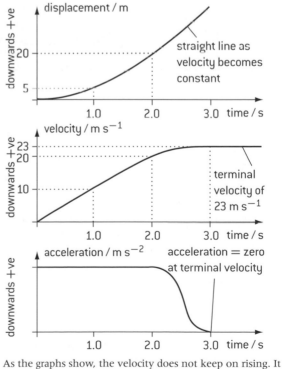

As the graphs show, the velocity does not keep on rising. It eventually reaches a maximum or **terminal velocity**. A piece of falling paper will reach its terminal velocity in a much shorter time than a falling book.

Projectile motion

COMPONENTS OF PROJECTILE MOTION

If two children are throwing and catching a tennis ball between them, the path of the ball is always the same shape. This motion is known as **projectile motion** and the shape is called a **parabola**.

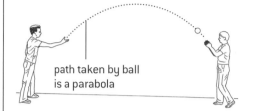

path taken by ball is a parabola

The only forces acting during its flight are gravity and friction. In many situations, air resistance can be ignored.

It is moving horizontally and vertically **at the same time** but the horizontal and vertical components of the motion are **independent** of one another. Assuming the gravitional force is constant, this is always true.

Horizontal component

There are no forces in the horizontal direction, so there is no horizontal acceleration. This means that the horizontal velocity must be constant.

ball travels at a constant horizontal velocity

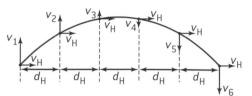

Vertical component

There is a constant vertical force acting down, so there is a constant vertical acceleration. The value of the vertical acceleration is 10 m s^{-2}, which is the acceleration due to gravity.

vertical velocity changes

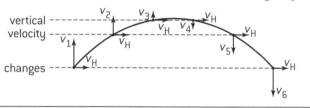

MATHEMATICS OF PARABOLIC MOTION

The graphs of the components of parabolic motion are shown below.

motion in the x-direction **motion in the y-direction**

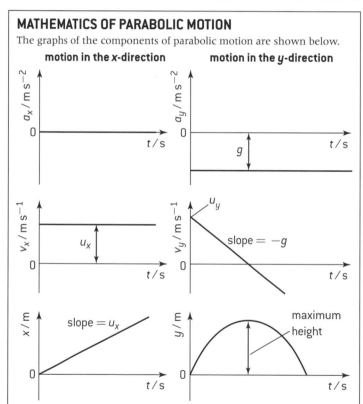

Once the components have been worked out, the actual velocities (or displacements) at any time can be worked out by vector addition.

The solution of any problem involving projectile motion is as follows:
- use the angle of launch to resolve the initial velocity into components.
- the time of flight will be determined by the vertical component of velocity.
- the range will be determined by the horizontal component (and the time of flight).
- the velocity at any point can be found by vector addition.

Useful 'short-cuts' in calculations include the following facts:
- for a given speed, the greatest range is achieved if the launch angle is 45°.
- if two objects are released together, one with a horizontal velocity and one from rest, they will both hit the ground together.

EXAMPLE

A projectile is launched horizontally from the top of a cliff.

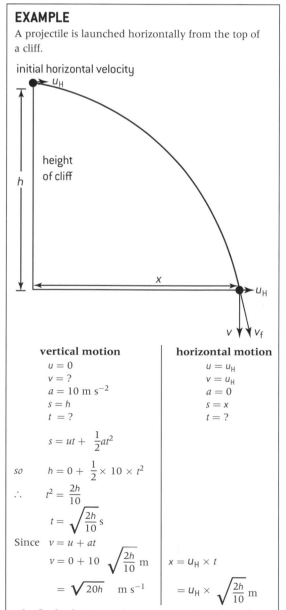

vertical motion	horizontal motion
$u = 0$	$u = u_H$
$v = ?$	$v = u_H$
$a = 10$ m s^{-2}	$a = 0$
$s = h$	$s = x$
$t = ?$	$t = ?$

$$s = ut + \frac{1}{2}at^2$$

so $\quad h = 0 + \frac{1}{2} \times 10 \times t^2$

$\therefore \quad t^2 = \frac{2h}{10}$

$\quad t = \sqrt{\frac{2h}{10}}$ s

Since $\quad v = u + at$

$\quad v = 0 + 10 \sqrt{\frac{2h}{10}}$ m $\qquad x = u_H \times t$

$\quad = \sqrt{20h}$ m s^{-1} $\qquad = u_H \times \sqrt{\frac{2h}{10}}$ m

The final velocity v_f is the vector addition of v and u_H.

Fluid resistance and free-fall

FLUID RESISTANCE

When an object moves through a fluid (a liquid or a gas), there will be a frictional fluid resistance that affects the object's motion. An example of this effect is the terminal velocity that is reached by a free-falling object, e.g. a spherical mass falling through a liquid or a parachutist falling towards the Earth. See page 11 for how the motion graphs will be altered in these situations.

Modelling the precise effect of fluid resistance on moving objects is complex but simple predictions are possible. The Engineering Physics option (see page 167) introduces a mathematical analysis of the frictional drag force that acts on a perfect sphere when it moves through a fluid. Key points to note are that:

- Viscous drag acts to oppose motion through a fluid
- The drag force is dependent on:
  - Relative velocity of the object with respect to the fluid
  - The shape and size of the object (whether the object is aerodynamic or not)
  - The fluid used (and a property called its viscosity).

For example page 12 shows how, in the absence of fluid resistance, an object that is in projectile motion will follow a parabolic path. When fluid resistance is taken into account, the vertical and the horizontal components of velocity will both be reduced. The effect will be a reduced range and, in the extreme, the horizontal velocity can be reduced to near zero.

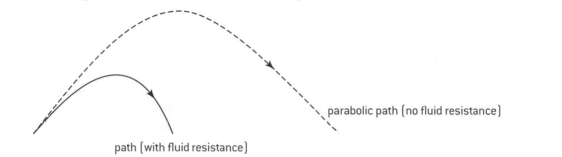

parabolic path (no fluid resistance)

path (with fluid resistance)

EXPERIMENT TO DETERMINE FREE-FALL ACCELERATION

All experiments to determine the free-fall acceleration for an object are based on the use of a constant acceleration equation with recorded measurements of displacement and time. Some experimental set-ups will be more sophisticated and use more equipment than others. This increased use of technology potentially brings greater precision but can introduce more complications. Simple equipment often means that, with a limited time available for experimentation, it is easier for many repetitions to be attempted.

If an object free-falls a height, h, from rest in a time, t, the acceleration, g, can be calculated using $s = ut + \frac{1}{2}at^2$ which rearranges to give $= \frac{2h}{t^2}$. Rather than just calculating a single value, a more reliable value comes from taking a series of measurement of the different times of fall for different heights $h = \frac{1}{2}gt^2$. A graph of h on the y-axis against t^2 on the x-axis will give a straight line graph that goes through the origin with a gradient equal to $\frac{1}{2}g$, making g twice the gradient.

Possible set-ups include:

Set-up	Comments
Direct measurement of a falling object, e.g. ball bearing with a stop watch and a metre ruler	Very simple set-up meaning many repetitions easily achieved so random error can be eliminated. If height of fall is carefully controlled, great precision is possible even though equipment is standard. For a simple everyday object such as a ball bearing, the effect of air resistance will be negligible in the laboratory whereas the effect of air resistance on a Ping-Pong ball will be significant.
Electromagnet release and electronic timing version of the above	The increased precision of the timing can improve accuracy but set-up will take longer. Introduction of technology can mean that systematic errors are harder to identify.
Motion of falling object automatically recording on ticker-tape attached to falling object	Physical record allows detailed analysis of motion and thus allows the object's whole fall to be considered (not just the overall time taken) and for the data to be graphically analysed. Addition of moving paper tape introduces friction to the motion, however.
Distance sensor and data logger	All measurements can be automated and very precise. Software can be programmed to perform all the calculations and to plot appropriate graphs. Experimenter needs to understand how to operate the data logger and associated software.
Video analysis of falling object	Capturing a visual record of the object's fall against a known scale, allows detailed measurements to be taken. Timing information from the video recording needed, which often involves ICT.

Forces and free-body diagrams

FORCES – WHAT THEY ARE AND WHAT THEY DO

In the examples below, a force (the kick) can cause deformation (the ball changes shape) or a change in motion (the ball gains a velocity). There are many different types of forces, but in general terms one can describe any force as 'the cause of a deformation or a velocity change'. The SI unit for the measurement of forces is the newton (N).

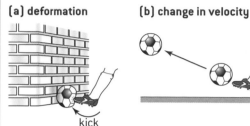

(a) deformation

kick causes
deformation of football

(b) change in velocity

kick

kick causes a change in
motion of football

Effect of a force on a football

- A (resultant) force causes a CHANGE in velocity. If the (resultant) force is zero then the velocity is constant. Remember a change in velocity is called an acceleration, so we can say that **a force causes an acceleration**. A (resultant) force is NOT needed for a constant velocity (see page 16).

- The fact that a force can cause deformation is also important, but the deformation of the ball was, in fact, not caused by just one force – there was another one from the wall.

- One force can act on only one object. To be absolutely precise the description of a force should include
  - its magnitude
  - its direction
  - the object on which it acts (or the part of a large object)
  - the object that exerts the force
  - the nature of the force

A description of the force shown in the example would thus be 'a 50 N push at 20° to the horizontal acting ON the football FROM the boot'.

DIFFERENT TYPES OF FORCES

The following words all describe the forces (the pushes or pulls) that exist in nature.

Gravitational force	**Normal reaction**	**Compression**
Electrostatic force	**Friction**	**Upthrust**
Magnetic force	**Tension**	**Lift**

One way of categorizing these forces is whether they result from the contact between two surfaces or whether the force exists even if a distance separates the objects.

The origin of all these everyday forces is either gravitational or electromagnetic. The vast majority of everyday effects that we observe are due to electromagnetic forces.

MEASURING FORCES

The simplest experimental method for measuring the size of a force is to use the **extension** of a spring. When a spring is in tension it increases in length. The difference between the natural length and stretched length is called the extension of a spring.

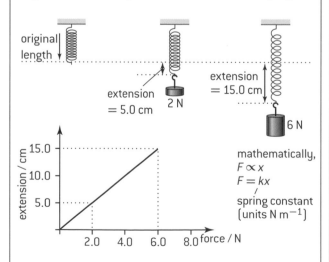

original
length

extension
= 5.0 cm

2 N

extension
= 15.0 cm

6 N

mathematically,
$F \propto x$
$F = kx$

spring constant
(units N m^{-1})

Hooke's law

Hooke's law states that up to the elastic limit, the extension, x, of a spring is proportional to the tension force, F. The constant of proportionality k is called the **spring constant**. The SI units for the spring constant are N m^{-1}. Thus by measuring the extension, we can calculate the force.

FORCES AS VECTORS

Since forces are vectors, vector mathematics must be used to find the resultant force from two or more other forces. A force can also be split into its components. See page 7 for more details.

(a) by vector mathematics

example: block being pushed on rough surface

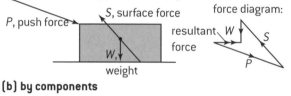

P, push force

S, surface force

force diagram:

resultant force

W, weight

(b) by components

example: block sliding down a smooth slope

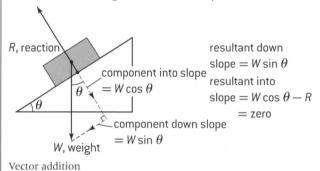

R, reaction

component into slope
$= W \cos \theta$

component down slope
$= W \sin \theta$

W, weight

resultant down
slope $= W \sin \theta$
resultant into
slope $= W \cos \theta - R$
$=$ zero

Vector addition

FREE-BODY DIAGRAMS

In a **free-body diagram**

- one object (and ONLY one object) is chosen
- all the forces on that object are shown and labelled.

For example, if we considered the simple situation of a book resting on a table, we can construct free-body diagrams for either the book or the table.

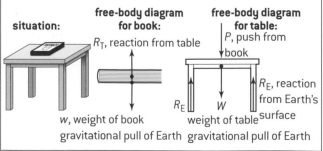

situation:

**free-body diagram
for book:**

R_T, reaction from table

w, weight of book
gravitational pull of Earth

**free-body diagram
for table:**

P, push from
book

R_E, reaction
from Earth's
surface

R_E W

weight of table
gravitational pull of Earth

Newton's first law

NEWTON'S FIRST LAW

Newton's first law of motion states that 'an object continues in uniform motion in a straight line or at rest unless a resultant external force acts'. On first reading, this can sound complicated but it does not really add anything to the description of a force given on page 14. All it says is that a resultant force causes acceleration. No resultant force means no acceleration – i.e. 'uniform motion in a straight line'.

Book on a table at rest

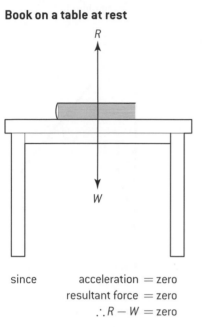

since   acceleration = zero
resultant force = zero
$\therefore R - W$ = zero

Parachutist in free fall

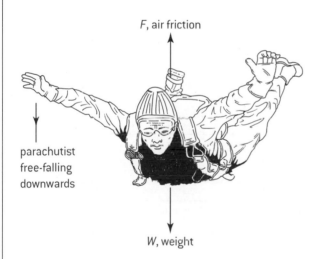

F, air friction

parachutist free-falling downwards

W, weight

If $W > F$ the parachutist accelerates downwards.
As the parachutist gets faster, the air friction increases until $W = F$
The parachutist is at constant velocity (the *acceleration* is zero).

Lifting a heavy suitcase

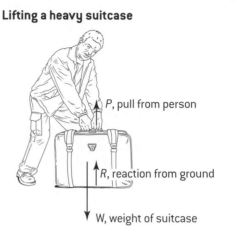

P, pull from person

R, reaction from ground

W, weight of suitcase

If the suitcase is too heavy to lift, it is not moving:
$\therefore$ acceleration = zero
$\therefore P + R = W$

Car travelling in a straight line

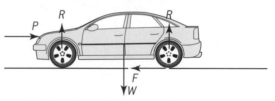

F is force forwards, due to engine
P is force backwards due to air resistance

At all times force up ($2R$) = force down (W).
If $F > P$ the car accelerates forwards.
If $F = P$ the car is at constant velocity (zero acceleration).
If $F < P$ the car decelerates (i.e. there is negative acceleration and the car slows down).

Person in a lift that is moving upwards

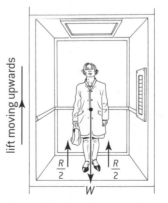

lift moving upwards

$\frac{R}{2}$ $\frac{R}{2}$

W

The total force up from the floor of the lift $= R$.
The total force down due to gravity $= W$.
If $R > W$ the person is accelerating upwards.
If $R = W$ the person is at constant velocity (acceleration = zero).
If $R < W$ the person is decelerating (acceleration is negative).

Equilibrium

EQUILIBRIUM

If the resultant force on an object is zero then it is said to be in **translational equilibrium** (or just in equilibrium). Mathematically this is expressed as follows:

$$\Sigma F = zero$$

From Newton's first law, we know that the objects in the following situations must be in equilibrium.

1. An object that is constantly at rest.

2. An object that is moving with constant (uniform) velocity in a straight line.

Since forces are vector quantities, a zero resultant force means no force IN ANY DIRECTION.

For 2-dimensional problems it is sufficient to show that the forces balance in any two non-parallel directions. If this is the case then the object is in equilibrium.

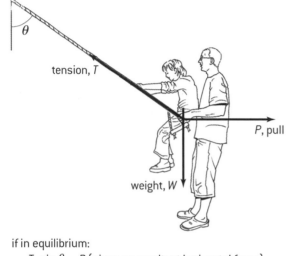

if in equilibrium:
  $T \sin \theta = P$ (since no resultant horizontal force)
  $T \cos \theta = W$ (since no resultant vertical force)

Translational equilibrium does NOT mean the same thing as being at rest. For example if the child in the previous example is allowed to swing back and forth, there are times when she is instantaneously at rest but he is never in equilibrium.

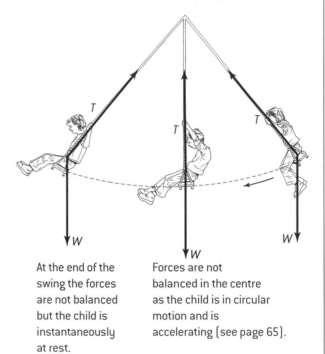

At the end of the swing the forces are not balanced but the child is instantaneously at rest.

Forces are not balanced in the centre as the child is in circular motion and is accelerating (see page 65).

DIFFERENT TYPES OF FORCES

Name of force	Description
Gravitational force	The force between objects as a result of their masses. This is sometimes referred to as the **weight** of the object but this term is, unfortunately, ambiguous – see page 19.
Electrostatic force	The force between objects as a result of their electric charges.
Magnetic force	The force between magnets and/or electric currents.
Normal reaction	The force between two surfaces that acts at right angles to the surfaces. If two surfaces are smooth then this is the only force that acts between them.
Friction	The force that opposes the relative motion of two surfaces and acts along the surfaces. **Air resistance** or **drag** can be thought of as a frictional force – technically this is known as **fluid friction**.
Tension	When a string (or a spring) is stretched, it has equal and opposite forces on its ends pulling outwards. The tension force is the force that the end of the string applies to another object.
Compression	When a rod is compressed (squashed), it has equal and opposite forces on its ends pushing inwards. The compression force is the force that the ends of the rod applies to another object. This is the opposite of the tension force.
Upthrust	This is the upward force that acts on an object when it is submerged in a fluid. It is the buoyancy force that causes some objects to float in water (see page 164).
Lift	This force can be exerted on an object when a fluid flows over it in an asymmetrical way. The shape of the wing of an aircraft causes the aerodynamic lift that enables the aircraft to fly (see page 166).

Newton's second law

NEWTON'S SECOND LAW OF MOTION

Newton's first law states that a resultant force causes an acceleration. His second law provides a means of calculating the value of this acceleration. The best way of stating the second law is use the concept of the **momentum** of an object. This concept is explained on page 23.

A correct statement of Newton's second law using momentum would be 'the resultant force is proportional to the rate of change of momentum'. If we use SI units (and you always should) then the law is even easier to state – 'the resultant force is equal to the rate of change of momentum'. In symbols, this is expressed as follows

$$\text{In SI units, } F = \frac{\Delta p}{\Delta t}$$

$$\text{or, in full calculus notation, } F = \frac{dp}{dt}$$

p is the symbol for the momentum of a body.

Until you have studied what this means this will not make much sense, but this version of the law is given here for completeness.

An equivalent (but more common) way of stating Newton's second law applies when we consider the action of a force on a single mass. If the amount of mass stays constant we can state the law as follows. 'The resultant force is proportional to the acceleration.' If we also use SI units then 'the resultant force is equal to the product of the mass and the acceleration'.

In symbols, in SI units,

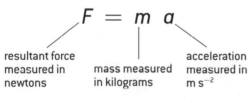

$$F = m\,a$$

resultant force measured in newtons | mass measured in kilograms | acceleration measured in m s⁻²

Note:

- The '$F = ma$' version of the law only applies if we use SI units – for the equation to work the mass must be in **kilograms** rather than in grams.

- F is the resultant force. If there are several forces acting on an object (and this is usually true) then one needs to work out the resultant force before applying the law.

- This is an experimental law.

- There are no exceptions – Newton's laws apply throughout the Universe. (To be absolutely precise, Einstein's theory of relativity takes over at very large values of speed and mass.)

The $F = ma$ version of the law can be used whenever the situation is simple – for example, a constant force acting on a constant mass giving a constant acceleration. If the situation is more difficult (e.g. a changing force or a changing mass) then one needs to use the $F = \frac{dp}{dt}$ version.

EXAMPLES OF NEWTON'S SECOND LAW

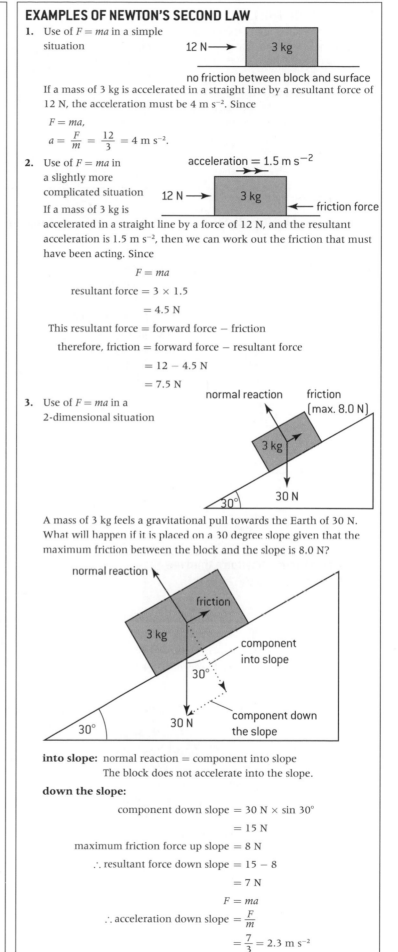

1. Use of $F = ma$ in a simple situation

If a mass of 3 kg is accelerated in a straight line by a resultant force of 12 N, the acceleration must be 4 m s⁻². Since

$$F = ma,$$
$$a = \frac{F}{m} = \frac{12}{3} = 4 \text{ m s}^{-2}.$$

2. Use of $F = ma$ in a slightly more complicated situation

If a mass of 3 kg is accelerated in a straight line by a force of 12 N, and the resultant acceleration is 1.5 m s⁻², then we can work out the friction that must have been acting. Since

$$F = ma$$
$$\text{resultant force} = 3 \times 1.5$$
$$= 4.5 \text{ N}$$

This resultant force = forward force − friction

therefore, friction = forward force − resultant force

$$= 12 - 4.5 \text{ N}$$
$$= 7.5 \text{ N}$$

3. Use of $F = ma$ in a 2-dimensional situation

A mass of 3 kg feels a gravitational pull towards the Earth of 30 N. What will happen if it is placed on a 30 degree slope given that the maximum friction between the block and the slope is 8.0 N?

into slope: normal reaction = component into slope
The block does not accelerate into the slope.

down the slope:

$$\text{component down slope} = 30 \text{ N} \times \sin 30°$$
$$= 15 \text{ N}$$
$$\text{maximum friction force up slope} = 8 \text{ N}$$
$$\therefore \text{ resultant force down slope} = 15 - 8$$
$$= 7 \text{ N}$$
$$F = ma$$
$$\therefore \text{ acceleration down slope} = \frac{F}{m}$$
$$= \frac{7}{3} = 2.3 \text{ m s}^{-2}$$

Newton's third law

STATEMENT OF THE LAW

Newton's second law is an experimental law that allows us to calculate the effect that a force has. Newton's third law highlights the fact that forces always come in pairs. It provides a way of checking to see if we have remembered all the forces involved.

It is very easy to state. 'When two bodies A and B interact, the force that A exerts on B is equal and opposite to the force that B exerts on A'. Another way of saying the same thing is that 'for every action on one object there is an equal but opposite reaction on another object'.

In symbols,

$$F_{AB} = -F_{BA}$$

Key points to notice include

- The two forces in the pair act on different objects – this means that equal and opposite forces that act on the same object are NOT Newton's third law pairs.

- Not only are the forces equal and opposite, but they must be of the same type. In other words, if the force that A exerts on B is a gravitational force, then the equal and opposite force exerted by B on A is also a gravitational force.

EXAMPLES OF THE LAW

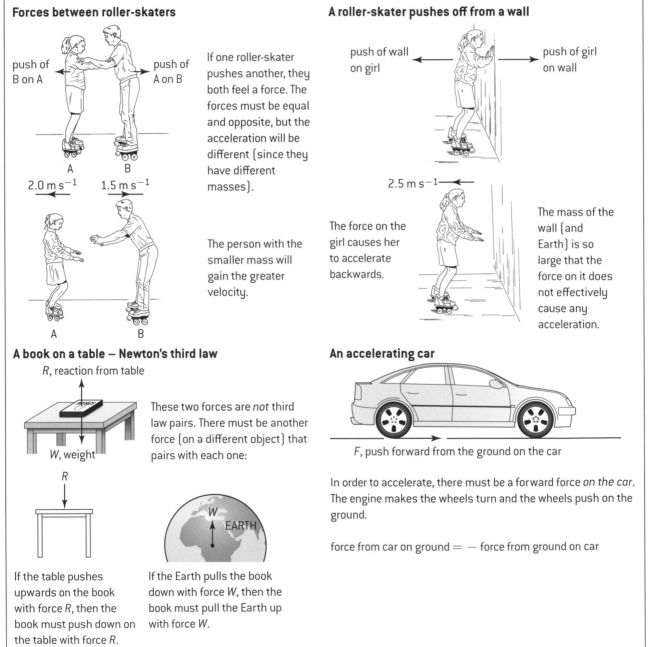

Forces between roller-skaters

push of B on A

push of A on B

A   B

$2.0\,\text{m s}^{-1}$ $1.5\,\text{m s}^{-1}$

A   B

If one roller-skater pushes another, they both feel a force. The forces must be equal and opposite, but the acceleration will be different (since they have different masses).

The person with the smaller mass will gain the greater velocity.

A roller-skater pushes off from a wall

push of wall on girl

push of girl on wall

$2.5\,\text{m s}^{-1}$

The force on the girl causes her to accelerate backwards.

The mass of the wall (and Earth) is so large that the force on it does not effectively cause any acceleration.

A book on a table – Newton's third law

R, reaction from table

W, weight

R

These two forces are *not* third law pairs. There must be another force (on a different object) that pairs with each one:

W
EARTH

If the table pushes upwards on the book with force R, then the book must push down on the table with force R.

If the Earth pulls the book down with force W, then the book must pull the Earth up with force W.

An accelerating car

F, push forward from the ground on the car

In order to accelerate, there must be a forward force *on the car*. The engine makes the wheels turn and the wheels push on the ground.

force from car on ground $= -$ force from ground on car

Mass and weight

WEIGHT

Mass and **weight** are two very different things. Unfortunately their meanings have become muddled in everyday language. Mass is the amount of matter contained in an object (measured in kg) whereas the weight of an object is a force (measured in N).

If an object is taken to the Moon, its mass would be the same, but its weight would be less (the gravitational forces on the Moon are less than on the Earth). On the Earth the two terms are often muddled because they are proportional. People talk about wanting to gain or lose weight – what they are actually worried about is gaining or losing mass.

Double the mass means double the weight

To make things worse, the term 'weight' can be ambiguous even to physicists. Some people choose to define weight as the gravitational force on an object. Other people define it to be the reading on a supporting scale. Whichever definition you use, you weigh less at the top of a building compared with at the bottom – the pull of gravity is slightly less!

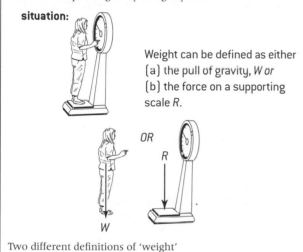

Weight can be defined as either
(a) the pull of gravity, W or
(b) the force on a supporting scale R.

Two different definitions of 'weight'

Although these two definitions are the same if the object is in equilibrium, they are very different in non-equilibrium situations. For example, if both the object and the scale were put into a lift and the lift accelerated upwards then the definitions would give different values.

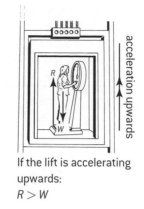

If the lift is accelerating upwards:
$R > W$

The safe thing to do is to avoid using the term weight if at all possible! Stick to the phrase 'gravitational force' or force of gravity and you cannot go wrong.

Gravitational force $= m\,g$

On the surface of the Earth, g is approximately 10 N kg^{-1}, whereas on the surface of the moon, $g \approx 1.6 \text{ N kg}^{-1}$

Solid friction

FACTORS AFFECTING FRICTION – STATIC AND DYNAMIC

Friction is the force that opposes the relative motion of two surfaces. It arises because the surfaces involved are not perfectly smooth on the microscopic scale. If the surfaces are prevented from relative motion (they are at rest) then this is an example of **static friction**. If the surfaces are moving, then it is called **dynamic friction** or **kinetic friction**.

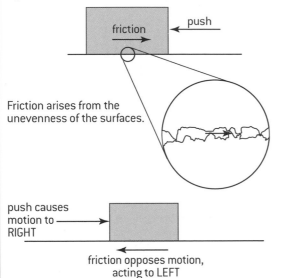

Friction arises from the unevenness of the surfaces.

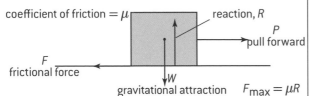

push causes motion to RIGHT

friction opposes motion, acting to LEFT

A key experimental fact is that the value of static friction changes depending on the applied force. Up to a certain maximum force, F_{max}, the resultant force is zero. For example, if we try to get a heavy block to move, any value of pushing force below F_{max} would fail to get the block to accelerate.

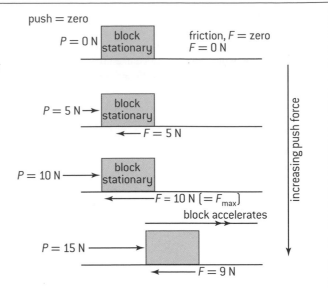

push = zero

$P = 0\,\text{N}$ block stationary — friction, F = zero $F = 0\,\text{N}$

$P = 5\,\text{N} \rightarrow$ block stationary — $\leftarrow F = 5\,\text{N}$

$P = 10\,\text{N} \rightarrow$ block stationary — $\leftarrow F = 10\,\text{N}\ (= F_{max})$

block accelerates

$P = 15\,\text{N} \rightarrow$ — $\leftarrow F = 9\,\text{N}$

increasing push force

The value of F_{max} depends upon

- the nature of the two surfaces in contact.
- the normal reaction force between the two surfaces. The maximum frictional force and the normal reaction force are proportional.

If the two surfaces are kept in contact by gravity, the value of F_{max} does NOT depend upon the area of contact

Once the object has started moving, the maximum value of friction slightly reduces. In other words,

$$F_k < F_{max}$$

For two surfaces moving over one another, the dynamic frictional force remains roughly constant even if the speed changes slightly.

COEFFICIENT OF FRICTION

Experimentally, the maximum frictional force and the normal reaction force are proportional. We use this to define the **coefficient of friction, μ**.

coefficient of friction = μ

reaction, R

P pull forward

F frictional force

W gravitational attraction $F_{max} = \mu R$

The coefficient of friction is defined from the maximum value that friction can take

$$F_{max} = \mu R$$

where R = normal reaction force

It should be noted that

- since the maximum value for dynamic friction is less than the maximum value for static friction, the values for the coefficients of friction will be different
$$\mu_d < \mu_s$$
- the coefficient of friction is a ratio between two forces – it has no units.
- if the surfaces are smooth then the maximum friction is zero i.e. $\mu = 0$.
- the coefficient of friction is less than 1 unless the surfaces are stuck together.
$$F_f \leq \mu_s R \text{ and } F_f = \mu_d R$$

EXAMPLE

If a block is placed on a slope, the angle of the slope can be increased until the block just begins to slide down the slope. This turns out to be an easy experimental way to measure the coefficient of static friction.

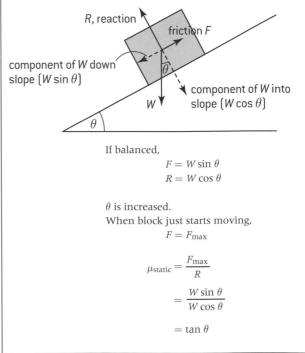

R, reaction

friction F

component of W down slope ($W \sin \theta$)

W

component of W into slope ($W \cos \theta$)

θ

If balanced,
$$F = W \sin \theta$$
$$R = W \cos \theta$$

θ is increased.
When block just starts moving,
$$F = F_{max}$$

$$\mu_{static} = \frac{F_{max}}{R}$$

$$= \frac{W \sin \theta}{W \cos \theta}$$

$$= \tan \theta$$

Work

WHEN IS WORK DONE?

Work is done when a force moves its point of application in the direction of the force. If the force moves at right angles to the direction of the force, then no work has been done.

1) before — at rest / **after** — block now moving – work has been done

force → (block at rest) / force → (block moving, v)

distance

- In 1) the force has made the object move faster.

2) before / **after** — block now higher up – work has been done

↑ force

distance ↑ force

3) before

force → (block with spring)

after — spring has been compressed – work has been done

force → (block compressing spring)

distance

4) before / **after** — book supported by shelf – no work is done

PHYSICS / PHYSICS

5) before — constant velocity v / **after** — v

friction-free surface / friction-free surface

object continues at constant velocity – no work is done

In the examples above the work done has had different results.

- In 1) the force has made the object move faster.
- In 2) the object has been lifted higher in the gravitational field.
- In 3) the spring has been compressed.
- In 4) and 5), NO work is done. Note that even though the object is moving in the last example, there is no force moving along its direction of action so no work is done.

DEFINITION OF WORK

Work is a scalar quantity. Its definition is as follows.

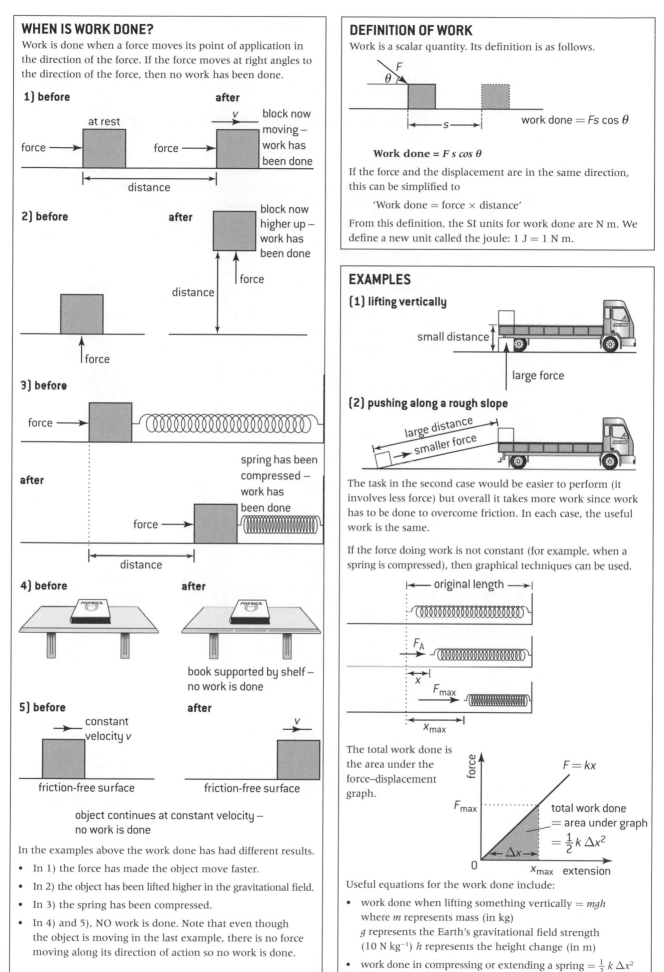

work done = $Fs \cos \theta$

Work done = $F s \cos \theta$

If the force and the displacement are in the same direction, this can be simplified to

'Work done = force × distance'

From this definition, the SI units for work done are N m. We define a new unit called the joule: $1 \text{ J} = 1 \text{ N m}$.

EXAMPLES

(1) lifting vertically

small distance

large force

(2) pushing along a rough slope

large distance

smaller force

The task in the second case would be easier to perform (it involves less force) but overall it takes more work since work has to be done to overcome friction. In each case, the useful work is the same.

If the force doing work is not constant (for example, when a spring is compressed), then graphical techniques can be used.

← original length →

F_A

x

F_{max}

x_{max}

The total work done is the area under the force–displacement graph.

F_{max} ... $F = kx$

total work done = area under graph = $\frac{1}{2} k \Delta x^2$

Δx

0 ... x_{max} extension

Useful equations for the work done include:

- work done when lifting something vertically = mgh
  where m represents mass (in kg)
  g represents the Earth's gravitational field strength (10 N kg^{-1}) h represents the height change (in m)
- work done in compressing or extending a spring = $\frac{1}{2} k \Delta x^2$

Energy and power

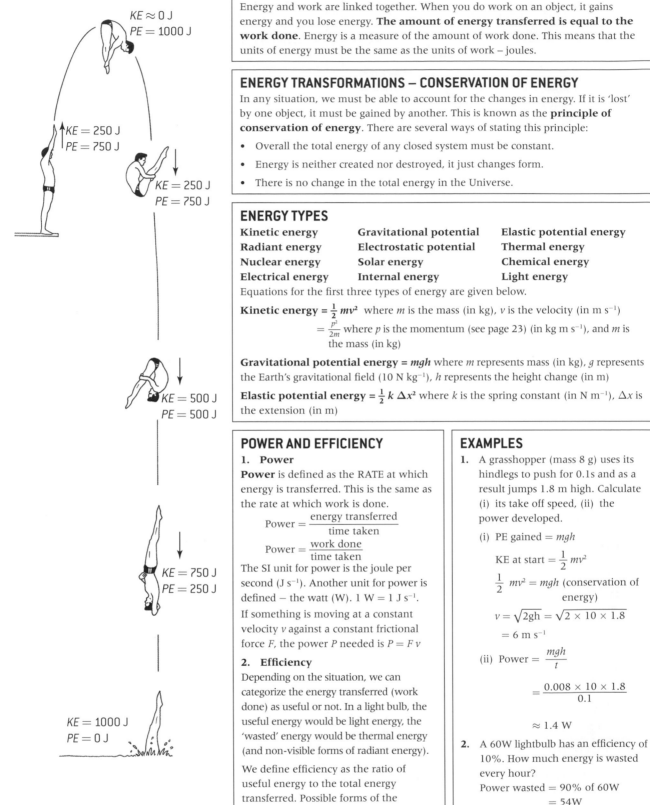

KE ≈ 0 J
PE = 1000 J

KE = 250 J
PE = 750 J

KE = 250 J
PE = 750 J

KE = 500 J
PE = 500 J

KE = 750 J
PE = 250 J

KE = 1000 J
PE = 0 J

CONCEPTS OF ENERGY AND WORK

Energy and work are linked together. When you do work on an object, it gains energy and you lose energy. **The amount of energy transferred is equal to the work done.** Energy is a measure of the amount of work done. This means that the units of energy must be the same as the units of work – joules.

ENERGY TRANSFORMATIONS – CONSERVATION OF ENERGY

In any situation, we must be able to account for the changes in energy. If it is 'lost' by one object, it must be gained by another. This is known as the **principle of conservation of energy**. There are several ways of stating this principle:

- Overall the total energy of any closed system must be constant.
- Energy is neither created nor destroyed, it just changes form.
- There is no change in the total energy in the Universe.

ENERGY TYPES

Kinetic energy	**Gravitational potential**	**Elastic potential energy**
Radiant energy	**Electrostatic potential**	**Thermal energy**
Nuclear energy	**Solar energy**	**Chemical energy**
Electrical energy	**Internal energy**	**Light energy**

Equations for the first three types of energy are given below.

Kinetic energy $= \frac{1}{2} mv^2$ where m is the mass (in kg), v is the velocity (in m s^{-1})

$\quad\quad = \frac{p^2}{2m}$ where p is the momentum (see page 23) (in kg m s^{-1}), and m is the mass (in kg)

Gravitational potential energy $= mgh$ where m represents mass (in kg), g represents the Earth's gravitational field (10 N kg^{-1}), h represents the height change (in m)

Elastic potential energy $= \frac{1}{2} k \, \Delta x^2$ where k is the spring constant (in N m^{-1}), Δx is the extension (in m)

POWER AND EFFICIENCY

1. Power

Power is defined as the RATE at which energy is transferred. This is the same as the rate at which work is done.

$$\text{Power} = \frac{\text{energy transferred}}{\text{time taken}}$$

$$\text{Power} = \frac{\text{work done}}{\text{time taken}}$$

The SI unit for power is the joule per second (J s^{-1}). Another unit for power is defined – the watt (W). 1 W = 1 J s^{-1}.

If something is moving at a constant velocity v against a constant frictional force F, the power P needed is $P = F v$

2. Efficiency

Depending on the situation, we can categorize the energy transferred (work done) as useful or not. In a light bulb, the useful energy would be light energy, the 'wasted' energy would be thermal energy (and non-visible forms of radiant energy).

We define efficiency as the ratio of useful energy to the total energy transferred. Possible forms of the equation include:

$$\text{Efficiency} = \frac{\text{useful work OUT}}{\text{total work IN}}$$

$$\text{Efficiency} = \frac{\text{useful energy OUT}}{\text{total energy IN}}$$

$$\text{Efficiency} = \frac{\text{useful power OUT}}{\text{total power IN}}$$

Since this is a ratio it does not have any units. Often it is expressed as a percentage.

EXAMPLES

1. A grasshopper (mass 8 g) uses its hindlegs to push for 0.1s and as a result jumps 1.8 m high. Calculate (i) its take off speed, (ii) the power developed.

   (i) PE gained $= mgh$

   $\quad$ KE at start $= \frac{1}{2} mv^2$

   $\quad \frac{1}{2} mv^2 = mgh$ (conservation of energy)

   $\quad v = \sqrt{2gh} = \sqrt{2 \times 10 \times 1.8}$

   $\quad\quad = 6 \text{ m s}^{-1}$

   (ii) Power $= \dfrac{mgh}{t}$

   $\quad\quad = \dfrac{0.008 \times 10 \times 1.8}{0.1}$

   $\quad\quad \approx 1.4 \text{ W}$

2. A 60W lightbulb has an efficiency of 10%. How much energy is wasted every hour?

   Power wasted = 90% of 60W
   $\quad\quad$ = 54W
   Energy wasted = 54 × 60 × 60J
   $\quad\quad$ = 190 kJ

Momentum and impulse

DEFINITIONS – LINEAR MOMENTUM AND IMPULSE

Linear momentum (always given the symbol p) is defined as the product of mass and velocity.

Momentum = mass × velocity

$$p = m v$$

The SI units for momentum must be kg m s^{-1}. Alternative units of N s can also be used (see below). Since velocity is a vector, momentum must be a vector. In any situation, particularly if it happens quickly, the change of momentum Δp is called the **impulse** ($\Delta p = F \, \Delta t$).

USE OF MOMENTUM IN NEWTON'S SECOND LAW

Newton's second law states that the resultant force is proportional to the rate of change of momentum. Mathematically we can write this as

$$F = \frac{(\text{final momentum} - \text{initial momentum})}{\text{time taken}} = \frac{\Delta p}{\Delta t}$$

Example 1

A jet of water leaves a hose and hits a wall where its velocity is brought to rest. If the hose cross-sectional area is 25 cm^2, the velocity of the water is 50 m s^{-1} and the density of the water is 1000 kg m^{-3}, what is the force acting on the wall?

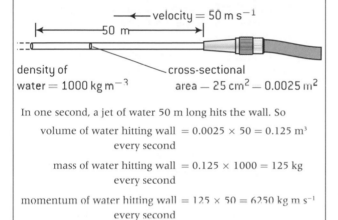

density of water = 1000 kg m^{-3}

cross-sectional area = 25 cm^2 = 0.0025 m^2

In one second, a jet of water 50 m long hits the wall. So

volume of water hitting wall = $0.0025 \times 50 = 0.125$ m^3 every second

mass of water hitting wall = $0.125 \times 1000 = 125$ kg every second

momentum of water hitting wall = $125 \times 50 = 6250$ kg m s^{-1} every second

This water is all brought to rest,

∴ change in momentum, $\Delta p = 6250$ kg m s^{-1}

$$\therefore \text{force} = \frac{\Delta p}{\Delta t} = \frac{6250}{1} = 6250 \text{ N}$$

Example 2

The graph below shows the variation with time of the force on a football of mass 500 g. Calculate the final velocity of the ball.

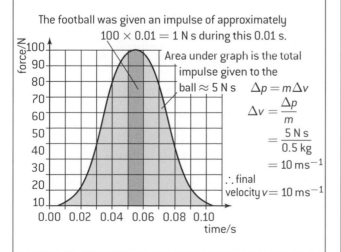

The football was given an impulse of approximately $100 \times 0.01 = 1$ N s during this 0.01 s.

Area under graph is the total impulse given to the ball ≈ 5 N s

$\Delta p = m \Delta v$

$\Delta v = \dfrac{\Delta p}{m}$

$= \dfrac{5 \text{ N s}}{0.5 \text{ kg}}$

$= 10 \text{ ms}^{-1}$

∴ final velocity $v = 10$ ms^{-1}

CONSERVATION OF MOMENTUM

The law of conservation of linear momentum states that 'the total linear momentum of a system of interacting particles remains constant **provided there is no resultant external force**'.

To see why, we start by imagining two isolated particles A and B that collide with one another.

- The force from A onto B, F_{AB} will cause B's momentum to change by a certain amount.
- If the time taken was Δt, then the momentum change (the impulse) given to B will be given by $\Delta p_B = F_{AB} \, \Delta t$
- By Newton's third law, the force from B onto A, F_{BA} will be equal and opposite to the force from A onto B, $F_{AB} = - F_{BA}$.
- Since the time of contact for A and B is the same, then the momentum change for A is equal and opposite to the momentum change for B, $\Delta p_A = - F_{AB} \, \Delta t$.
- This means that the total momentum (momentum of A plus the momentum of B) will remain the same. Total momentum is conserved.

This argument can be extended up to any number of interacting particles so long as the system of particles is still isolated. If this is the case, the momentum is still conserved.

ELASTIC AND INELASTIC COLLISIONS

The law of conservation of linear momentum is not enough to always predict the outcome after a collision (or an explosion). This depends on the nature of the colliding bodies. For example, a moving railway truck, m_A, velocity v, collides with an identical stationary truck m_B. Possible outcomes are:

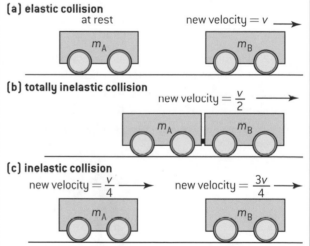

(a) elastic collision
at rest   new velocity = v →

(b) totally inelastic collision
new velocity = $\dfrac{v}{2}$ →

(c) inelastic collision
new velocity = $\dfrac{v}{4}$ →   new velocity = $\dfrac{3v}{4}$ →

In (a), the trucks would have to have elastic bumpers. If this were the case then no mechanical energy at all would be lost in the collision. A collision in which no mechanical energy is lost is called an **elastic collision**. In reality, collisions between everyday objects always lose some energy – the only real example of elastic collisions is the collision between molecules. For an elastic collision, the relative velocity of approach always equals the relative velocity of separation.

In (b), the railway trucks stick together during the collision (the relative velocity of separation is zero). This collision is what is known as a **totally inelastic collision**. A large amount of mechanical energy is lost (as heat and sound), but the total momentum is still conserved.

In energy terms, (c) is somewhere between (a) and (b). Some energy is lost, but the railway trucks do not join together. This is an example of an **inelastic collision**. Once again the total momentum is conserved.

Linear momentum is also conserved in explosions.

IB Questions – mechanics

1. Two identical objects A and B fall from rest from different heights. If B takes twice as long as A to reach the ground, what is the ratio of the heights from which A and B fell? Neglect air resistance.

   A. $1 : \sqrt{2}$ **B.** 1:2   **C.** 1:4   **D.** 1:8

2. A trolley is given an initial push along a horizontal floor to get it moving. The trolley then travels forward along the floor, gradually slowing. What is true of the horizontal force(s) on the trolley while it is slowing?

   A. There is a forward force and a backward force, but the forward force is larger.

   B. There is a forward force and a backward force, but the backward force is larger.

   C. There is only a forward force, which diminishes with time.

   D. There is only a backward force.

3. A mass is suspended by cord from a ring which is attached by two further cords to the ceiling and the wall as shown. The cord from the ceiling makes an angle of less than 45° with the vertical as shown. The tensions in the three cords are labelled R, S and T in the diagram.

   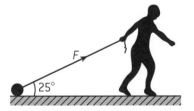

   How do the tensions R, S and T in the three cords compare in magnitude?

   A. $R > T > S$ **B.** $S > R > T$

   C. $R = S = T$ **D.** $R = S > T$

4. A 24 N force causes a 2.0 kg mass to accelerate at 8.0 m s⁻² along a horizontal surface. The coefficient of dynamic friction is:

   A. 0.0   **B.** 0.4

   C. 0.6   **D.** 0.8

5. An athlete trains by dragging a heavy load across a rough horizontal surface.

   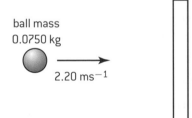

   The athlete exerts a force of magnitude F on the load at an angle of 25° to the horizontal.

   a) Once the load is moving at a steady speed, the average horizontal frictional force acting on the load is 470 N.

   Calculate the average value of F that will enable the load to move at constant speed. [2]

   b) The load is moved a horizontal distance of 2.5 km in 1.2 hours.

   Calculate

   (i)   the work done on the load by the force F. [2]

   (ii) the minimum average power required to move the load. [2]

   c) The athlete pulls the load uphill at the same speed as in part (a).

   Explain, in terms of energy changes, why the minimum average power required is greater than in (b)(ii). [2]

6. A car and a truck are both travelling at the speed limit of 60 km h⁻¹ but in opposite directions as shown. The truck has **twice** the mass of the car.

   The vehicles collide head-on and become entangled together.

   a) During the collision, how does the force exerted by the car on the truck compare with the force exerted by the truck on the car? Explain. [2]

   b) In what direction will the entangled vehicles move after collision, or will they be stationary? Support your answer, referring to a physics principle. [2]

   c) Determine the speed (in km h⁻¹) of the combined wreck immediately after the collision. [3]

   d) How does the acceleration of the car compare with the acceleration of the truck during the collision? Explain. [2]

   e) Both the car and truck drivers are wearing seat belts. Which driver is likely to be more severely jolted in the collision? Explain. [2]

   f) The total kinetic energy of the system decreases as a result of the collision. Is the principle of conservation of energy violated? Explain. [1]

7. a) A net force of magnitude F acts on a body. Define the *impulse I* of the force. [1]

   b) A ball of mass 0.0750 kg is travelling horizontally with a speed of 2.20 m s⁻¹. It strikes a vertical wall and rebounds horizontally.

   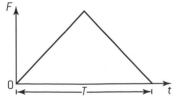

   ball mass
   0.0750 kg

   2.20 ms⁻¹

   Due to the collision with the wall, 20 % of the ball's initial kinetic energy is dissipated.

   (i)   Show that the ball rebounds from the wall with a speed of 1.97 m s⁻¹. [2]

   (ii) Show that the impulse given to the ball by the wall is 0.313 N s. [2]

   c) The ball strikes the wall at time $t = 0$ and leaves the wall at time $t = T$.

   The sketch graph shows how the force F that the wall exerts on the ball is assumed to vary with time t.

   The time T is measured electronically to equal 0.0894 s.

   Use the impulse given in (b)(ii) to estimate the average value of F. [4]

Thermal concepts

TEMPERATURE AND HEAT FLOW

Hot and cold are just labels that identify the direction in which thermal energy (sometimes known as heat) will be naturally transferred when two objects are placed in thermal contact. This leads to the concept of the 'hotness' of an object. The direction of the natural flow of thermal energy between two objects is determined by the 'hotness' of each object. Thermal energy naturally flows from hot to cold.

The temperature of an object is a measure of how hot it is. In other words, if two objects are placed in thermal contact, then the temperature difference between the two objects will determine the direction of the natural transfer of thermal energy. Thermal energy is naturally transferred 'down' the temperature difference – from high temperature to low temperature. Eventually, the two objects would be expected to reach the same temperature. When this happens, they are said to be in **thermal equilibrium**.

Heat is not a substance that flows from one object to another. What has happened is that thermal energy has been transferred. Thermal energy (heat) refers to the non-mechanical transfer of energy between a system and its surroundings.

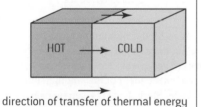

direction of transfer of thermal energy

KELVIN AND CELSIUS

Most of the time, there are only two sensible temperature scales to chose between – the Kelvin scale and the Celsius scale.

In order to use them, you do not need to understand the details of how either of these scales has been defined, but you do need to know the relation between them. Most everyday thermometers are marked with the Celsius scale and temperature is quoted in degrees Celsius (°C).

There is an easy relationship between a temperature T as measured on the Kelvin scale and the corresponding temperature t as measured on the Celsius scale. The approximate relationship is

$$T \text{ (K)} = t \text{ (°C)} + 273$$

This means that the 'size' of the units used on each scale is identical, but they have different zero points.

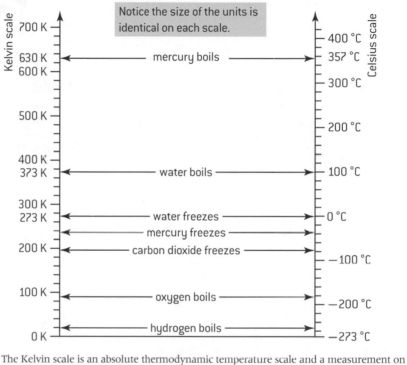

The Kelvin scale is an absolute thermodynamic temperature scale and a measurement on this scale is also called the *absolute temperature*.

Zero Kelvin is called *absolute zero* (see page 29).

EXAMPLES: GASES

For a given sample of a gas, the *pressure*, the *volume* and the *temperature* are all related to one another.

- The pressure, P, is the force per unit area from the gas acting at 90° on the container wall.

$$p = \frac{F}{A}$$

  The SI units of pressure are N m⁻² or Pa (Pascals).
  $1 \text{ Pa} = 1 \text{ N m}^{-2}$

  Gas pressure can also be measured in atmospheres
  $(1 \text{ atm} \approx 10^5 \text{ Pa})$

- The volume, V, of the gas is measured in m³ or cm³
  $(1 \text{ m}^3 = 10^6 \text{ cm}^3)$

- The temperature, t, of the gas is measured in °C or K

In order to investigate how these quantities are interrelated, we choose:

- one quantity to be the independent variable (the thing we alter and measure)

- another quantity to be the dependent variable (the second thing we measure).

- The third quantity needs to be controlled (i.e. kept constant). The specific values that will be recorded also depend on the mass of gas being investigated and the type of gas being used so these need to be controlled as well.

Heat and internal energy

MICROSCOPIC VS MACROSCOPIC

When analysing something physical, we have a choice.

- The **macroscopic** point of view considers the system as a whole and sees how it interacts with its surroundings.

- The **microscopic** point of view looks inside the system to see how its component parts interact with each other.

So far we have looked at the temperature of a system in a macroscopic way, but all objects are made up of **atoms** and **molecules**.

According to **kinetic theory** these particles are constantly in random motion – hence the name. See below for more details. Although atoms and molecules are different things (a molecule is a combination of atoms), the difference is not important at this stage. The particles can be thought of as little 'points' of mass with velocities that are continually changing.

INTERNAL ENERGY

If the temperature of an object changes then it must have gained (or lost) energy. From the microscopic point of view, the molecules must have gained (or lost) this energy.

The two possible forms are kinetic energy and potential energy.

speed in a random direction ⟶ v
∴ molecule has KE

equilibrium position

resultant force back towards equilibrium position due to neighbouring molecules
∴ molecule has PE

- The molecules have kinetic energy because they are moving. To be absolutely precise, a molecule can have either translational kinetic energy (the whole molecule is moving in a certain direction) or rotational kinetic energy (the molecule is rotating about one or more axes).

- The molecules have potential energy because of the **intermolecular** forces. If we imagine pulling two molecules further apart, this would require work against the intermolecular forces.

The total energy that the molecules possess (random kinetic plus inter molecule potential) is called the **internal energy** of a substance. Whenever we heat a substance, we increase its internal energy.

Temperature is a measure of the average kinetic energy of the molecules in a substance.

If two substances have the same temperature, then their molecules have the same average kinetic energy.

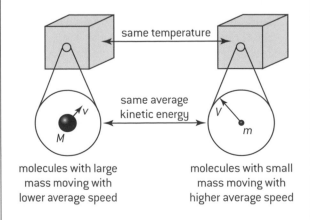

same temperature

same average kinetic energy

molecules with large mass moving with lower average speed

molecules with small mass moving with higher average speed

KINETIC THEORY

Molecules are arranged in different ways depending on the **phase** of the substance (i.e. solid, liquid or gas).

SOLIDS

Macroscopically, solids have a fixed volume and a fixed shape. This is because the molecules are held in position by bonds. However the bonds are not absolutely rigid. The molecules vibrate around a mean (average) position. The higher the temperature, the greater the vibrations.

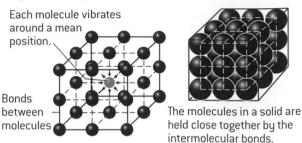

Each molecule vibrates around a mean position.

Bonds between molecules

The molecules in a solid are held close together by the intermolecular bonds.

LIQUIDS

A liquid also has a fixed volume but its shape can change. The molecules are also vibrating, but they are not completely fixed in position. There are still strong forces between the molecules. This keeps the molecules close to one another, but they are free to move around each other.

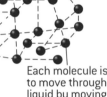

Bonds between neighbouring molecules; these can be made and broken, allowing a molecule to move.

Each molecule is free to move throughout the liquid by moving around its neighbours.

GASES

A gas will always expand to fill the container in which it is put. The molecules are not fixed in position, and any forces between the molecules are very weak. This means that the molecules are essentially independent of one another, but they do occasionally collide. More detail is given on page 31.

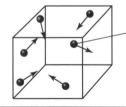

Molecules in random motion; no fixed bonds between molecules so they are free to move

HEAT AND WORK

Many people have confused ideas about heat and work. In answers to examination questions it is very common to read, for example, that 'heat rises' – when what is meant is that the transfer of thermal energy is upwards.

- When a force moves through a distance, we say that work is done. Work is the energy that has been transmitted from one system to another from the macroscopic point of view.

- When work is done on a microscopic level (i.e. on individual molecules), we say that heating has taken place. Heat is the energy that has been transmitted. It can either increase the kinetic energy of the molecules or their potential energy or, of course, both.

In both cases energy is being transferred.

Specific heat capacity

DEFINITIONS AND MICROSCOPIC EXPLANATION

In theory, if an object could be heated up with no energy loss, then the increase in temperature ΔT depends on three things:

- the energy given to the object Q,
- the mass, m, and
- the substance from which the object is made.

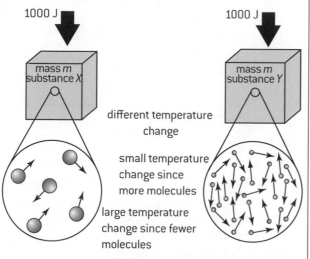

Two different blocks with the same mass and same energy input will have a different temperature change.

We define the **thermal capacity** C of an object as the energy required to raise its temperature by 1 K. Different objects (even different samples of the same substance) will have different values of heat capacity. **Specific heat capacity** is the energy required to raise a unit mass of a substance by 1 K. 'Specific' here just means 'per unit mass'.

In symbols,

Thermal capacity $\quad C = \dfrac{Q}{\Delta T}$ (J K^{-1} or J °C^{-1})

Specific heat capacity $\quad c = \dfrac{Q}{(m\,\Delta T)}$ (J kg^{-1} K^{-1} or J kg^{-1} °C^{-1})

$$Q = mc\Delta T$$

Note

- A particular gas can have many different values of specific heat capacity – it depends on the conditions used – see page 161.

- These equations refer to the **temperature difference** resulting from the addition of a certain amount of energy. In other words, it generally takes the same amount of energy to raise the temperature of an object from 25 °C to 35 °C as it does for the same object to go from 402 °C to 412 °C. This is only true so long as energy is not lost from the object.

- If an object is raised above room temperature, it starts to lose energy. The hotter it becomes, the greater the rate at which it loses energy.

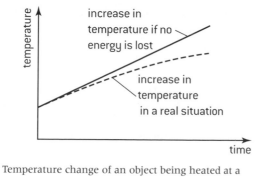

Temperature change of an object being heated at a constant rate

METHODS OF MEASURING HEAT CAPACITIES AND SPECIFIC HEAT CAPACITIES

The are two basic ways to measure heat capacity.

1. Electrical method

The experiment would be set up as below:

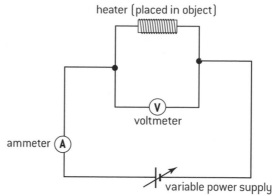

- the specific heat capacity $c = \dfrac{I\,t\,V}{m(T_2 - T_1)}$.

Sources of experimental error

- loss of thermal energy from the apparatus.
- the container for the substance and the heater will also be warmed up.
- it will take some time for the energy to be shared uniformly through the substance.

2. Method of mixtures

The known specific heat capacity of one substance can be used to find the specific heat capacity of another substance.

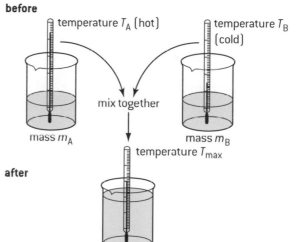

Procedure:

- measure the masses of the liquids m_A and m_B.
- measure the two starting temperatures T_A and T_B.
- mix the two liquids together.
- record the maximum temperature of the mixture T_{max}.

If no energy is lost from the system then,

energy lost by hot substance cooling down = energy gained by cold substance heating up

$$m_A\, c_A\, (T_A - T_{max}) = m_B\, c_B\, (T_{max} - T_B)$$

Again, the main source of experimental error is the loss of thermal energy from the apparatus; particularly while the liquids are being transferred. The changes of temperature of the container also need to be taken into consideration for a more accurate result.

Phases (states) of matter and latent heat

DEFINITIONS AND MICROSCOPIC VIEW

When a substance changes phase, the temperature remains constant even though thermal energy is still being transferred.

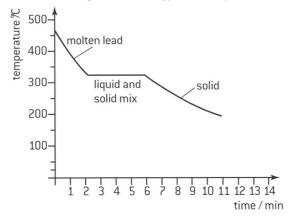

Cooling curve for molten lead (idealized)

The amount of energy associated with the phase change is called the **latent heat**. The technical term for the change of phase from solid to liquid is **fusion** and the term for the change from liquid to gas is **vaporization**.

The energy given to the molecules does not increase their kinetic energy so it must be increasing their potential energy. Intermolecular bonds are being broken and this takes energy. When the substance freezes bonds are created and this process releases energy.

It is a very common mistake to think that the molecules must speed up during a phase change. The molecules in water vapour at 100 °C must be moving with the same average speed as the molecules in liquid water at 100 °C.

The **specific latent heat** of a substance is defined as the amount of energy per unit mass absorbed or released during a change of phase.

In symbols,

$$\text{Specific latent heat } L = \frac{Q}{M} \quad (\text{J kg}^{-1}) \quad Q = ML$$

In the idealized situation of no energy loss, a constant rate of energy transfer into a solid substance would result in a constant rate of increase in temperature until the melting point is reached:

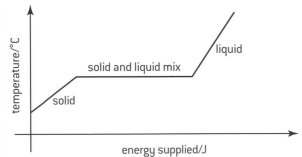

Phase-change graph with temperature vs energy

In the example above, the specific heat capacity of the liquid is less than the specific heat capacity of the solid as the gradient of the line that corresponds to the liquid phase is greater than the gradient of the line that corresponds to the solid phase. A given amount of energy will cause a greater increase in temperature for the liquid when compared with the solid.

METHODS OF MEASURING

The two possible methods for measuring latent heats shown below are very similar in principle to the methods for measuring specific heat capacities (see previous page).

1. **A method for measuring the specific latent heat of vaporization of water**

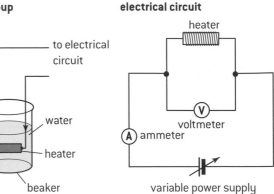

The amount of thermal energy provided to water at its boiling point is calculated using electrical energy $= I\,t\,V$. The mass vaporized needs to be recorded.

- The specific latent heat $L = \dfrac{I\,t\,V}{(m_1 - m_2)}$.

Sources of experimental error

- Loss of thermal energy from the apparatus.
- Some water vapour will be lost before and after timing.

2. **A method for measuring the specific latent heat of fusion of water**

Providing we know the specific heat capacity of water, we can calculate the specific latent heat of fusion for water. In the example below, ice (at 0 °C) is added to warm water and the temperature of the resulting mix is measured.

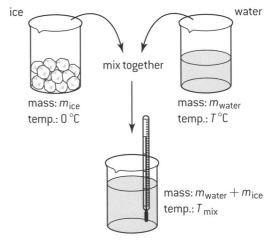

If no energy is lost from the system then,

energy lost by water cooling down = energy gained by ice

$$m_{\text{water}}\,c_{\text{water}}\,(T_{\text{water}} - T_{\text{mix}}) = m_{\text{ice}}\,L_{\text{fusion}} + m_{\text{ice}}\,c_{\text{water}}\,T_{\text{mix}}$$

Sources of experimental error

- Loss (or gain) of thermal energy from the apparatus.
- If the ice had not started at exactly zero, then there would be an additional term in the equation in order to account for the energy needed to warm the ice up to 0 °C.
- Water clinging to the ice before the transfer.

The gas laws 1

GAS LAWS

For the experimental methods shown below, the graphs below outline what might be observed.

(a) constant volume

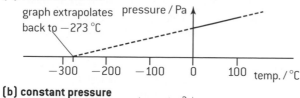

graph extrapolates back to −273 °C

(b) constant pressure

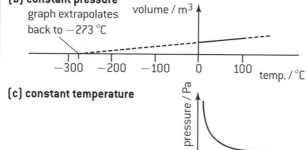

graph extrapolates back to −273 °C

(c) constant temperature

Points to note:

- Although pressure and volume both vary linearly with Celsius temperature, neither pressure nor volume is proportional to Celsius temperature.

- A different sample of gas would produce a different straight-line variation for pressure (or volume) against temperature but both graphs would extrapolate back to the same low temperature, −273 °C. This temperature is known as **absolute zero**.

- As pressure increases, the volume decreases. In fact they are inversely proportional.

The trends can be seen more clearly if this information is presented in a slightly different way.

(1) constant volume

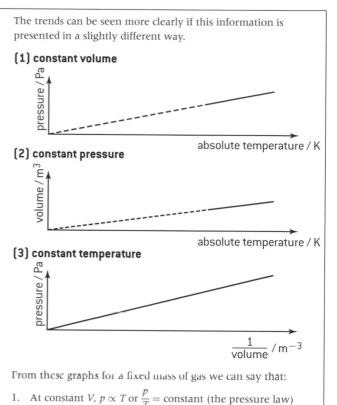

(2) constant pressure

(3) constant temperature

From these graphs for a fixed mass of gas we can say that:

1. At constant V, $p \propto T$ or $\frac{p}{T} = $ constant (the pressure law)

2. At constant p, $V \propto T$ or $\frac{V}{T} = $ constant (Charles's law)

3. At constant T, $p \propto \frac{1}{V}$ or $pV = $ constant (Boyle's law)

These relationships are known as the **ideal gas laws**. The temperature is always expressed in Kelvin (see page 25). These laws do not always apply to experiments done with real gases. A real gas is said to 'deviate' from ideal behaviour under certain conditions (e.g. high pressure).

EXPERIMENTAL INVESTIGATIONS

1. Temperature t as the independent variable; P as the dependent variable; V as the control.

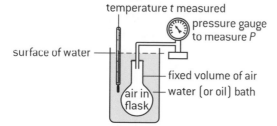

temperature t measured

pressure gauge to measure P

surface of water

fixed volume of air

air in flask

water (or oil) bath

- Fixed volume of gas is trapped in the flask. Pressure is measured by a pressure gauge.

- Temperature of gas altered by temperature of bath – time is needed to ensure bath and gas at same temperature.

2. Temperature t as the independent variable; V as the dependent variable; P as the control.

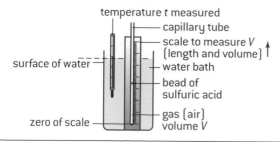

temperature t measured

capillary tube

scale to measure V (length and volume)

surface of water

water bath

bead of sulfuric acid

gas (air) volume V

zero of scale

- Volume of gas is trapped in capillary tube by bead of concentrated sulfuric acid.

- Concentrated sulfuric acid is used to ensure gas remains dry.

- Heating gas causes it to expand moving bead.

- Pressure remains equal to atmospheric.

- Temperature of gas altered by temperature of bath; time is needed to ensure bath and gas at same temperature.

3. P as the independent variable; V as the dependent variable; t as the control.

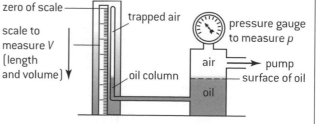

zero of scale

scale to measure V (length and volume)

trapped air

pressure gauge to measure p

air

pump

oil column

surface of oil

oil

- Volume of gas measured against calibrated scale.

- Increase of pressure forces oil column to compress gas.

- Temperature of gas will be altered when volume is changed; time is needed to ensure gas is always at room temperature.

The gas laws 2

EQUATION OF STATE

The three ideal gas laws can be combined together to produce one mathematical relationship.

$$\frac{pV}{T} = \text{constant}$$

This constant will depend on the mass and type of gas.

If we compare the value of this constant for different masses of different gases, it turns out to depend on the number of molecules that are in the gas – not their type. In this case we use the definition of the mole to state that for n moles of ideal gas

$$\frac{pV}{nT} = \text{a universal constant}.$$

The universal constant is called the **molar gas constant** R.

The SI unit for R is J mol^{-1} K^{-1}

$$R = 8.314 \text{ J mol}^{-1} \text{ K}^{-1}$$

Summary: $\frac{pV}{nT} = R$ Or $pV = nRT$

EXAMPLE

a) What volume will be occupied by 8 g of helium (mass number 4) at room temperature (20 °C) and atmospheric pressure (1.0×10^5 Pa)

$$n = \frac{8}{4} = 2 \text{ moles}$$

$$T = 20 + 273 = 293 \text{ K}$$

$$V = \frac{nRT}{p} = \frac{2 \times 8.314 \times 293}{1.0 \times 10^5} = 0.049 \text{ m}^3$$

b) How many atoms are there in 8 g of helium (mass number 4)?

$$n = \frac{8}{4} = 2 \text{ moles}$$

$$\begin{aligned}
\text{number of atoms} &= 2 \times 6.02 \times 10^{23} \\
&= 1.2 \times 10^{24}
\end{aligned}$$

DEFINITIONS

The concepts of the **mole**, **molar mass** and the **Avogadro constant** are all introduced so as to be able to relate the mass of a gas (an easily measurable quantity) to the number of molecules that are present in the gas.

Ideal gas	An ideal gas is one that follows the gas laws for all values of of P, V and T (see page 29).
Mole	The mole is the basic SI unit for 'amount of substance'. One mole of any substance is equal to the amount of that substance that contains the same number of particles as 0.012 kg of carbon–12 (^{12}C). When writing the unit it is (slightly) shortened to the mol.
Avogadro constant, N_A	This is the number of atoms in 0.012 kg of carbon–12 (^{12}C). It is 6.02×10^{23}.
Molar mass	The mass of one mole of a substance is called the molar mass. A simple rule applies. If an element has a certain mass number, A, then the molar mass will be A grams.

$$n = \frac{N}{N_A}$$

$$\text{number of moles} = \frac{\text{number of atoms}}{\text{Avogadro constant}}$$

IDEAL GASES AND REAL GASES

An ideal gas is a one that follows the gas laws for all values of p, V and T and thus ideal gases cannot be liquefied. The microscopic description of an ideal gas is given on page 31. Real gases, however, can approximate to ideal behaviour providing that the intermolecular forces are small enough to be ignored. For this to apply, the pressure/density of the gas must be low and the temperature must be moderate.

LINK BETWEEN THE MACROSCOPIC AND MICROSCOPIC

The equation of state for an ideal gas, $pV = nRT$, links the three macroscopic properties of a gas (p, V and T). Kinetic theory (page 26) describes a gas as being composed of molecules in random motion and for this theory to be valid, each of these macroscopic properties must be linked to the microscopic behaviour of molecules.

A detailed analysis of how a large number of randomly moving molecules interact beautifully predicts another formula that allows the links between the macroscopic and the microscopic to be identified. The derivation of the formula only uses Newton's laws and a handful of assumptions. These assumptions describe from the microscopic perspective what we mean by an ideal gas.

The detail of this derivation is not required by the IB syllabus but the assumptions and the approach are outlined on the following page. The result of this derivation is that the pressure and volume of the idealized gas are related to just two quantities:

$$pV = \frac{2}{3} N \overline{E}_K$$

- The number of molecules present, N
- The average random kinetic energy per molecule, $\overline{E}_K$.

Equating the right-hand side of this formula with the right-hand side of the macroscopic equation of state for an ideal gas shows that:

$$nRT = \frac{2}{3} N \overline{E}_K$$

But $n = \frac{N}{N_A}$, so

$$\frac{N}{N_A} RT = \frac{2}{3} N \overline{E}_K$$

$$\therefore \quad \overline{E}_K = \frac{3}{2} \frac{R}{N_A} T$$

R (the molar gas constant) and N_A (Avogadro constant) are fixed numbers so this equation shows that the absolute temperature is proportional to the average KE per molecule

$$T \propto \overline{E}_K$$

The ratio $\frac{R}{N_A}$ is called the Boltzmann's constant k_B. $k_B = \frac{R}{N_A}$

$$\overline{E}_K = \frac{3}{2} k_B T = \frac{3}{2} \frac{R}{N_A} T$$

Molecular model of an ideal gas

KINETIC MODEL OF AN IDEAL GAS

Assumptions:

- Newton's laws apply to molecular behaviour
- there are no intermolecular forces except during a collision
- the molecules are treated as points
- the molecules are in random motion
- the collisions between the molecules are elastic (no energy is lost)
- there is no time spent in these collisions.

The pressure of a gas is explained as follows:

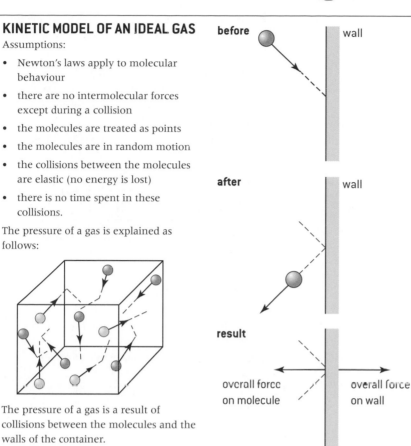

The pressure of a gas is a result of collisions between the molecules and the walls of the container.

A single molecule hitting the walls of the container.

- When a molecule bounces off the walls of a container its momentum changes (due to the change in direction – momentum is a vector).
- There must have been a force on the molecule from the wall (Newton II).
- There must have been an equal and opposite force on the wall from the molecule (Newton III).
- Each time there is a collision between a molecule and the wall, a force is exerted on the wall.
- The average of all the microscopic forces on the wall over a period of time means that there is effectively a constant force on the wall from the gas.
- This force per unit area of the wall is what we call pressure.

$$P = \frac{F}{A}$$

Since the temperature of a gas is a measure of the average kinetic energy of the molecules, as we lower the temperature of a gas the molecules will move slower. At absolute zero, we imagine the molecules to have zero kinetic energy. We cannot go any lower because we cannot reduce their kinetic energy any further!

PRESSURE LAW

Macroscopically, at a constant volume the pressure of a gas is proportional to its temperature in kelvin (see page 29). Microscopically this can be analysed as follows

- If the temperature of a gas goes up, the molecules have more average kinetic energy – they are moving faster on average.
- Fast moving molecules will have a greater change of momentum when they hit the walls of the container.
- Thus the microscopic force from each molecule will be greater.
- The molecules are moving faster so they hit the walls more often.
- For both these reasons, the total force on the wall goes up.
- Thus the pressure goes up.

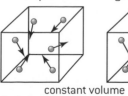

Microscopic justification of the pressure law

CHARLES'S LAW

Macroscopically, at a constant pressure, the volume of a gas is proportional to its temperature in kelvin (see page 29). Microscopically this can be analysed as follows

- A higher temperature means faster moving molecules (see left).
- Faster moving molecules hit the walls with a greater microscopic force (see left).
- If the volume of the gas increases, then the rate at which these collisions take place on a unit area of the wall must go down.
- The average force on a unit area of the wall can thus be the same.
- Thus the pressure remains the same.

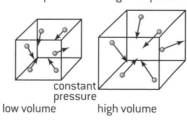

Microscopic justification of Charles's law

BOYLE'S LAW

Macroscopically, at a constant temperature, the pressure of a gas is inversely proportional to its volume (see page 29). Microscopically this can be seen to be correct.

- The constant temperature of gas means that the molecules have a constant average speed.
- The microscopic force that each molecule exerts on the wall will remain constant.
- Increasing the volume of the container decreases the rate with which the molecules hit the wall – average total force decreases.
- If the average total force decreases the pressure decreases.

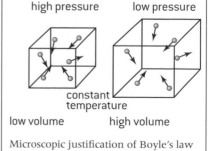

Microscopic justification of Boyle's law

IB Questions – thermal physics

The following information relates to questions 1 and 2 below.

A substance is heated at a constant rate of energy transfer. A graph of its temperature against time is shown below.

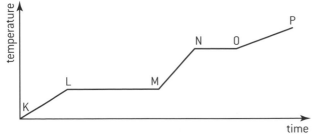

1. Which regions of the graph correspond to the substance existing in a mixture of two phases?

    A. KL, MN and OP

    B. LM and NO

    C. All regions

    D. No regions

2. In which region of the graph is the specific heat capacity of the substance greatest?

    A. KL

    B. LM

    C. MN

    D. OP

3. When the volume of a gas is isothermally compressed to a smaller volume, the pressure exerted by the gas on the container walls increases. The best microscopic explanation for this pressure increase is that at the smaller volume

    A. the individual gas molecules are compressed

    B. the gas molecules repel each other more strongly

    C. the average velocity of gas molecules hitting the wall is greater

    D. the frequency of collisions with gas molecules with the walls is greater

4. A lead bullet is fired into an iron plate, where it deforms and stops. As a result, the temperature of the lead increases by an amount ΔT. For an identical bullet hitting the plate with twice the speed, what is the best estimate of the temperature increase?

    A. ΔT

    B. $2 \Delta T$

    C. $2 \Delta T$

    D. $4 \Delta T$

5. In winter, in some countries, the water in a swimming pool needs to be heated.

    a) Estimate the cost of heating the water in a typical swimming pool from 5 °C to a suitable temperature for swimming. You may choose to consider any reasonable size of pool.

    Clearly show any estimated values. The following information will be useful:

Specific heat capacity of water	4186 J kg^{-1} K^{-1}
Density of water	1000 kg m^{-3}
Cost per kW h of electrical energy	$0.10

    (i) Estimated values [4]

    (ii) Calculations [7]

    b) An electrical heater for swimming pools has the following information written on its side:

    50 Hz   2.3 kW

    (i) Estimate how many days it would take this heater to heat the water in the swimming pool. [4]

    (ii) Suggest two reasons why this can only be an approximation. [2]

6. a) A cylinder fitted with a piston contains 0.23 mol of helium gas.

    The following data are available for the helium with the piston in the position shown.

    Volume $= 5.2 \times 10^{-3}$ m^3

    Pressure $= 1.0 \times 10^5$ Pa

    Temperature $= 290$ K

    (i) Use the data to calculate a value for the universal gas constant. (2)

    (ii) State the assumption made in the calculation in (a)(i). (1)

7. This question is about determining the specific latent heat of fusion of ice.

    A student determines the specific latent heat of fusion of ice at home. She takes some ice from the freezer, measures its mass and mixes it with a known mass of water in an insulating jug. She stirs until all the ice has melted and measures the final temperature of the mixture. She also measured the temperature in the freezer and the initial temperature of the water.

    She records her measurements as follows:

Mass of ice used	m_i	0.12 kg
Initial temperature of ice	T_i	−12 °C
Initial mass of water	m_w	0.40 kg
Initial temperature of water	T_w	22 °C
Final temperature of mixture	T_f	15 °C

    The specific heat capacities of water and ice are $c_w = 4.2$ kJ kg^{-1} °C^{-1} and $c_i = 2.1$ kJ kg^{-1} °C^{-1}

    a) Set up the appropriate equation, representing energy transfers during the process of coming to thermal equilibrium, that will enable her to solve for the specific latent heat L_i of ice. Insert values into the equation from the data above, **but do not solve the equation**. [5]

    b) Explain the physical meaning of each *energy transfer term* in your equation (but not each symbol). [4]

    c) State an assumption you have made about the experiment, in setting up your equation in (a). [1]

    d) Why should she take the temperature of the mixture *immediately* after all the ice has melted? [1]

    e) Explain from the microscopic point of view, in terms of molecular behaviour, why the temperature of the ice does not increase while it is melting. [4]

Oscillations

DEFINITIONS

Many systems involve vibrations or oscillations; an object continually moves to-and-fro about a fixed average point (the **mean position**) retracing the same path through space taking a fixed time between repeats. Oscillations involve the interchange of energy between kinetic and potential.

	Kinetic energy	Potential energy store
Mass moving between two horizontal springs	Moving mass	Elastic potential energy in the springs
Mass moving on a vertical spring	Moving mass	Elastic potential energy in the springs and gravitational potential energy
Simple pendulum	Moving pendulum bob	Gravitational potential energy of bob
Buoy bouncing up and down in water	Moving buoy	Gravitational PE of buoy and water
An oscillating ruler as a result of one end being displaced while the other is fixed	Moving sections of the ruler	Elastic PE of the bent ruler

	Definition
Displacement, x	The instantaneous distance (SI measurement: m) of the moving object from its mean position (in a specified direction)
Amplitude, A	The maximum displacement (SI measurement: m) from the mean position
Frequency, f	The number of oscillations completed per unit time. The SI measurement is the number of cycles per second or Hertz (Hz).
Period, T	The time taken (SI measurement: s) for one complete oscillation. $T = \frac{1}{f}$
Phase difference, ø	This is a measure of how 'in step' different particles are. If moving together they are **in phase**. ø is measured in either degrees (°) or radians (rad). 360° or 2π rad is one complete cycle so 180° or π rad is completely **out of phase** by half a cycle. A phase difference of 90° or $\pi/2$ rad is a quarter of a cycle.

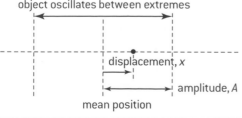

object oscillates between extremes

displacement, x

amplitude, A

mean position

SIMPLE HARMONIC MOTION (SHM)

Simple harmonic motion is defined as the motion that takes place when the acceleration, a, of an object is always directed towards, and is proportional to, its displacement from a fixed point. This acceleration is caused by a **restoring force** that must always be pointed towards the mean position and also proportional to the displacement from the mean position.

$$F \propto -x \text{ or } F = - \text{(constant)} \times x$$

Since $F = ma$

$$a \propto -x \text{ or } a = - \text{(constant)} \times x$$

The negative sign signifies that the acceleration is always pointing back towards the mean position.

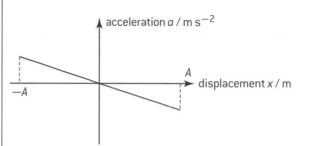

Points to note about SHM:

- The time period T does not depend on the amplitude A. It is **isochronous**.

- Not all oscillations are SHM, but there are many everyday examples of natural SHM oscillations.

EXAMPLE OF SHM: MASS BETWEEN TWO SPRINGS

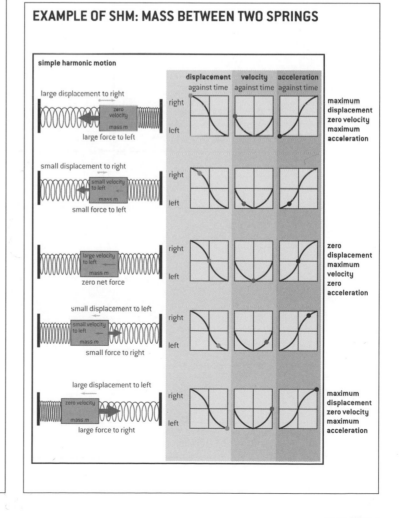

Graphs of simple harmonic motion

ACCELERATION, VELOCITY AND DISPLACEMENT DURING SHM

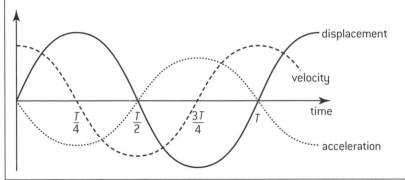

- acceleration leads velocity by 90°
- velocity leads displacement by 90°
- acceleration and displacement are 180° out of phase
- displacement lags velocity by 90°
- velocity lags acceleration by 90°

ENERGY CHANGES DURING SIMPLE HARMONIC MOTION

During SHM, energy is interchanged between KE and PE. Providing there are no resistive forces which dissipate this energy, the total energy must remain constant. The oscillation is said to be **undamped**.

Energy in SHM is proportional to:

- the mass m
- the (amplitude)^2
- the (frequency)^2

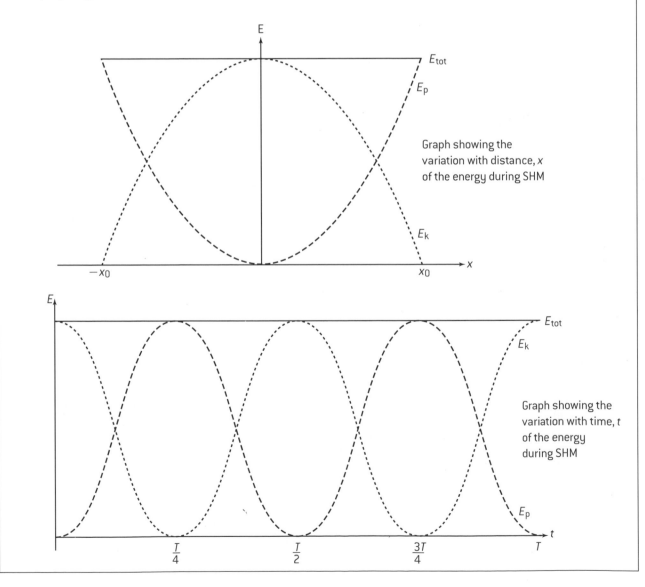

Graph showing the variation with distance, x of the energy during SHM

Graph showing the variation with time, t of the energy during SHM

Travelling waves

INTRODUCTION – RAYS AND WAVE FRONTS

Light, sound and ripples on the surface of a pond are all examples of wave motion.

- They all transfer energy from one place to another.
- They do so without a net motion of the medium through which they travel.
- They all involve oscillations (vibrations) of one sort or another. The oscillations are SHM.

A **continuous wave** involves a succession of individual oscillations. A **wave pulse** involves just one oscillation. Two important categories of wave are **transverse** and **longitudinal** (see below). The table gives some examples.

The following pages analyse some of the properties that are common to all waves.

	Example of energy transfer
Water ripples (Transverse)	A floating object gains an 'up and down' motion.
Sound waves (Longitudinal)	The sound received at an ear makes the eardrum vibrate.
Light wave (Transverse)	The back of the eye (the retina) is stimulated when light is received.
Earthquake waves (Both T and L)	Buildings collapse during an earthquake.
Waves along a stretched rope (Transverse)	A 'sideways pulse' will travel down a rope that is held taut between two people.
Compression waves down a spring (Longitudinal)	A compression pulse will travel down a spring that is is held taut between two people.

LONGITUDINAL WAVES

Sound is a longitudinal wave. This is because the oscillations are **parallel** to the direction of energy transfer.

situation

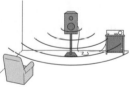

view from above

[1] wave front diagram

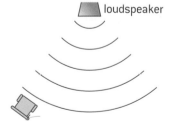

loudspeaker

[2] ray diagram

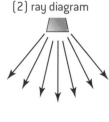

cross-section through wave at one instant of time

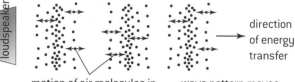

direction of energy transfer

motion of air molecules in same direction as energy transfer

wave pattern moves out from loudspeaker

TRANSVERSE WAVES

Suppose a stone is thrown into a pond. Waves spread out as shown below.

situation

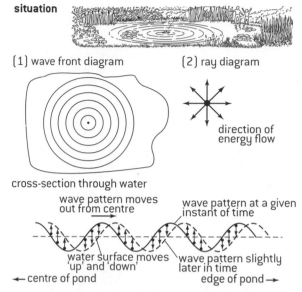

[1] wave front diagram

[2] ray diagram

direction of energy flow

cross-section through water

wave pattern moves out from centre

wave pattern at a given instant of time

water surface moves 'up' and 'down'

wave pattern slightly later in time

← centre of pond

edge of pond →

The top of the wave is known as the **crest**, whereas the bottom of the wave is known as the **trough**.

Note that there are several aspects to this wave that can be studied. These aspects are important to all waves.

- The movement of the wave pattern. The **wave fronts** highlight the parts of the wave that are moving together.
- The direction of energy transfer. The **rays** highlight the direction of energy transfer.
- The oscillations of the medium.

It should be noted that the rays are at right angles to the wave fronts in the above diagrams. This is always the case.

This wave is an example of a transverse wave because the oscillations are **at right angles** to the direction of energy transfer.

Transverse mechanical waves cannot be propagated through fluids (liquids or gases).

A point on the wave where everything is 'bunched together' (high pressure) is known as a **compression**. A point where everything is 'far apart' (low pressure) is known as a **rarefaction**.

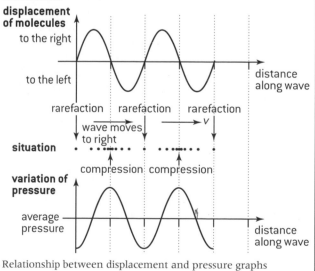

displacement of molecules

to the right

to the left

distance along wave

rarefaction   rarefaction   rarefaction

wave moves to right

v

situation

compression   compression

variation of pressure

average pressure

distance along wave

Relationship between displacement and pressure graphs

Wave characteristics

DEFINITIONS

There are some useful terms that need to be defined in order to analyse wave motion in more detail. The table below attempts to explain these terms and they are also shown on the graphs.

Because the graphs seem to be identical, you need to look at the axes of the graphs carefully.

- The displacement–time graph on the left represents the oscillations for one point on the wave. All the other points on the wave will oscillate in a similar manner, but they will not start their oscillations at exactly the same time.

- The displacement–position graph on the right represents a 'snapshot' of all the points along the wave at one instant of time. At a later time, the wave will have moved on but it will retain the same shape.

- The graphs can be used to represent longitudinal AND transverse waves because the y-axis records only the value of the displacement. It does NOT specify the direction of this displacement. So, if this displacement were parallel to the direction of the wave energy, the wave would be a longitudinal wave. If this displacement were at right angles to the direction of the wave energy, the wave would be a transverse wave.

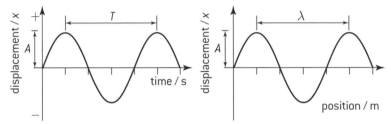

Term	Symbol	Definition
Displacement	x	This measures the change that has taken place as a result of a wave passing a particular point. Zero displacement refers to the mean (or average) position. For mechanical waves the displacement is the distance (in metres) that the particle moves from its undisturbed position.
Amplitude	A	This is the maximum displacement from the mean position. If the wave does not lose any of its energy its amplitude is constant.
Period	T	This is the time taken (in seconds) for one complete oscillation. It is the time taken for one complete wave to pass any given point.
Frequency	f	This is the number of oscillations that take place in one second. The unit used is the hertz (Hz). A frequency of 50 Hz means that 50 cycles are completed every second.
Wavelength	λ	This is the shortest distance (in metres) along the wave between two points that are **in phase** with one another. 'In phase' means that the two points are moving exactly in step with one another. For example, the distance from one crest to the next crest on a water ripple or the distance from one compression to the next one on a sound wave.
Wave speed	c	This is the speed (in m s⁻¹) at which the wave fronts pass a stationary observer.
Intensity	I	The intensity of a wave is the power per unit area that is received by the observer. The unit is W m⁻². The intensity of a wave is proportional to the square of its amplitude: $I \propto A^2$.

The period and the frequency of any wave are inversely related. For example, if the frequency of a wave is 100 Hz, then its period must be exactly $\frac{1}{100}$ of a second.

In symbols,

$$T = \frac{1}{f}$$

WAVE EQUATIONS

There is a very simple relationship that links wave speed, wavelength and frequency. It applies to all waves.

The time taken for one complete oscillation is the period of the wave, T.

In this time, the wave pattern will have moved on by one wavelength, λ.

This means that the speed of the wave must be given by

$$c = \frac{\text{distance}}{\text{time}} = \frac{\lambda}{T}$$

Since $\frac{1}{T} = f$

$$c = f\lambda$$

In words,

velocity = frequency × wavelength

EXAMPLE

A stone is thrown onto a still water surface and creates a wave. A small floating cork 1.0 m away from the impact point has the following displacement–time graph (time is measured from the instant the stone hits the water):

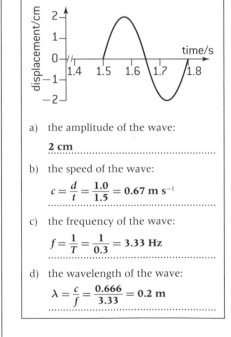

a) the amplitude of the wave:

    2 cm

b) the speed of the wave:

$$c = \frac{d}{t} = \frac{1.0}{1.5} = 0.67 \text{ m s}^{-1}$$

c) the frequency of the wave:

$$f = \frac{1}{T} = \frac{1}{0.3} = 3.33 \text{ Hz}$$

d) the wavelength of the wave:

$$\lambda = \frac{c}{f} = \frac{0.666}{3.33} = 0.2 \text{ m}$$

Electromagnetic spectrum

ELECTROMAGNETIC WAVES

Visible light is one part of a much larger spectrum of similar waves that are all electromagnetic.

Charges that are accelerating generate electromagnetic fields. If an electric charge oscillates, it will produce a varying electric and magnetic field at right angles to one another.

These oscillating fields propagate (move) as a transverse wave through space. Since no physical matter is involved in this propagation, they can travel through a vacuum. The speed of this wave can be calculated from basic electric and magnetic constants and it is the same for all electromagnetic waves, 3.0×10^8 m s^{-1}.

Although all electromagnetic waves are identical in their nature, they have very different properties. This is because of the huge range of frequencies (and thus energies) involved in the electromagnetic spectrum.

See page 132 (option A) for more details.

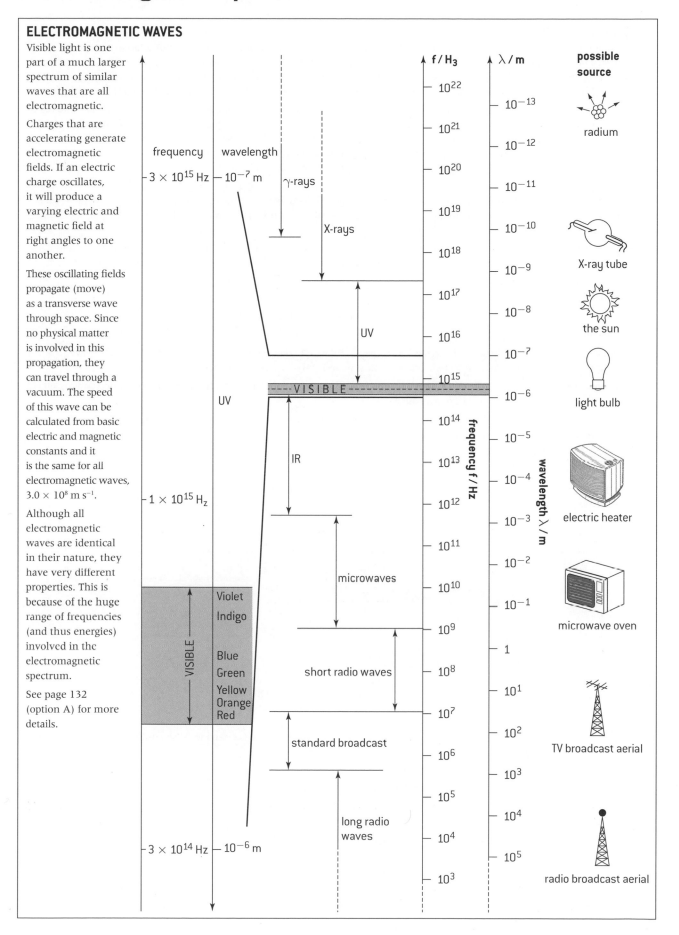

Investigating speed of sound experimentally

1. DIRECT METHODS

The most direct method to measure the speed of sound is to record the time taken t for sound to cover a known distance d: speed $c = \frac{d}{t}$. In air at normal pressures and temperatures, sound travels at approximately 330 m s^{-1}. Given the much larger speed of light (3×10^8 m s^{-1}), a possible experiment would be to use a stop watch to time the difference between seeing an event (e.g. the firing of a starting pistol for a race or seeing two wooden planks being hit together) and hearing the same event some distance away (100 m or more).

Echoes can be used to put the source and observer of the sound in the same place. Standing a distance d in front of a tall wall (e.g. the side of a building that is not surrounded by other buildings) can allow the echo from a pulse of sound (e.g. a single clap of the hands) to be heard. With practice, it is possible for an experimenter to adjust the frequency of clapping to synchronize the sound of the claps with their echoes. When this is achieved, the frequency of clapping f can be recorded (counting the number of claps in a given time) and the time period T between claps is just $T = \frac{1}{f}$. In this time, the sound travels to the wall **and back.** The speed of sound is thus $c = 2df$.

In either of the above situations a more reliable result will be achieved if a range of distances, rather than one single value is used. A graph of distance against time will allow the speed of sound to be calculated from the gradient of the best-fit straight line (which should go through the origin).

Timing pulses of sound over smaller distances requires small time intervals to be recorded with precision. It is possible to automate the process using electronic timers and / or data loggers. This equipment would allow, for example, the speed of a sound wave along a metal rod or through water to be investigated.

2. INDIRECT METHODS

Since $c = f\lambda$, the speed of sound can be calculated if we measure a sound's frequency and wavelength.

Frequency measurement

a) A microphone and a cathode ray oscilloscope (CRO) [page 116] can display a graph of the oscillations of a sound wave. Appropriate measurements from the graph allow the time period and hence the frequency to be calculated.

b) **Stroboscopic** techniques (e.g. flashing light of known frequency) can be used to measure the frequency of the vibrating object (e.g. a tuning fork) that is the source of the sound.

c) Frequency of sound can be controlled at source using a known frequency source (e.g. a standard tuning fork) or a calibrated electronic frequency generator.

d) Comparisons can also be made between the unknown frequency and a known frequency.

Wavelength measurement

a) The interference of waves (see page 40) can be employed to find the path difference between consecutive positions of destructive interference. The path difference between these two situations will be λ.

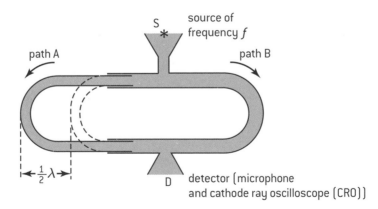

b) Standing waves (see page 48) in a gas can be employed to find the location of adjacent nodes. The positions in an enclosed tube can be revealed either:

- in the period pattern made by dust in the tube
- electronically using a small movable microphone.

c) A resonance tube (see page 49) allows the column length for different maxima to be recorded. The length distance between adjacent maxima will be $\frac{\lambda}{2}$.

3. FACTORS THAT AFFECT THE SPEED OF SOUND

Factors include:

- Nature of material
- Density
- Temperature (for an ideal gas, $c \propto \sqrt{T}$)
- Humidity (for air).

Intensity

INTENSITY

The **sound intensity**, I, is the amount of energy that a sound wave brings to a unit area every second. The units of sound intensity are W m⁻².

It depends on the amplitude of the sound. A more intense sound (one that is louder) must have a larger amplitude.

Intensity ∝ (amplitude)²

This relationship between intensity and amplitude is true for all waves.

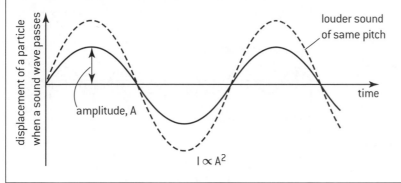

INVERSE SQUARE LAW OF RADIATION

As the distance of an observer from a point source of light increases, the power received by the observer will decrease as the energy spreads out over a larger area. A doubling of distance will result in the reduction of the power received to a quarter of the original value.

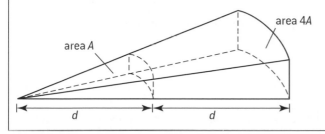

The surface area A of a sphere of radius r is calculated using:

$$A = 4\pi r^2$$

If the point source radiates a total power P in all directions, then the power received per unit area (the **intensity** I) at a distance r away from the point source is:

$$I = \frac{P}{4\pi r^2}$$

For a given area of receiver, the intensity of the received radiation is inversely proportional to the square of the distance from the point source to the receiver. This is known as the **inverse square law** and applies to all waves.

$$I \propto x^{-2}$$

WAVEFRONTS AND RAYS

As introduced on page 35, waves can be described in terms of the motion of a wavefront and/or in terms of rays.

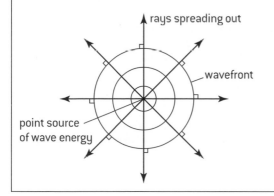

A **ray** is the path taken by the wave energy as it travels out from the source.

A **wavefront** is a surface joining neighbouring points where the oscillations are in phase with one another. In two dimensions, the wavefront is a line and in one dimension, the wavefront is a point.

Superposition

INTERFERENCE OF WAVES

When two waves of the same type meet, they **interfere** and we can work out the resulting wave using the principle of superposition. The overall disturbance at any point and at any time where the waves meet is the vector sum of the disturbances that would have been produced by each of the individual waves. This is shown below.

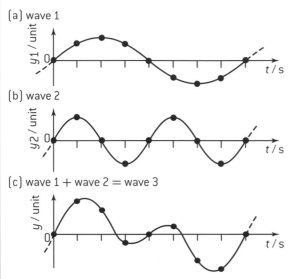

(a) wave 1

(b) wave 2

(c) wave 1 + wave 2 = wave 3

Wave superposition

If the waves have the same amplitude and the same frequency then the interference **at a particular point** can be **constructive** or **destructive**.

graphs

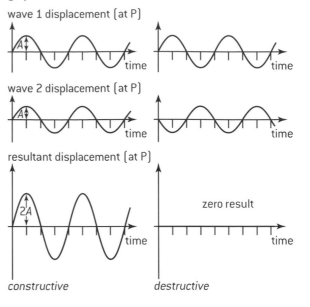

wave 1 displacement (at P)

wave 2 displacement (at P)

resultant displacement (at P)

constructive   *destructive*

TECHNICAL LANGUAGE

Constructive interference takes place when the two waves are 'in step' with one another – they are said to be **in phase**. There is a zero **phase difference** between them. Destructive interference takes place when the waves are exactly 'out of step' – they are said to be **out of phase**. There are several different ways of saying this. One could say that the phase difference is equal to 'half a cycle' or '180 degrees' or 'π radians'.

Interference can take place if there are two possible routes for a ray to travel from source to observer. If the path difference between the two rays is a whole number of wavelengths, then constructive interference will take place.

path difference $= n\lambda \rightarrow$ constructive

path difference $= (n + \frac{1}{2})\lambda \rightarrow$ destructive

$$n = 0, 1, 2, 3 \ldots$$

For constructive or destructive interference to take place, the sources of the waves must be phase linked or **coherent**.

EXAMPLES OF INTERFERENCE

Water waves

A ripple tank can be used to view the interference of water waves. Regions of large-amplitude waves are constructive interference. Regions of still water are destructive interference.

Sound

It is possible to analyse any noise in terms of the component frequencies that make it up. A computer can then generate exactly the same frequencies but of different phase. This 'antisound' will interfere with the original sound. An observer in a particular position in space could have the overall noise level reduced if the waves superimposed destructively at that position.

Light

The colours seen on the surface of a soap bubble are a result of constructive and destructive interference of two light rays. One ray is reflected off the outer surface of the bubble whereas the other is reflected off the inner surface.

SUPERPOSITION OF WAVE PULSES

Whenever wave pulses meet, the principle of superposition applies: At any instant in time, the net displacement that results from different waves meeting at the same point in space is just the vector sum of the displacements that would have been produced by each individual wave. $y_{\text{overall}} = y_1 + y_2 + y_3$ etc.

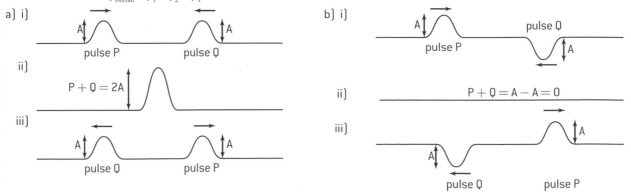

a) i)
pulse P   pulse Q

ii)
P + Q = 2A

iii)
pulse Q   pulse P

b) i)
pulse P   pulse Q

ii)
P + Q = A − A = 0

iii)
pulse Q   pulse P

Polarization

POLARIZED LIGHT

Light is part of the electromagnetic spectrum. It is made up of oscillating electric and magnetic fields that are at right angles to one another (for more details see page 132). They are transverse waves; both fields are at right angles to the direction of propagation. The **plane of vibration** of electromagnetic waves is defined to be the plane that contains the electric field and the direction of propagation.

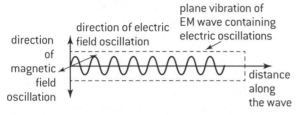

There are an infinite number of ways for the fields to be oriented. Light (or any EM wave) is said to be **unpolarized** if the plane of vibration varies randomly whereas **plane-polarized** light has a fixed plane of vibration. The diagrams below represent the electric fields of light when being viewed 'head on'.

polarized light: over a period of time, the electric field only oscillates in one direction

unpolarized light: over a period of time, the electric field oscillates in random directions

A mixture of polarized light and unpolarized light is **partially plane-polarized**. If the plane of polarization rotates uniformly the light is said to be **circularly polarized**.

Most light sources emit unpolarized light whereas radio waves, radar and laboratory microwaves are often plane-polarized as a result of the processes that produce the waves. Light can be polarized as a result of reflection or selective absorption. In addition, some crystals exhibit **double refraction** or **birefringence** where an unpolarized ray that enters a crystal is split into two plane-polarized beams that have mutually perpendicular planes of polarization.

BREWSTER'S LAW

A ray of light incident on the boundary between two media will, in general, be reflected and refracted. The reflected ray is always partially plane-polarized. If the reflected ray and the refracted ray are at right angles to one another, then the reflected ray is totally plane-polarized. The angle of incidence for this condition is known as the **polarizing angle**.

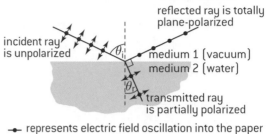

● represents electric field oscillation into the paper
↕ represents electric field oscillation in the plane of the paper

$$\theta_i + \theta_r = 90°$$

Brewster's law relates the refractive index of medium 2, n, to the incident angle θ_i:

$$n = \frac{\sin \theta_i}{\sin \theta_r} = \frac{\sin \theta_i}{\cos \theta_i} = tan\,\theta_i$$

A **polarizer** is any device that produces plane-polarized light from an unpolarized beam. An **analyser** is a polarizer used to detect polarized light.

Polaroid is a material which preferentially absorbs any light in one particular plane of polarization allowing transmission only in the plane at 90° to this.

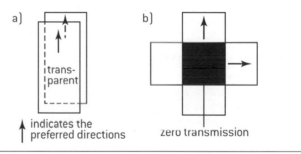

↑ indicates the preferred directions

zero transmission

MALUS'S LAW

When plane-polarized light is incident on an analyser, its preferred direction will allow a component of the light to be transmitted:

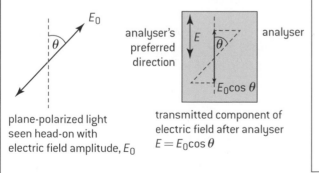

plane-polarized light seen head-on with electric field amplitude, E_0

transmitted component of electric field after analyser $E = E_0 \cos \theta$

The intensity of light is proportional to the (amplitude)².

Transmitted intensity $I \propto E^2$

$\therefore\ I \propto E_0^{\ 2} \cos^2\theta$ as expressed by **Malus's law:**

$$I = I_0 \cos^2\theta$$

OPTICALLY ACTIVE SUBSTANCES

An **optically active** substance is one that rotates the plane of polarization of light that passes through it. Many solutions (e.g. sugar solutions of different concentrations) are optically active.

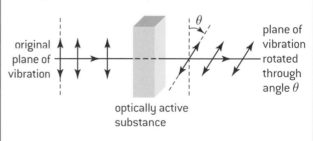

original plane of vibration

optically active substance

plane of vibration rotated through angle θ

I is transmitted intensity of light in W m⁻²

I_0 is incident intensity of light in W m⁻²

θ is the angle between the plane of vibration and the analyser's preferred direction

Uses of polarization

POLAROID SUNGLASSES

Polaroid is a material containing long chain molecules. The molecules selectively absorb light that have electric fields aligned with the molecules in the same way that a grid of wires will selectively absorb microwaves.

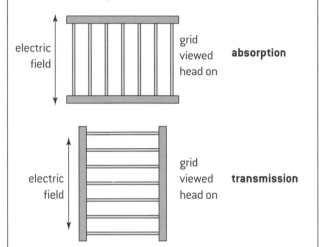

When worn normally by a person standing up, Polaroid dark glasses allow light with vertically oscillating electric fields to be transmitted and absorb light with horizontally oscillating electric fields.

- The absorption will mean that the overall light intensity is reduced.
- Light that has reflected from horizontal surfaces will be horizontally plane-polarized to some extent.
- Polaroid sunglasses will preferentially absorb reflected light, reducing 'glare' from horizontal surfaces.

CONCENTRATION OF SOLUTIONS

For a given optically active solution, the angle θ through which the plane of polarization is rotated is proportional to:

- The length of the solution through which the plane-polarized light passes.
- The concentration of the solution.

A polarimeter is a device that measures θ for a given solution. It consists of two polarizers (a polarizer and an analyser) that are initially aligned. The optically active solution is introduced between the two and the analyser is rotated to find the maximum transmitted light.

STRESS ANALYSIS

Glass and some plastics become birefringent (see page 41) when placed under stress. When polarized white light is passed through stressed plastics and then analysed, bright coloured lines are observed in the regions of maximum stress.

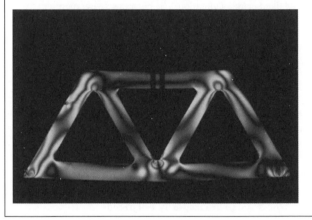

LIQUID-CRYSTAL DISPLAYS (LCDS)

LCDs are used in a wide variety of different applications that include calculator displays and computer monitors. The liquid crystal is sandwiched between two glass electrodes and is birefringent. One possible arrangement with crossed polarizers surrounding the liquid crystal is shown below:

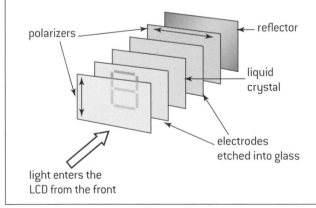

- With no liquid crystal between the electrodes, the second polarizer would absorb all the light that passed through the first polarizer. The screen would appear black.
- The liquid crystal has a twisted structure and, in the absence of a potential difference, causes the plane of polarization to rotate through 90°.
- This means that light can pass through the second polarizer, reach the reflecting surface and be transmitted back along its original direction.
- With no pd between the electrodes, the LCD appears light.
- A pd across the liquid crystal causes the molecules to align with the electric field. This means less light will be transmitted and this section of the LCD will appear darker.
- The extent to which the screen appears grey or black can be controlled by the pd
- Coloured filters can be used to create a colour image.
- A picture can be built up from individual picture elements.

FURTHER POLARIZATION EXAMPLES

Only transverse waves can be polarized. Page 41 has concentrated on the polarization of light but all EM waves that are transverse are able, in principle, to be polarized.

- Sound waves, being longitudinal waves, cannot be polarized.
- The nature of radio and TV broadcasts means that the signal is often polarized and aerials need to be properly aligned if they are to receive the maximum possible signal strength.

- Microwave radiation (with a typical wavelength of a few cm) can be used to demonstrate wave characteristics in the laboratory. Polarization can be demonstrated using a grid of conducting wires. If the grid wires are aligned parallel to the plane of vibration of the electric field the microwaves will be absorbed. Rotation of the grid through 90° will allow the microwaves to be transmitted.

Wave behaviour – reflection

REFLECTION AND TRANSMISSION

In general, when any wave meets the boundary between two different media it is partially reflected and partially transmitted.

incident ray
normal
reflected ray
medium (1)
medium (2)
transmitted ray
medium (2) is optically denser than medium (1)

REFLECTION OF TWO-DIMENSIONAL PLANE WAVES

The diagram below shows what happens when plane waves are reflected at a boundary. When working with rays, by convention we always measure the angles between the rays and the **normal**. The normal is a construction line that is drawn at right angles to the surface.

incident ray
incident angle i
normal
reflected angle r
reflected ray
surface
Law of reflection: $i = r$

TYPES OF REFLECTION

When a single ray of light strikes a smooth mirror it produces a single reflected ray. This type of 'perfect' reflection is very different to the reflection that takes place from an uneven surface such as the walls of a room. In this situation, a single incident ray is generally scattered in all directions. This is an example of a **diffuse** reflection.

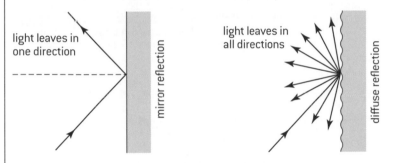

light leaves in one direction

mirror reflection

light leaves in all directions

diffuse reflection

We see objects by receiving light that has come from them. Most objects do not give out light by themselves so we cannot see them in the dark. Objects become visible with a source of light (e.g. the Sun or a light bulb) because diffuse reflections have taken place that scatter light from the source towards our eyes.

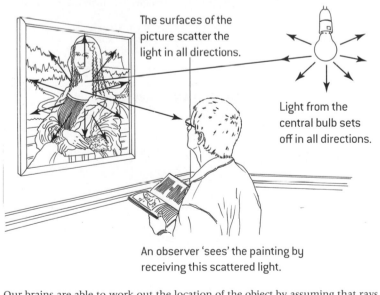

The surfaces of the picture scatter the light in all directions.

Light from the central bulb sets off in all directions.

An observer 'sees' the painting by receiving this scattered light.

Our brains are able to work out the location of the object by assuming that rays travel in straight lines.

LAW OF REFLECTION

The location and nature of optical images can be worked out using ray **diagrams** and the principles of **geometric optics**. A ray is a line showing the direction in which light energy is propagated. The ray must always be at right angles to the wavefront. The study of geometric optics ignores the wave and particle nature of light.

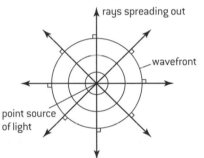

rays spreading out

wavefront

point source of light

When a mirror reflection takes place, the direction of the reflected ray can be predicted using the laws of reflection. In order to specify the ray directions involved, it is usual to measure all angles with respect to an imaginary construction line called the **normal**. For example, the incident angle is always taken as the angle between the incident ray and the normal. The normal to a surface is the line at right angles to the surface as shown below.

normal
incident ray
i
r
reflected ray

The laws of reflection are that:

- the incident angle is equal to the reflected angle

- the incident ray, the reflected ray and the normal all lie in the same plane (as shown in the diagram).

The second statement is only included in order to be precise and is often omitted. It should be obvious that a ray arriving at a mirror (such as the one represented above) is not suddenly reflected in an odd direction (e.g. out of the plane of the page).

Snell's law and refractive index

REFRACTIVE INDEX AND SNELL'S LAW

Refraction takes place at the boundary between two media. In general, a wave that crosses the boundary will undergo a change of direction. The reason for this change in direction is the change in wave speed that has taken place.

As with reflection, the ray directions are always specified by considering the angles between the ray and **the normal**. If a ray travels into an optically denser medium (e.g. from air into water), then the ray of light is refracted **towards** the normal. If the ray travels into an optically less dense medium then the ray of light is refracted **away from** the normal.

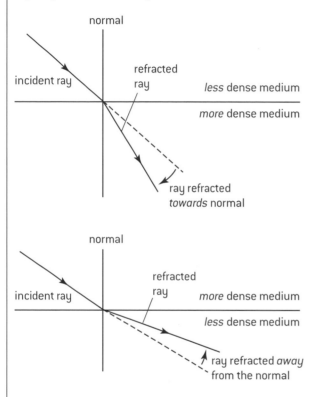

Snell's law allows us to work out the angles involved. When a ray is refracted between two different media, the ratio $\frac{\sin(\text{angle of incidence})}{\sin(\text{angle of refraction})}$ is a constant.

The constant is called the refractive index n between the two media. This ratio is equal to the ratio of the speeds of the waves in the two media.

$$\frac{\sin i}{\sin r} = n$$

If the refractive index for a particular substance is given as a particular number and the other medium is not mentioned then you can assume that the other medium is air (or to be absolutely correct, a vacuum). Another way of expressing this is to say that the refractive index of air can be taken to be 1.0.

For example the refractive index for a type of glass might be given as

$$n_{\text{glass}} = 1.34$$

This means that a ray entering the glass from air with an incident angle of 40° would have a refracted angle given by

$$\sin r = \frac{\sin 40°}{1.34} = 0.4797$$

$$\therefore r = 28.7°$$

$$n_{\text{glass}} = \frac{n_{\text{glass}}}{n_{\text{air}}} = \frac{\sin \theta_{\text{air}}}{\sin \theta_{\text{glass}}} = \frac{V_{\text{air}}}{V_{\text{glass}}}$$

EXAMPLES

1. Parallel-sided block

A ray will always leave a parallel-sided block travelling in a parallel direction to the one with which it entered the block. The overall effect of the block has been to move the ray sideways. An example of this is shown below.

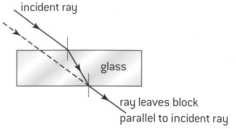

2. Ray travelling between two media

If a ray goes between two different media, the two individual refractive indices can be used to calculate the overall refraction using the following equation

$$n_1 \sin \theta_1 = n_2 \sin \theta_2 \text{ or } \frac{n_1}{n_2} = \frac{\sin \theta_2}{\sin \theta_1}$$

n_1 refractive index of medium 1

θ_1 angle in medium 1

n_2 refractive index of medium 2

θ_2 angle in medium 2

Suppose a ray of light is shone into a fish tank that contains water. The refraction that takes place would be calculated as shown below:

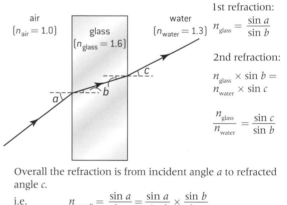

1st refraction:

$$n_{\text{glass}} = \frac{\sin a}{\sin b}$$

2nd refraction:

$$n_{\text{glass}} \times \sin b = n_{\text{water}} \times \sin c$$

$$\frac{n_{\text{glass}}}{n_{\text{water}}} = \frac{\sin c}{\sin b}$$

Overall the refraction is from incident angle a to refracted angle c.

i.e.
$$n_{\text{overall}} = \frac{\sin a}{\sin c} = \frac{\sin a}{\sin b} \times \frac{\sin b}{\sin c}$$
$$= n_{\text{water}}$$

REFRACTION OF PLANE WAVES

The reason for the change in direction in refraction is the change in speed of the wave.

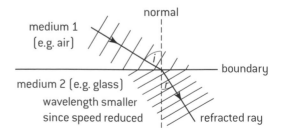

Snell's law (an experimental law of refraction) states that

the ratio $\frac{\sin i}{\sin r}$ = constant, for a given frequency.

The ratio is equal to the ratio of the speeds in the different media

$$\frac{n_1}{n_2} = \frac{\sin \theta_2}{\sin \theta_1} = \frac{V_2}{V_1} \leftarrow \text{speed of wave in medium 2}$$
$$\leftarrow \text{speed of wave in medium 1}$$

Refraction and critical angle

TOTAL INTERNAL REFLECTION AND CRITICAL ANGLE

In general, both reflection and refraction can happen at the boundary between two media.

It is, under certain circumstances, possible to guarantee complete (total) reflection with no transmission at all. This can happen when a ray meets the boundary and it is travelling in the denser medium.

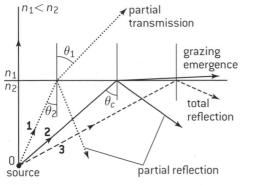

Ray1 This ray is partially reflected and partially refracted.

Ray2 This ray has a refracted angle of nearly 90°. The **critical ray** is the name given to the ray that has a refracted angle of 90°. The **critical angle** is the angle of incidence θ_c for the critical ray.

Ray3 This ray has an angle of incidence **greater** than the critical angle. Refraction cannot occur so the ray must be totally reflected at the boundary and stay inside medium 2. The ray is said to be **totally internally reflected**.

The critical angle can be worked out as follows. For the critical ray,

$$n_1 \sin \theta_1 = n_2 \sin \theta_2$$

$$\theta_1 = 90°$$

$$\theta_2 = \theta_c$$

$$\therefore \sin \theta_c = \frac{1}{n_2}$$

METHODS FOR DETERMINING REFRACTIVE INDEX EXPERIMENTALLY

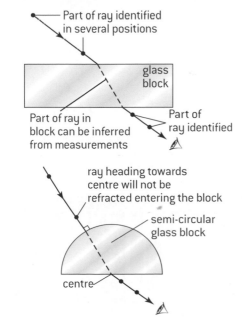

EXAMPLES

1. What a fish sees under water

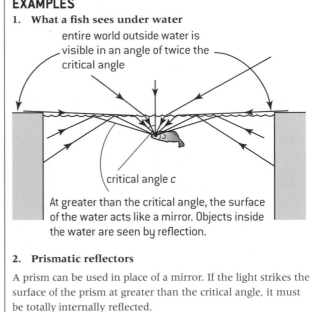

entire world outside water is visible in an angle of twice the critical angle

critical angle c

At greater than the critical angle, the surface of the water acts like a mirror. Objects inside the water are seen by reflection.

2. Prismatic reflectors

A prism can be used in place of a mirror. If the light strikes the surface of the prism at greater than the critical angle, it must be totally internally reflected.

Prisms are used in many optical devices. Examples include:

- periscopes – the double reflection allows the user to see over a crowd.
- binoculars – the double reflection means that the binoculars do not have to be too long
- SLR cameras – the view through the lens is reflected up to the eyepiece.

binoculars
The prism arrangement delivers the image to the eyepiece the right way up. By sending the light along the instrument three times, it also allows the binoculars to be shorter.

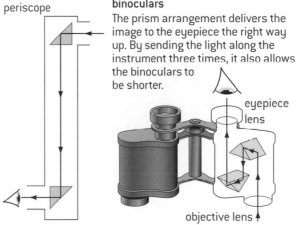

1. Locate paths taken by different rays either by sending a ray through a solid and measuring its position or aligning objects by eye. Uncertainties in angle measurement are dependent on protractor measurements. (See diagrams on left)

2. Use a travelling microscope to measure real and apparent depth and apply following formula:

$$n = \frac{\text{real depth of object}}{\text{apparent depth of object}}$$

3. Very accurate measurements of angles of refraction can be achieved using a prism of the substance and a custom piece of equipment call a *spectrometer*.

Diffraction

DIFFRACTION

When waves pass through apertures they tend to spread out. Waves also spread around obstacles. This wave property is called **diffraction**.

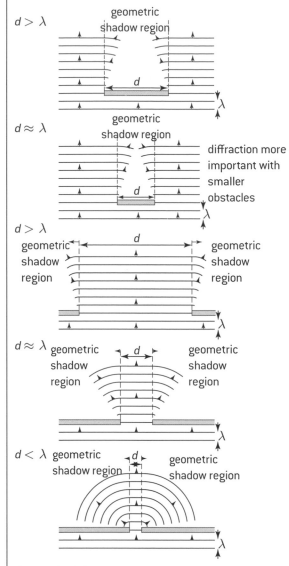

$d > \lambda$

$d \approx \lambda$

diffraction more important with smaller obstacles

$d > \lambda$

$d \approx \lambda$

$d < \lambda$

d = width of obstacle/gap
Diffraction – wave energy is received in geometric shadow region.

BASIC OBSERVATIONS

Diffraction is a wave effect. The objects involved (slits, apertures, etc.) have a size that is of the same order of magnitude as the wavelength.

intensity

There is a central maximum intensity.
Other maxima occur roughly halfway between the minima.

As the angle increases, the intensity of the maxima decreases.

1st minimum

angle

Diffraction of a single slit.

There are some important points to note from these diagrams.

- Diffraction becomes relatively more important when the wavelength is large in comparison to the size of the aperture (or the object).
- The wavelength needs to be of the same order of magnitude as the aperture for diffraction to be noticeable.

PRACTICAL SIGNIFICANCE OF DIFFRACTION

Whenever an observer receives information from a source of electromagnetic waves, diffraction causes the energy to spread out. This spreading takes place as a result of any obstacle in the way and the width of the device receiving the electromagnetic radiation. Two sources of electromagnetic waves that are angularly close to one another will both spread out and interfere with one another. This can affect whether or not they can be resolved (see page 101).

Diffraction effects mean that it is impossible ever to see atoms because they are smaller than the wavelength of visible light, meaning that light will diffract around the atoms. It is, however, possible to image atoms using smaller wavelengths. Practical devices where diffraction needs to be considered include:

- CDs and DVDs – the maximum amount of information that can be stored depends on the size and the method used for recording information.
- The electron microscope – resolves items that cannot be resolved using a light microscope. The electrons have an effective wavelength that is much smaller than the wavelength of visible light (see page 127).
- Radio telescopes – the size of the dish limits the maximum resolution possible. Several radio telescopes can be linked together in an array to create a virtual radio telescope with a greater diameter and with a greater ability to resolve astronomical objects. (See page 181)

EXAMPLES OF DIFFRACTION

Diffraction provides the reason why we can hear something even if we can not see it.

If you look at a distant street light at night and then squint your eyes the light spreads sideways – this is as a result of diffraction taking place around your eyelashes! (Needless to say, this explanation is a simplification.)

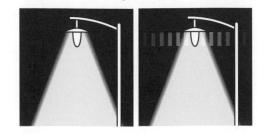

Two-source interference of waves

PRINCIPLES OF THE TWO-SOURCE INTERFERENCE PATTERN

Two-source interference is simply another application of the principle of superposition, for two coherent sources having roughly the same amplitude.

Two sources are coherent if:
- they have the same frequency
- there is a constant phase relationship between the two sources.

regions where waves are in phase:
constructive interference

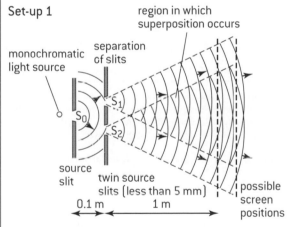

destructive interference

Two dippers in water moving together are coherent sources. This forms regions of water ripples and other regions with no waves.

Two loudspeakers both connected to the same signal generator are coherent sources. This forms regions of loud and soft sound.

A set-up for viewing two-source interference with light is shown below. It is known as **Young's double slit** experiment. A **monochromatic** source of light is one that gives out only one frequency. Light from the twin slits (the sources) interferes and patterns of light and dark regions, called **fringes**, can be seen on the screen.

Set-up 1

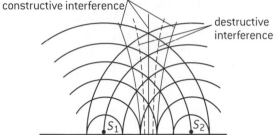

region in which superposition occurs

monochromatic light source

separation of slits

source slit

twin source slits (less than 5 mm)

0.1 m   1 m

possible screen positions

Set-up 2

The use of a laser makes the set-up easier.

laser   double slit   screen

The experiment results in a regular pattern of light and dark strips across the screen as represented below.

intensity distribution   **view seen**

fringe width, d

intensity

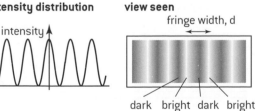

dark bright dark bright

MATHEMATICS

The location of the light and dark fringes can be mathematically derived in one of two ways. The derivations do not need to be recalled.

Method 1

The simplest way is to consider two parallel rays setting off from the slits as shown below.

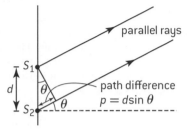

parallel rays

path difference
$p = d\sin\theta$

If these two rays result in a bright patch, then the two rays must arrive in phase. The two rays of light started out in phase but the light from source 2 travels an extra distance. This extra distance is called the **path difference**.

Constructive interference can only happen if the path difference is a whole number of wavelengths. Mathematically,

Path difference $= n\lambda$

[where n is an integer – e.g. 1, 2, 3 etc.]

From the geometry of the situation

Path difference $= d\sin\theta$

In other words $n\lambda = d\sin\theta$

Method 2

If a screen is used to make the fringes visible, then the rays from the two slits cannot be absolutely parallel, but the physical set-up means that this is effectively true.

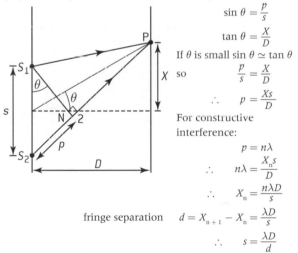

$$\sin\theta = \frac{p}{s}$$

$$\tan\theta = \frac{X}{D}$$

If θ is small $\sin\theta \simeq \tan\theta$

so $$\frac{p}{s} = \frac{X}{D}$$

$$\therefore \quad p = \frac{Xs}{D}$$

For constructive interference:

$$p = n\lambda$$

$$\therefore \quad n\lambda = \frac{X_n s}{D}$$

$$\therefore \quad X_n = \frac{n\lambda D}{s}$$

fringe separation $$d = X_{n+1} - X_n = \frac{\lambda D}{s}$$

$$\therefore \quad s = \frac{\lambda D}{d}$$

This equation only applies when the angle is small.

Example

Laser light of wavelength 450 nm is shone on two slits that are 0.1 mm apart. How far apart are the fringes on a screen placed 5.0 m away?

$$d = \frac{\lambda D}{s} = \frac{4.5 \times 10^{-7} \times 5}{1.0 \times 10^{-4}} = 0.0225 \text{ m} = 2.25 \text{ cm}$$

Nature and production of standing (stationary) waves

STANDING WAVES

A special case of interference occurs when two waves meet that are:

- of the same amplitude
- of the same frequency
- travelling in opposite directions.

In these conditions a **standing wave** will be formed.

The conditions needed to form standing waves seem quite specialized, but standing waves are in fact quite common. They often occur when a wave reflects back from a boundary along the route that it came. Since the reflected wave and the incident wave are of (nearly) equal amplitude, these two waves can interfere and produce a standing wave.

Perhaps the simplest way of picturing a standing wave would be to consider two transverse waves travelling in opposite directions along a stretched rope. The series of diagrams below shows what happens.

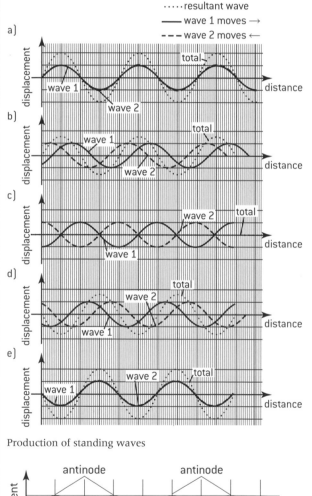

Production of standing waves

A standing wave – the pattern remains fixed

There are some points on the rope that are always at rest. These are called the **nodes**. The points where the maximum movement takes place are called **antinodes**. The resulting standing wave is so called because the wave pattern remains fixed in space – it is its amplitude that changes over time. A comparison with a normal (travelling) wave is given below.

	Stationary wave	Normal (travelling) wave
Amplitude	All points on the wave have different amplitudes. The maximum amplitude is 2A at the antinodes. It is zero at the nodes.	All points on the wave have the same amplitude.
Frequency	All points oscillate with the same frequency.	All points oscillate with the same frequency.
Wavelength	This is **twice** the distance from one node (or antinode) to the next node (or antinode).	This is the shortest distance (in metres) along the wave between two points that are in phase with one another.
Phase	All points between one node and the next node are moving in phase.	All points along a wavelength have different phases.
Energy	Energy is not transmitted by the wave, but it does have an energy associated with it.	Energy is transmitted by the wave.

Although the example left involved transverse waves on a rope, a standing wave can also be created using sound or light waves. All musical instruments involve the creation of a standing sound wave inside the instrument. The production of laser light involves a standing light wave. Even electrons in hydrogen atoms can be explained in terms of standing waves.

A standing longitudinal wave can be particularly hard to imagine. The diagram below attempts to represent one example – a standing sound wave.

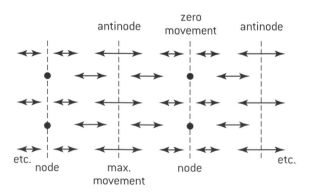

A longitudinal standing wave

Boundary conditions

BOUNDARY CONDITIONS

The boundary conditions of the system specify the conditions that must be met at the edges (the boundaries) of the system when standing waves are taking place. Any standing wave that meets these boundary conditions will be a possible resonant mode of the system.

1. Transverse waves on a string

If the string is fixed at each end, the ends of the string cannot oscillate. Both ends of the string would reflect a travelling wave and thus a standing wave is possible. The only standing waves that fit these boundary conditions are ones that have nodes at each end. The diagrams below show the possible resonant modes.

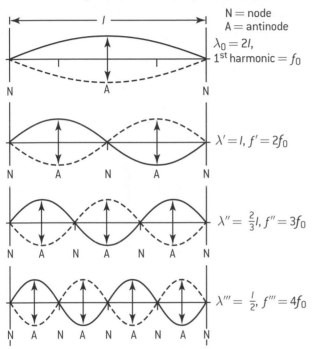

N = node
A = antinode
$\lambda_0 = 2l$,
1st harmonic = f_0

$\lambda' = l$, $f' = 2f_0$

$\lambda'' = \frac{2}{3}l$, $f'' = 3f_0$

$\lambda''' = \frac{l}{2}$, $f''' = 4f_0$

Harmonic modes for a string

The resonant mode that has the lowest frequency is called the fundamental or the **first harmonic**. Higher resonant modes are called **harmonics**. Many musical instruments (e.g. piano, violin, guitar etc.) involve similar oscillations of metal 'strings'.

2. Longitudinal sound waves in a pipe

A longitudinal standing wave can be set up in the column of air enclosed in a pipe. As in the example above, this results from the reflections that take place at both ends.

As before, the boundary conditions determine the standing waves that can exist in the tube. A closed end must be a displacement node. An open end must be an antinode. Possible standing waves are shown for a pipe open at both ends and a pipe closed at one end.

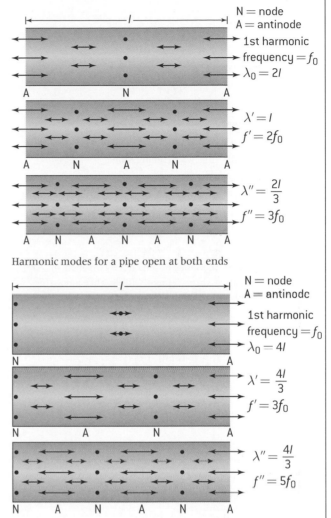

N = node
A = antinode
1st harmonic frequency = f_0
$\lambda_0 = 2l$

$\lambda' = l$
$f' = 2f_0$

$\lambda'' = \frac{2l}{3}$
$f'' = 3f_0$

Harmonic modes for a pipe open at both ends

N = node
A = antinode
1st harmonic frequency = f_0
$\lambda_0 = 4l$

$\lambda' = \frac{4l}{3}$
$f' = 3f_0$

$\lambda'' = \frac{4l}{3}$
$f'' = 5f_0$

Harmonic modes for a pipe closed at one end

Musical instruments that involve a standing wave in a column of air include the flute, the trumpet, the recorder and organ pipes.

EXAMPLE

An organ pipe (open at one end) is 1.2 m long.

Calculate its fundamental frequency.
The speed of sound is 330 m s⁻¹.

$l = 1.2$ m   ∴ $\frac{\lambda}{4} = 1.2$ m (first harmonic)

∴ $\lambda = 4.8$ m

$v = f\lambda$

$f = \frac{330}{4.8} \simeq 69$ Hz

RESONANCE TUBE

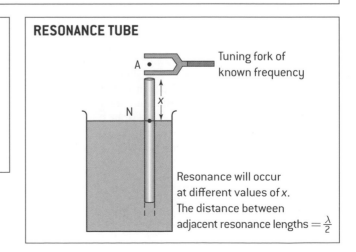

Tuning fork of known frequency

Resonance will occur at different values of x. The distance between adjacent resonance lengths = $\frac{\lambda}{2}$

IB Questions – waves

1. A surfer is out beyond the breaking surf in a deep-water region where the ocean waves are sinusoidal in shape. The crests are 20 m apart and the surfer rises a vertical distance of 4.0 m from wave **trough** to **crest**, in a time of 2.0 s. What is the speed of the waves?

   A. 1.0 m s⁻¹   B. 2.0 m s⁻¹

   C. 5.0 m s⁻¹   D. 10.0 m s⁻¹

2. A standing wave is established in air in a pipe with one closed and one open end.

   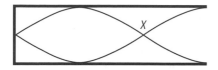

   The air molecules near X are

   A. always at the centre of a compression.

   B. always at the centre of a rarefaction.

   C. sometimes at the centre of a compression and sometimes at the centre of a rarefaction.

   D. never at the centre of a compression or a rarefaction.

3. This question is about sound waves.

   A sound wave of frequency 660 Hz passes through air. The variation of particle displacement with distance along the wave at one instant of time is shown below.

   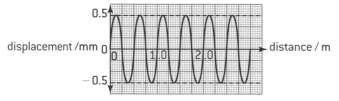

   a) State whether this wave is an example of a longitudinal or a transverse wave. [1]

   b) Using data from the above graph, deduce for this sound wave,

      (i) the wavelength. [1]

      (ii) the amplitude. [1]

      (iii) the speed. [2]

4. The diagram below represents the direction of oscillation of a disturbance that gives rise to a wave.

   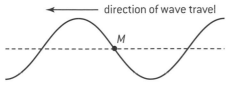

   a) By redrawing the diagram, add arrows to show the direction of wave energy transfer to illustrate the difference between

      (i) a transverse wave and [1]

      (ii) a longitudinal wave. [1]

   A wave travels along a stretched string. The diagram below shows the variation with distance along the string of the displacement of the string at a particular instant in time. A small marker is attached to the string at the point labelled M. The undisturbed position of the string is shown as a dotted line.

   ← direction of wave travel

   M

---

b) On the diagram above

   (i) draw an arrow to indicate the direction in which the marker is moving. [1]

   (ii) indicate, with the letter A, the amplitude of the wave. [1]

   (iii) indicate, with the letter λ, the wavelength of the wave. [1]

   (iv) draw the displacement of the string a time $\frac{T}{4}$ later, where T is the period of oscillation of the wave. Indicate, with the letter N, the new position of the marker. [2]

The wavelength of the wave is 5.0 cm and its speed is 10 cm s⁻¹.

c) Determine

   (i) the frequency of the wave. [1]

   (ii) how far the wave has moved in $\frac{T}{4}$ s. [2]

Interference of waves

d) By reference to the principle of superposition, explain what is meant by constructive interference. [4]

The diagram below (not drawn to scale) shows an arrangement for observing the interference pattern produced by the light from two narrow slits S_1 and S_2.

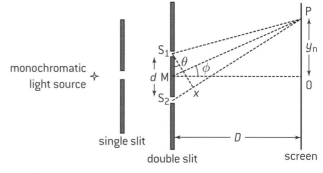

The distance S_1S_2 is d, the distance between the double slit and screen is D and $D \gg d$ such that the angles θ and ϕ shown on the diagram are small. M is the mid-point of S_1S_2 and it is observed that there is a bright fringe at point P on the screen, a distance y_n from point O on the screen. Light from S_2 travels a distance S_2X further to point P than light from S_1.

e) (i) State the condition in terms of the distance S_2X and the wavelength of the light λ, for there to be a bright fringe at P. [2]

   (ii) Deduce an expression for θ in terms of S_2X and d. [2]

   (iii) Deduce an expression for ϕ in terms of D and y_n. [1]

For a particular arrangement, the separation of the slits is 1.40 mm and the distance from the slits to the screen is 1.50 m. The distance y_n is the distance of the eighth bright fringe from O and the angle $\theta = 2.70 \times 10^{-3}$ rad.

f) Using your answers to (e) to determine

   (i) the wavelength of the light. [2]

   (ii) the separation of the fringes on the screen. [3]

5. A bright source of light is viewed through two polarisers whose preferred directions are initially parallel. Calculate the angle through which one sheet should be turned to reduce the transmitted intensity to half its original value.

Electric charge and Coulomb's law

CONSERVATION OF CHARGE

Two types of charge exist – positive and negative. Equal amounts of positive and negative charge cancel each other. Matter that contains no charge, or matter that contains equal amounts of positive and negative charge, is said to be electrically **neutral**.

Charges are known to exist because of the forces that exist between all charges, called the **electrostatic force**: like charges repel, unlike charges attract.

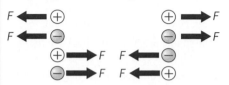

A very important experimental observation is that charge is always conserved.

Charged objects can be created by friction. In this process electrons are physically moved from one object to another. In order for the charge to remain on the object, it normally needs to be an insulator.

before

neutral comb
neutral hair

after

attraction
negative comb
positive hair

electrons have been transferred from hair to comb

The total charge before any process must be equal to the total charge afterwards. It is impossible to create a positive charge without an equal negative charge. This is the law of conservation of charge.

COULOMB'S LAW

The diagram shows the force between two point charges that are far away from the influence of any other charges.

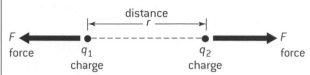

The directions of the forces are along the line joining the charges. If they are like charges, the forces are away from each other – they repel. If they are unlike charges, the forces are towards each other – they attract.

Each charge must feel a force of the same size as the force on the other one.

Experimentally, the force is proportional to the size of both charges and inversely proportional to the square of the distance between the charges.

$$F = \frac{kq_1q_2}{r^2} = k\frac{q_1q_2}{r^2}$$

This is known as Coulomb's law and the constant k is called the Coulomb constant. In fact, the law is often quoted in a slightly different form using a different constant for the medium called the permittivity, ε.

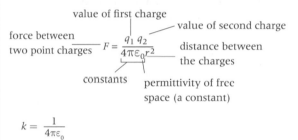

$$k = \frac{1}{4\pi\varepsilon_0}$$

If there are two or more charges near another charge, the overall force can be worked out using vector addition.

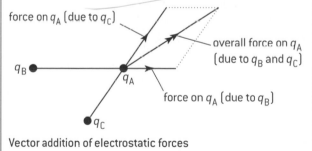

Vector addition of electrostatic forces

CONDUCTORS AND INSULATORS

A material that allows the flow of charge through it is called an electrical **conductor**. If charge cannot flow through a material it is called an electrical **insulator**. In solid conductors the flow of charge is always as a result of the flow of electrons from atom to atom.

Electrical conductors	Electrical insulators
all metals	plastics
e.g. copper	e.g. polythene
aluminium	nylon
brass	acetate
graphite	rubber
	dry wood
	glass
	ceramics

Electric fields

ELECTRIC FIELDS – DEFINITION

A charge, or combination of charges, is said to produce an **electric field** around it. If we place a **test charge** at any point in the field, the value of the force that it feels at any point will depend on the value of the test charge only.

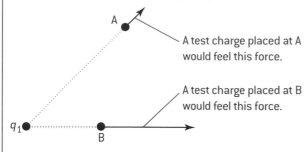

A test charge would feel a different force at different points around a charge q_1.

In practical situations, the test charge needs to be small so that it doesn't disturb the charge or charges that are being considered.

The definition of electric field, E, is

$$E = \frac{F}{q_2} = \text{force per unit positive point test charge.}$$

Coulomb's law can be used to relate the electric field around a point charge to the charge producing the field.

$$E = \frac{q_1}{4\pi\varepsilon_0 r^2}$$

When using these equations you have to be very careful:

- not to muddle up the charge producing the field and the charge sitting in the field (and thus feeling a force)
- not to use the mathematical equation for the field around a point charge for other situations (e.g. parallel plates).

REPRESENTATION OF ELECTRIC FIELDS

This is done using field lines.

At any point in a field:

- the direction of field is represented by the direction of the field lines closest to that point
- the magnitude of the field is represented by the number of field lines passing near that point.

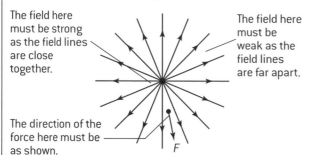

Field around a positive point charge

The resultant electric field at any position due to a collection of point charges is shown to the right.

The parallel field lines between two plates mean that the electric field is uniform.

Electric field lines:

- begin on positive charges and end on negative charges
- never cross
- are close together when the field is strong.

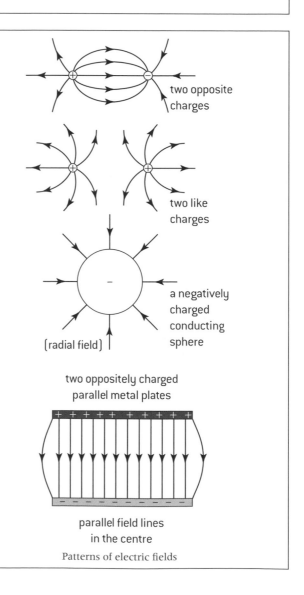

Patterns of electric fields

Electric potential energy and electric potential difference

ENERGY DIFFERENCE IN AN ELECTRIC FIELD

When placed in an electric field, a charge feels a force. This means that if it moves around in an electric field work will be done. As a result, the charge will either gain or lose electric potential energy. Electric potential energy is the energy that a charge has as a result of its position in an electric field. This is the same idea as a mass in a gravitational field. If we lift a mass up, its gravitational potential energy increases. If the mass falls, its gravitational potential energy decreases. In the example below a positive charge is moved from position A to position B. This results in an increase in electric potential energy. Since the field is uniform, the force is constant. This makes it very easy to calculate the work done.

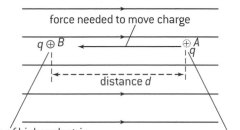

Charge moving in an electric field

Change in electric potential energy = force × distance
$$= E\,q \times d$$

See page 52 for a definition of electric field, E.

In the example above the electric potential energy at B is greater than the electric potential energy at A. We would have to put in this amount of work to push the charge from A to B. If we let go of the charge at B it would be pushed by the electric field. This push would accelerate it so that the loss in electrical potential energy would be the same as the gain in kinetic energy.

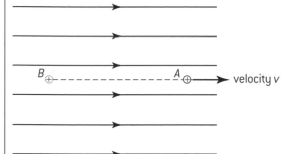

A positive charge released at B will be accelerated as it travels to point A.

gain in kinetic energy = loss in electric potential energy
$$\tfrac{1}{2}mv^2 = Eqd$$
$$mv^2 = 2Eqd$$
$$\therefore\ v = \sqrt{\frac{2Eqd}{m}}$$

ELECTRIC POTENTIAL DIFFERENCE

In the example on the left, the actual energy difference between A and B depended on the charge that was moved. If we doubled the charge we would double the energy difference. The quantity that remains fixed between A and B is the energy difference **per unit charge**. This is called the **potential difference**, or **pd**, between the points.

$$\begin{aligned} \text{Potential difference} \atop \text{between two points} &= \frac{\text{energy difference}}{\text{per unit charge moved}} \\[4pt] &= \frac{\text{energy difference}}{\text{charge}} = \frac{\text{work done}}{\text{charge}} \\[4pt] V &= \frac{W}{q} \end{aligned}$$

The basic unit for potential difference is the joule/coulomb, J C^{-1}. A very important point to note is that for a given electric field, the potential difference between any two points is a single fixed scalar quantity. The work done between these two points does not depend on the path taken by the test charge. A technical way of saying this is 'the electric field is **conservative**'.

UNITS

The smallest amount of negative charge available is the charge on an electron; the smallest amount of positive charge is the charge on a proton. In everyday situations this unit is far too small so we use the **coulomb**, **C**. One coulomb of negative charge is the charge carried by a total of 6.25×10^{18} electrons.

From its definition, the unit of potential difference (pd) is J C^{-1}. This is given a new name, the volt, V. Thus:

1 volt = 1 J C^{-1}

Voltage and potential difference are different words for the same thing. Potential difference is probably the better name to use as it reminds you that it is measuring the difference between two points.

When working at the atomic scale, the joule is far too big to use for a unit for energy. The everyday unit used by physicists for this situation is the electronvolt. As could be guessed from its name, the electronvolt is simply the energy that would be gained by an electron moving through a potential difference of 1 volt.

$$1 \text{ electronvolt} = 1 \text{ volt} \times 1.6 \times 10^{-19} \text{ C}$$
$$= 1.6 \times 10^{-19} \text{ J}$$

The normal SI prefixes also apply so one can measure energies in kiloelectronvolts (keV) or megaelectronvolts (MeV). The latter unit is very common in particle physics.

Example

Calculate the speed of an electron accelerated in a vacuum by a pd of 1000 V (energy = 1 KeV).

$$\begin{aligned} \text{KE of electron} &= V \times e = 1000 \times 1.6 \times 10^{-19} \\ &= 1.6 \times 10^{-16} \text{ J} \\ \tfrac{1}{2}mv^2 &= 1.6 \times 10^{-16} \text{ J} \\ v &= 1.87 \times 10^{7} \text{ m s}^{-1} \end{aligned}$$

Electric current

ELECTRICAL CONDUCTION IN A METAL

Whenever charges move we say that a **current** is flowing. A current is the name for moving charges and the path that they follow is called the **circuit**. Without a complete circuit, a current cannot be maintained for any length of time.

Current flows THROUGH an object when there is a potential difference ACROSS the object. A battery (or power supply) is the device that creates the potential difference.

By convention, currents are always represented as the flow of positive charge. Thus **conventional current**, as it is known, flows from positive to negative. Although currents can flow in solids, liquids and gases, in most everyday electrical circuits the currents flow through wires. In this case the things that actually move are the negative electrons – the **conduction electrons**. The direction in which they move is opposite to the direction of the representation of conventional current. As they move the interactions between the conduction electrons and the lattice ions means that work needs to be done. Therefore, when a current flows, the metal heats up. The speed of the electrons due to the current is called their **drift velocity**.

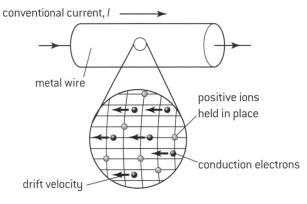

Electrical conduction in a metal

It is possible to estimate the drift velocity of electrons using the generalized **drift speed equation**. All currents are comprised of the movement of charge-carriers and these could be positive or negative; not all currents involve just the movement of electrons. Suppose that the number density of the charge-carriers (the number per unit volume that are available to move) is n, the charge on each carrier is q and their average speed is v.

In a time Δt,

the average distance moved by a charge-carrier $= v \times \Delta t$

so volume of charge moved past a point $= A \times v\Delta t$

so number of charge-carriers moved past a point $= n \times Av\Delta t$

so charge moved past a point, $\Delta Q = nAv\Delta t \times q$

$$\text{current } I = \frac{\Delta Q}{\Delta t}$$

$$I = nAvq$$

It is interesting to compare:

- A typical drift speed of an electron: 10^{-4} m s^{-1}
  (5A current in metal conductor of cross section 1 mm^2)

- The speeds of the electrons due to their random motion: 10^6 m s^{-1}

- The speed of an electrical signal down a conductor: approx. 3×10^8 m s^{-1}

CURRENT

Current is defined as the **rate of flow of electrical charge**. It is always given the symbol, I. Mathematically the definition for current is expressed as follows:

$$\text{Current} = \frac{\text{charge flowed}}{\text{time taken}}$$

$$I = \frac{\Delta Q}{\Delta t} \text{ or (in calculus notation) } I = \frac{\mathrm{d}Q}{\mathrm{d}t}$$

$$1 \text{ ampere} = \frac{\textbf{1 coulomb}}{\textbf{1 second}}$$

$$1 \text{ A} = 1 \text{ C s}^{-1}$$

If a current flows in just one direction it is known as a **direct current**. A current that constantly changes direction (first one way then the other) is known as an **alternating current** or ac.

In SI units, the ampere is the base unit and the coulomb is a derived unit

$$1 \text{ C} = 1 \text{ A s}$$

Electric circuits

OHM'S LAW – OHMIC AND NON-OHMIC BEHAVIOUR

The graphs below show how the current varies with potential difference for some typical devices.

(a) metal at constant temperature

(b) filament lamp

(c) diode

If current and potential difference are proportional (like the metal at constant temperature) the device is said to be **ohmic**. Devices where current and potential difference are not proportional (like the filament lamp or the diode) are said to be **non-ohmic**.

Ohm's law states that the current flowing through a piece of metal is proportional to the potential difference across it providing the temperature remains constant.

In symbols,

$V \propto I$ [if temperature is constant]

A device with constant resistance (in other words an ohmic device) is called a **resistor**.

RESISTANCE

Resistance is the mathematical ratio between potential difference and current. If something has a high resistance, it means that you would need a large potential difference across it in order to get a current to flow.

$$\text{Resistance} = \frac{\text{potential difference}}{\text{current}}$$

In symbols, $R = \frac{V}{I}$

We define a new unit, the ohm, Ω, to be equal to one volt per amp.

$$\textbf{1 ohm = 1 V A}^{-1}$$

POWER DISSIPATION

Since potential difference $= \frac{\text{energy difference}}{\text{charge flowed}}$

And current $= \frac{\text{charge flowed}}{\text{time taken}}$

This means that potential difference × current

$= \frac{(\text{energy difference})}{(\text{charge flowed})} \times \frac{(\text{charge flowed})}{(\text{time taken})} = \frac{\text{energy difference}}{\text{time}}$

This energy difference per time is the power dissipated by the resistor. All this energy is going into heating up the resistor. In symbols:

$$P = V \times I$$

Sometimes it is more useful to use this equation in a slightly different form, e.g.

$P = V \times I$ but $V = I \times R$ so

$P = (I \times R) \times I$

$P = I^2 R$

Similarly $P = \frac{V^2}{R}$

CIRCUITS – KIRCHOFF'S CIRCUIT LAWS

An electric circuit can contain many different devices or **components**. The mathematical relationship $V = IR$ can be applied to any component or groups of components in a circuit.

When analysing a circuit it is important to look at the circuit as a whole. The power supply is the device that is providing the energy, but it is the whole circuit that determines what current flows through the circuit.

Two fundamental conservation laws apply when analysing circuits: the conservation of electric charge and the conservation of energy. These laws are collectively known as Kirchoff's circuit laws and can be stated mathematically as:

First law: $\sum I = 0$ (junction)

Second law: $\sum V = 0$ (loop)

The first law states that the algebraic sum of the currents at any junction in the circuit is zero. The current flowing into a junction must be equal to the current flowing out of a junction. In the example (right) the unknown current $x = 5.5 + 2.7 - 3.4 = 4.8$ A

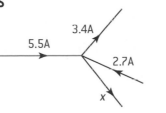

The second law states that around any loop, the total energy per unit charge must sum to zero. Any source of potential difference within the loop must be completely dissipated across the components in the loop (potential drop across the component). Care needs to be taken to get the sign of any pd correct.

- If the chosen loop direction is from the negative side of a battery to its positive side, this is an increase in potential and the value is positive when calculating the sum.

- If the direction around the loop is in the same direction as the current flowing through the component, this is a potential drop and the value is negative when calculating the sum.

EXAMPLE

A 1.2 kW electric kettle is plugged into the 250 V mains supply. Calculate

(i) the current drawn

(ii) its resistance

(i) $I = \frac{1200}{250} = 4.8$ A

(ii) $R = \frac{250}{4.8} = 52\ \Omega$

The example below shows one loop in a larger circuit. Anti-clockwise consideration of the loop means that:

$12.0 - 5.3 - x + 2.7 - 3.2 = 0$.

The potential difference across the bulb, $x = 6.2$ V

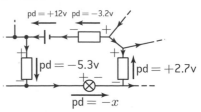

An example of the use of Kirchoff's circuit laws is shown on page 59.

Resistors in series and parallel

RESISTORS IN SERIES

A **series circuit** has components connected one after another in a continuous chain. The current must be the same everywhere in the circuit since charge is conserved. The total potential difference is shared among the components.

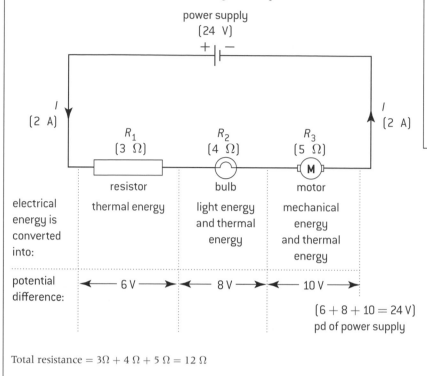

electrical energy is converted into:	thermal energy	light energy and thermal energy	mechanical energy and thermal energy
potential difference:	← 6 V →	← 8 V →	← 10 V →

(6 + 8 + 10 = 24 V)
pd of power supply

Total resistance = $3\,\Omega + 4\,\Omega + 5\,\Omega = 12\,\Omega$

ELECTRICAL METERS

A current-measuring meter is called an **ammeter**. It should be connected in series at the point where the current needs to be measured. A perfect ammeter would have zero resistance.

A meter that measures potential difference is called a **voltmeter**. It should be placed in parallel with the component or components being considered. A perfect voltmeter has infinite resistance.

Example of a series circuit

We can work out what share they take by looking at each component in turn, e.g.

The potential difference across the resistor $= I \times R_1$

The potential difference across the bulb $= I \times R_2$

$$R_{total} = R_1 + R_2 + R_3$$

This always applies to a series circuit. Note that $V = IR$ correctly calculates the potential difference across each individual component as well as calculating it across the total.

RESISTORS IN PARALLEL

A **parallel circuit** branches and allows the charges more than one possible route around the circuit.

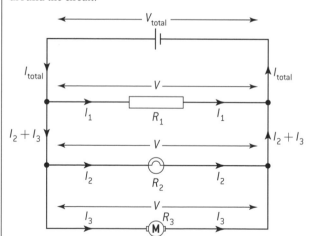

Example of a parallel circuit

Since the power supply fixes the potential difference, each component has the same potential difference across it. The total current is just the addition of the currents in each branch.

$$
\begin{aligned}
I_{total} &= I_1 + I_2 + I_3 \\
&= \frac{V}{R_1} + \frac{V}{R_2} + \frac{V}{R_3} \\
\frac{1}{R_{total}} &= \frac{1}{R_1} + \frac{1}{R_2} + \frac{1}{R_3}
\end{aligned}
$$

$$
\begin{aligned}
\frac{1}{R_{total}} &= \frac{1}{3} + \frac{1}{4} + \frac{1}{5}\,\Omega^{-1} \\
&= \frac{20 + 15 + 12}{60}\,\Omega^{-1} \\
&= \frac{47}{60}\,\Omega^{-1} \\
\therefore R_{total} &= \frac{60}{47}\,\Omega \\
&= 1.28\,\Omega
\end{aligned}
$$

Potential divider circuits and sensors

POTENTIAL DIVIDER CIRCUIT

The example on the right is an example of a circuit involving a **potential divider**. It is so called because the two resistors 'divide up' the potential difference of the battery. You can calculate the 'share' taken by one resistor from the ratio of the resistances but this approach does not work unless the voltmeter's resistance is also considered. An ammeter's internal resistance also needs to be considered. One of the most common mistakes when solving problems involving electrical circuits is to assume the current or potential difference remains constant after a change to the circuit. After a change, the only way to ensure your calculations are correct is to start again.

A variable potential divider (a **potentiometer**) is often the best way to produce a variable power supply. When designing the potential divider, the smallest resistor that is going to be connected needs to be taken into account: the potentiometer's resistance should be significantly smaller.

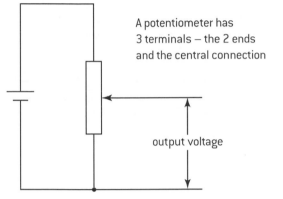

A potentiometer has 3 terminals – the 2 ends and the central connection

output voltage

In order to measure the V–I characteristics of an unknown resistor R, the two circuits (A and B) below are constructed. Both will both provide a range of readings for the potential difference, V, across and current, I, through R. Providing that $R \gg$ the resistance of the potentiometer, this circuit (circuit B) is preferred because the range of readings is greater.

- Circuit B allows the potential difference across R (and hence the current through R) to be reduced down to zero. Circuit A will not go below the minimum value achieved when the variable resistor is at its maximum value.

- Circuit B allows the potential difference across R (and hence the current through R) to be increased up to the maximum value V_{supply} that can be supplied by the power supply in regular intervals. The range of values obtainable by Circuit A depends on a maximum of resistance of the variable resistor.

Circuit A – variable resistor

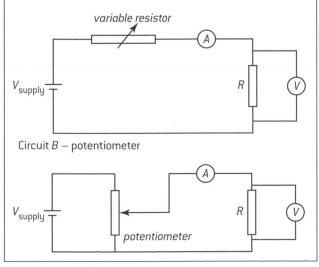

Circuit B – potentiometer

EXAMPLE

In the circuit below the voltmeter has a resistance of 20 kΩ. Calculate:

(a) the pd across the 20 kΩ resistor with the switch open

(b) the reading on the voltmeter with the switch closed.

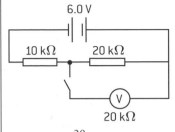
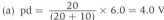

(a) pd $= \dfrac{20}{(20 + 10)} \times 6.0 = 4.0$ V

(b) resistance of 20 kΩ resistor and voltmeter combination, R, given by:

$$\frac{1}{R} = \frac{1}{20} + \frac{1}{20} \ \text{k}\Omega^{-1}$$

$$\therefore R = 10 \ \text{k}\Omega$$

$$\therefore \text{pd} = \frac{10}{(10 + 10)} \times 6.0 = 3.0 \ \text{V}$$

SENSORS

A **light-dependent resistor** (**LDR**), is a device whose resistance depends on the amount of light shining on its surface. An increase in light causes a decrease in resistance.

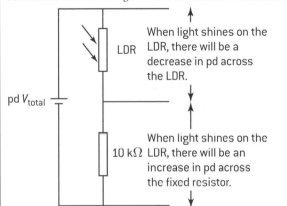

When light shines on the LDR, there will be a decrease in pd across the LDR.

When light shines on the LDR, there will be an increase in pd across the fixed resistor.

A **thermistor** is a resistor whose value of resistance depends on its temperature. Most are semi-conducting devices that have a **negative temperature coefficient** (**NTC**). This means that an increase in temperature causes a decrease in resistance. Both of these devices can be used in potential divider circuits to create sensor circuits. The output potential difference of a **sensor circuit** depends on an external factor.

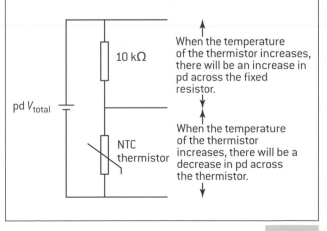

When the temperature of the thermistor increases, there will be an increase in pd across the fixed resistor.

When the temperature of the thermistor increases, there will be a decrease in pd across the thermistor.

Resistivity

RESISTIVITY

The resistivity, ρ, of a material is defined in terms of its resistance, R, its length l and its cross-sectional area A.

$$R = \rho \frac{l}{A}$$

The units of resistivity must be ohm metres (Ω m). Note that this is the ohm multiplied by the metre, not 'ohms per metre'.

Example

The resistivity of copper is 3.3×10^{-7} Ω m; the resistance of a 100 m length of wire of cross-sectional area 1.0 mm² is:

$$R = 3.3 \times 10^{-7} \times \frac{100}{10^{-4}} = 0.3 \ \Omega$$

INVESTIGATING RESISTANCE

The resistivity equation predicts that the resistance R of a substance will be:

a) Proportional to the length l of the substance

b) Inversely proportional to the cross-sectional area A of the substance.

These relationships can be predicted by considering resistors in series and in parallel:

a) Increasing l is like putting another resistor in series. Doubling l is the same as putting an identical resistor in series. R in series with R has an overall resistance of $2R$. Doubling l means doubling R. So $R \propto l$. A graph of R vs l will be a straight line going through the origin.

b) Increasing A is like putting another resistor in parallel. Doubling A is the same as putting an identical resistor in parallel. R in parallel with R has an overall resistance of $\frac{R}{2}$. Doubling A means halving R. So $R \propto \frac{1}{A}$. A graph of R vs $\frac{1}{A}$ will be a straight line going through the origin.

To practically investigate these relationships, we have:

Independent variable:	Either l or A
Control variables:	A or l (depending on above choice); Temperature; Substance.
Data collection:	For each value of independent variable: • a range of values for V and I should be recorded • R can be calculated from the gradient of a V vs I graph.
Data analysis	Values of R and the independent variable analysed graphically.

Possible sources of error/uncertainty include:

- Temperature variation of the substance (particularly if currents are high). Circuits should not be left connected.
- The cross-sectional area of the wire is calculated by measuring the wire's diameter, d, and using $A = \pi r^2 = \frac{\pi d^2}{4}$. Several sets of measurements should be taken along the length of the wire and the readings in a set should be mutually perpendicular.
- The small value of the wire's diameter will mean that the uncertainties generated using a ruler will be large. This will be improved using a **vernier calliper** or a **micrometer**.

Example of use of Kirchoff's laws

KIRCHOFF CIRCUIT LAWS EXAMPLE

Great care needs to be taken when applying Kirchoff's laws to ensure that every term in the equation is correctly identified as positive or negative. The concept of emf (see page 60) as sources of electrical energy can be used along with $V = IR$ to provide an alternative statement of the second law which may help avoid confusion: 'Round any closed circuit, the sum of the emfs is equal to the sum of the products of current and resistance'.

$$\sum(\text{emf}) = \sum(IR)$$

Process to follow

- Draw a full circuit diagram.
- It helps to set up the equations in symbols before substituting numbers and units.
- It helps to be as precise as possible. Potential difference V is a difference between two points in the circuit so specify which two points are being considered (use labels).
- Give the unknown currents symbols and mark their directions on the diagram. If you make a mistake and choose the wrong direction for a current, the solution to the equations will be negative.
- Use Kirchoff's first law to identify appropriate relationships between currents.
- Identify a loop to apply Kirchoff's second law. Go all around the loop in one direction (clockwise or anticlockwise) adding the emfs and $I \times R$ in senses shown below:

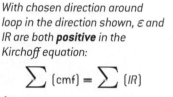

With chosen direction around loop in the direction shown, ε and IR are both **positive** in the Kirchoff equation:

$$\sum(\text{emf}) = \sum(IR)$$

(If chosen direction opposite to that shown, values are negative)

- The total number of different equations generated by Kirchoff's laws needs to be the same as the number of unknowns for the problem to be able to be solved.
- Use simultaneous equations to substitute and solve for the unknown values.
- A new loop can be identified to check that calculated values are correct.

Example

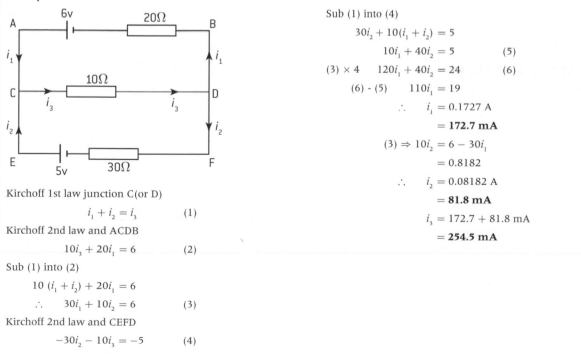

Kirchoff 1st law junction C(or D)

$$i_1 + i_2 = i_3 \qquad (1)$$

Kirchoff 2nd law and ACDB

$$10i_3 + 20i_1 = 6 \qquad (2)$$

Sub (1) into (2)

$$10(i_1 + i_2) + 20i_1 = 6$$
$$\therefore \quad 30i_1 + 10i_2 = 6 \qquad (3)$$

Kirchoff 2nd law and CEFD

$$-30i_2 - 10i_3 = -5 \qquad (4)$$

Sub (1) into (4)

$$30i_2 + 10(i_1 + i_2) = 5$$
$$10i_1 + 40i_2 = 5 \qquad (5)$$
$$(3) \times 4 \qquad 120i_1 + 40i_2 = 24 \qquad (6)$$
$$(6) - (5) \qquad 110i_1 = 19$$
$$\therefore \quad i_1 = 0.1727 \text{ A}$$
$$= \textbf{172.7 mA}$$
$$(3) \Rightarrow 10i_2 = 6 - 30i_1$$
$$= 0.8182$$
$$\therefore \quad i_2 = 0.08182 \text{ A}$$
$$= \textbf{81.8 mA}$$
$$i_3 = 172.7 + 81.8 \text{ mA}$$
$$= \textbf{254.5 mA}$$

Internal resistance and cells

ELECTROMOTIVE FORCE AND INTERNAL RESISTANCE

When a 6V battery is connected in a circuit some energy will be used up inside the battery itself. In other words, the battery has some **internal resistance**. The TOTAL energy difference per unit charge around the circuit is still 6 volts, but some of this energy is used up inside the battery. The energy difference per unit charge from one terminal of the battery to the other is less than the total made available by the chemical reaction in the battery.

For historical reasons, the TOTAL energy difference per unit charge around a circuit is called the **electromotive force** (**emf**). However, remember that it is not a force (measured in newtons) but an energy difference per charge (measured in volts).

In practical terms, emf is exactly the same as potential difference if no current flows.

$$\varepsilon = I\,(R + r)$$

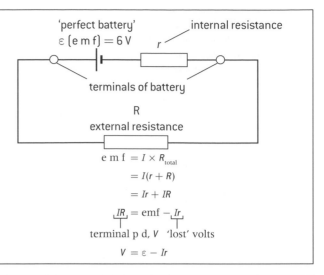

'perfect battery'
$\varepsilon\,(\text{e m f}) = 6\,\text{V}$ internal resistance r

terminals of battery

R
external resistance

$$\text{e m f} = I \times R_{\text{total}}$$
$$= I(r + R)$$
$$= Ir + IR$$
$$IR = \text{emf} - Ir$$

terminal p d, V 'lost' volts

$$V = \varepsilon - Ir$$

CELLS AND BATTERIES

An electric **battery** is a device consisting of one or more cells joined together. In a cell, a chemical reaction takes place, which converts stored chemical energy into electrical energy. There are two different types of cell: primary and secondary.

A **primary** cell cannot be recharged. During the lifetime of the cell, the chemicals in the cell get used in a non-reversible reaction. Once a primary cell is no longer able to provide electrical energy, it is thrown away. Common examples include zinc–carbon batteries and alkaline batteries.

A **secondary** cell is designed to be recharged. The chemical reaction that produces the electrical energy is reversible. A reverse electrical current charges the cell allowing it to be reused many times. Common examples include a lead–acid car battery, nickel–cadmium and lithium-ion batteries.

The **charge capacity** of a cell is how much charge can flow before the cells stops working. Typical batteries have charge capacities that are measured in Amp-hours (A h). 1 A h is the charge that flows when a current of 1 A flows for one hour i.e. 1 A h = 3600 C.

DISCHARGE CHARACTERISTICS

When current (and thus electrical energy) is drawn from a cell, the terminal potential difference varies with time. A perfect cell would maintain its terminal pd throughout its lifetime; real cells, however, do not. The terminal potential difference of a typical cell:

* loses its initial value quickly,
* has a stable and reasonably constant value for most of its lifetime.

This is followed by a rapid decrease to zero (cell discharges). The graph below shows the discharge characteristics for one particular type of lead–acid car battery.

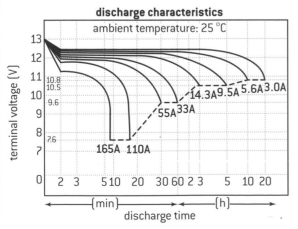

discharge characteristics
ambient temperature: 25 °C

terminal voltage (V)

14.3A 9.5A 5.6A 3.0A
55A 33A
165A 110A

{min}   {h}
discharge time

DETERMINING INTERNAL RESISTANCE EXPERIMENTALLY

To experimentally determine the internal resistance r of a cell (and its emf ε), the circuit below can be used:

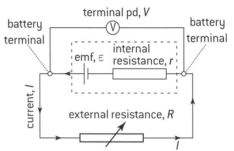

terminal pd, V
battery terminal   battery terminal
emf, ε internal resistance, r
current, I
external resistance, R

Procedure:

* Vary external resistance R to get a number (ideally 10 or more) of matching readings of V and I over as wide a range as possible.
* Repeat readings.
* Do not leave current running for too long (especially at high values of I).
* Take care that nothing overheats.

Data analysis:

* The relevant equation, $V = \varepsilon - Ir$ was introduced above.
* A plot of V on the y-axis and I on the x-axis gives a straight line graph with
  * gradient $= -r$
  * y-intercept $= \varepsilon$

RECHARGING SECONDARY CELLS

In order to recharge a secondary cell, it is connected to an external DC power source. The negative terminal of the secondary cell is connected to the negative terminal of the power source and the positive terminal of the power source with the positive terminal of the secondary cell. In order for a charging current, I, to flow, the voltage output of the power source must be slightly higher than that of the battery. A large difference between the power source and the cell's terminal potential difference means that the charging process will take less time but risks damaging the cell.

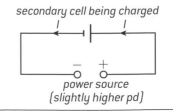

secondary cell being charged
I I
$-$ $+$
power source
(slightly higher pd)

Magnetic force and fields

MAGNETIC FIELD LINES

There are many similarities between the magnetic force and the electrostatic force. In fact, both forces have been shown to be two aspects of one force – the electromagnetic interaction (see page 78). It is, however, much easier to consider them as completely separate forces to avoid confusion.

Page 52 introduced the idea of electric fields. A similar concept is used for magnetic fields. A table of the comparisons between these two fields is shown below.

	Electric field	**Magnetic field**
Symbol	E	B
Caused by …	Charges	Magnets (or electric currents)
Affects …	Charges	Magnets (or electric currents)
Two types of …	Charge: positive and negative	Pole: North and South
Simple force rule:	Like charges repel, unlike charges attract	Like poles repel, unlike poles attract

In order to help visualize a magnetic field we, once again, use the concept of field lines. This time the field lines are lines of magnetic field – also called **flux** lines. If a 'test' magnetic North pole is placed in a magnetic field, it will feel a force.

- The direction of the force is shown by the direction of the field lines.

- The strength of the force is shown by how close the lines are to one another.

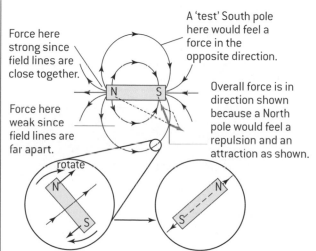

Force here strong since field lines are close together.

A 'test' South pole here would feel a force in the opposite direction.

Force here weak since field lines are far apart.

Overall force is in direction shown because a North pole would feel a repulsion and an attraction as shown.

rotate

A small magnet placed in the field would rotate until lined up with the field lines. This is how a compass works. Small pieces of iron (iron filings) will also line up with the field lines – they willbe induced to become little magnets.

Field pattern of an isolated bar magnet

Despite all the similarities between electric fields and magnetic fields, it should be remembered that they are very different. For example:

- A magnet does not feel a force when placed in an electric field.

- A positive charge does not feel a force when placed stationary in a magnetic field.

- Isolated charges exist whereas isolated poles do not.

- The Earth itself has a magnetic field. It turns out to be similar to that of a bar magnet with a magnetic South pole near the geographic North Pole as shown below.

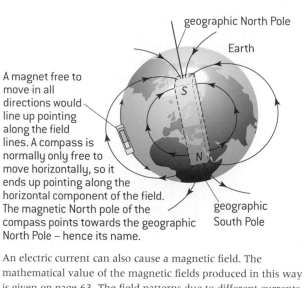

A magnet free to move in all directions would line up pointing along the field lines. A compass is normally only free to move horizontally, so it ends up pointing along the horizontal component of the field. The magnetic North pole of the compass points towards the geographic North Pole – hence its name.

An electric current can also cause a magnetic field. The mathematical value of the magnetic fields produced in this way is given on page 63. The field patterns due to different currents can be seen in the diagrams below.

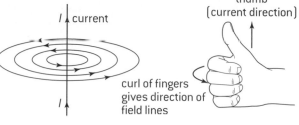

The field lines are circular around the current.

The direction of the field lines can be remembered with the right-hand grip rule. If the thumb of the right hand is arranged to point along the direction of a current, the way the fingers of the right hand naturally curl will give the direction of the field lines.

Field pattern of a straight wire carrying current

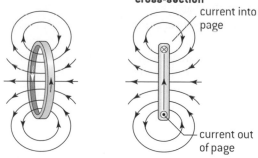

Field pattern of a flat circular coil

A long current-carrying coil is called a solenoid.

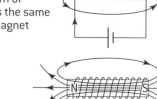

field pattern of solenoid is the same as a bar magnet

cross-section

poles of solenoid can be predicted using right-hand grip rule

Field pattern for a solenoid

Magnetic forces

MAGNETIC FORCE ON A CURRENT

When a current-carrying wire is placed in a magnetic field the magnetic interaction between the two results in a force. This is known as the **motor effect**. The direction of this force is at right angles to the plane that contains the field and the current as shown below.

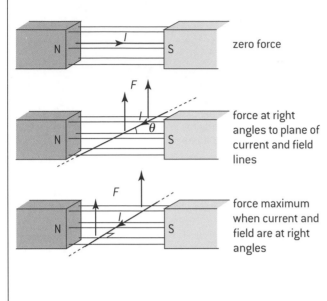

zero force

force at right angles to plane of current and field lines

force maximum when current and field are at right angles

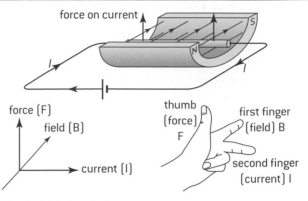

Fleming's left-hand rule

Experiments show that the force is proportional to:

- the magnitude of the magnetic field, B
- the magnitude of the current, I
- the length of the current, L, that is in the magnetic field
- the sine of the angle, θ, between the field and current.

The magnetic field strength, B is defined as follows:

$$F = BIL \sin \theta \quad \text{or}$$

$$B = \frac{F}{IL \sin \theta}$$

A new unit, the tesla, is introduced. 1 T is defined to be equal to $1 \text{ N A}^{-1} \text{ m}^{-1}$. Another possible unit for magnetic field strength is Wb m^{-2}. Another possible term is magnetic flux density.

MAGNETIC FORCE ON A MOVING CHARGE

A single charge moving through a magnetic field also feels a force in exactly the same way that a current feels a force.

In this case the force on a moving charge is proportional to:

- the magnitude of the magnetic field, B
- the magnitude of the charge, q
- the velocity of the charge, v
- the sine of the angle, θ, between the velocity of the charge and the field.

We can use these relationships to give an alternative definition of the magnetic field strength, B. This definition is exactly equivalent to the previous definition.

$$F = Bqv \sin \theta \quad \text{or} \quad B = \frac{F}{qv \sin \theta}$$

Since the force on a moving charge is always at right angles to the velocity of the charge the resultant motion can be circular. An example of this would be when an electron enters a region where the magnetic field is at right angles to its velocity as shown below.

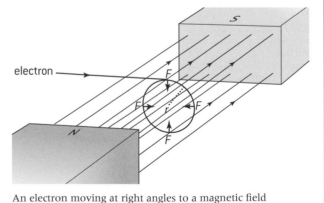

An electron moving at right angles to a magnetic field

Examples of the magnetic field due to currents

The formulae used on this page do not need to be remembered.

STRAIGHT WIRE

The field pattern around a long straight wire shows that as one moves away from the wire, the strength of the field gets weaker. Experimentally the field is proportional to:

- the value of the current, I
- the inverse of the distance away from the wire, r. If the distance away is doubled, the magnetic field will halve.
- The field also depends on the medium around the wire. These factors are summarized in the equation:

$$B = \frac{\mu I}{2\pi r}$$

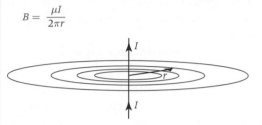

Magnetic field of a straight current

The constant μ is called the permeability and changes if the medium around the wire changes. Most of the time we consider the field around a wire when there is nothing there – so we use the value for the permeability of a vacuum, μ_0. There is almost no difference between the permeability of air and the permeability of a vacuum. There are many possible units for this constant, but it is common to use $N\ A^{-2}$ or $T\ m\ A^{-1}$.

Permeability and permittivity are related constants. In other words, if you know one constant you can calculate the other. In the SI system of units, the permeability of a vacuum is defined to have a value of exactly $4\pi \times 10^{-7}\ N\ A^{-2}$. See the definition of the ampere (right) for more detail.

MAGNETIC FIELD IN A SOLENOID

The magnetic field of a solenoid is very similar to the magnetic field of a bar magnet. As shown by the parallel field lines, the magnetic field inside the solenoid is constant. It might seem surprising that the field does not vary at all inside the solenoid, but this can be experimentally verified near the centre of a long solenoid. It does tend to decrease near the ends of the solenoid as shown in the graph below.

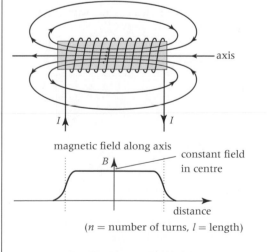

(n = number of turns, l = length)

Variation of magnetic field in a solenoid

TWO PARALLEL WIRES – DEFINITION OF THE AMPERE

Two parallel current-carrying wires provide a good example of the concepts of magnetic field and magnetic force. Because there is a current flowing down the wire, each wire is producing a magnetic field. The other wire is in this field so it feels a force. The forces on the wires are an example of a Newton's third law pair of forces.

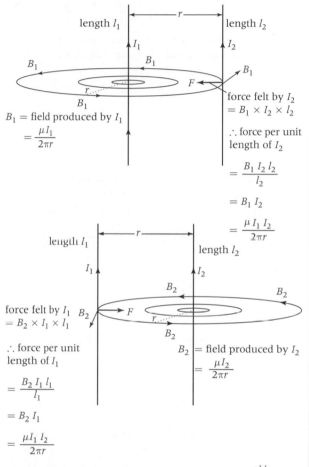

B_1 = field produced by I_1

$$= \frac{\mu I_1}{2\pi r}$$

force felt by I_2
$= B_1 \times I_2 \times l_2$

$\therefore$ force per unit length of I_2

$$= \frac{B_1 I_2 l_2}{l_2}$$

$$= B_1 I_2$$

$$= \frac{\mu I_1 I_2}{2\pi r}$$

force felt by I_1
$= B_2 \times I_1 \times l_1$

$\therefore$ force per unit length of I_1

$$= \frac{B_2 I_1 l_1}{l_1}$$

$$= B_2 I_1$$

$$= \frac{\mu I_1 I_2}{2\pi r}$$

B_2 = field produced by I_2

$$= \frac{\mu I_2}{2\pi r}$$

Magnitude of force per unit length on either wire $= \frac{\mu I_1 I_2}{2\pi r}$

This equation is experimentally used to define the ampere. The coulomb is then defined to be one ampere second. If we imagine two infinitely long wires carrying a current of one amp separated by a distance of one metre, the equation would predict the force per unit length to be $2 \times 10^{-7}\ N$. Although it is not possible to have infinitely long wires, an experimental set-up can be arranged with very long wires indeed. This allows the forces to be measured and ammeters to be properly calibrated.

The mathematical equation for this constant field at the centre of a long solenoid is

$$B = \mu \left(\frac{n}{l}\right) I$$

Thus the field only depends on:

- the current, I
- the number of turns per unit length, $\frac{n}{l}$
- the nature of the solenoid core, μ

It is independent of the cross-sectional area of the solenoid.

IB Questions – electricity and magnetism

1. Which **one** of the field patterns below could be produced by two point charges?

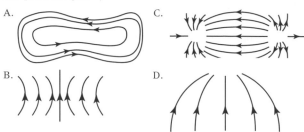

A.

B.

C.

D.

2. Two long, vertical wires **X** and **Y** carry currents in the same direction and pass through a horizontal sheet of card.

Iron filings are scattered on the card. Which **one** of the following diagrams best shows the pattern formed by the iron filings? (The dots show where the wires **X** and **Y** enter the card.)

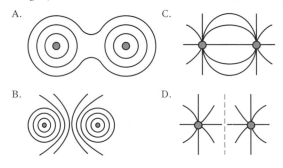

A.

B.

C.

D.

3. This question is about the electric field due to a charged sphere and the motion of electrons in that field.

The diagram below shows an isolated metal sphere in a vacuum that carries a negative electric charge of 9.0 nC.

$$\bigodot$$

a) On the diagram draw arrows to represent the electric field pattern due to the charged sphere. [3]

b) The electric field strength at the surface of the sphere and at points outside the sphere can be determined by assuming that the sphere acts as though a point charge of magnitude 9.0 nC is situated at its centre. The radius of the sphere is 4.5×10^{-2} m. Deduce that the magnitude of the field strength at the surface of the sphere is 4.0×10^4 V m^{-1}. [1]

An electron is initially at rest on the surface of the sphere.

c) (i) Describe the path followed by the electron as it leaves the surface of the sphere. [1]

(ii) Calculate the initial acceleration of the electron. [3]

(iii) State and explain whether the acceleration of the electron remains constant, increases or decreases as it moves away from the sphere. [2]

(iv) At a certain point P, the speed of the electron is 6.0×10^6 m s^{-1}. Determine the potential difference between the point P and the surface of the sphere. [2]

4. In order to measure the voltage-current (V-I) characteristics of a lamp, a student sets up the following electrical circuit.

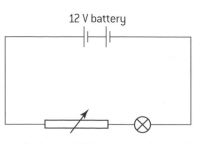

a) On the circuit above, add circuit symbols showing the correct positions of an ideal ammeter **and** an ideal voltmeter that would allow the V-I characteristics of this lamp to be measured. [2]

The voltmeter and the ammeter are connected correctly in the circuit above.

b) Explain why the potential difference across the lamp

(i) cannot be increased to 12 V. [2]

(ii) cannot be reduced to zero. [2]

An alternative circuit for measuring the V-I characteristic uses a *potential divider*. [3]

c) (i) Draw a circuit that uses a potential divider to enable the V-I characteristics of the filament to be found. [3]

(ii) Explain why this circuit enables the potential difference across the lamp to be reduced to zero volts. [2]

The graph below shows the V-I characteristic for two 12 V filament lamps A and B.

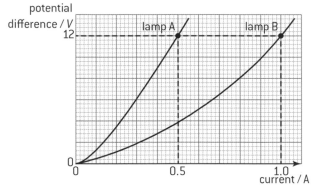

d) State and explain which lamp has the greater power dissipation for a potential difference of 12 V. [3]

The two lamps are now connected in series with a 12 V battery as shown below.

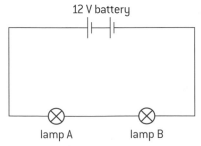

e) (i) State how the current in lamp A compares with that in lamp B. [1]

(ii) Use the V-I characteristics of the lamps to deduce the total current from the battery. [4]

(iii) Compare the power dissipated by the two lamps. [2]

Uniform circular motion

MECHANICS OF CIRCULAR MOTION

The phrase 'uniform circular motion' is used to describe an object that is going around a circle at constant speed. Most of the time this also means that the circle is horizontal. An example of uniform circular motion would be the motion of a small mass on the end of a string as shown below.

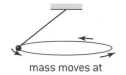

mass moves at
constant speed

Example of uniform circular motion

It is important to remember that even though the speed of the object is constant, its direction is changing all the time.

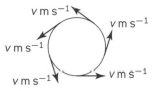

speed is constant but the direction is constantly changing

Circular motion – the direction of motion is changing all the time

This constantly changing direction means that the velocity of the object is constantly changing. The word 'acceleration' is used whenever an object's velocity changes. This means that an object in uniform circular motion MUST be accelerating even if the speed is constant.

The acceleration of a particle travelling in circular motion is called the **centripetal acceleration**. The force needed to cause the centripetal acceleration is called the **centripetal force**.

MATHEMATICS OF CIRCULAR MOTION

The diagram below allows us to work out the direction of the centripetal acceleration – which must also be the direction of the centripetal force. This direction is constantly changing.

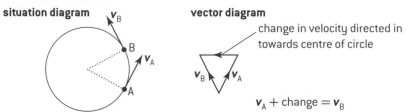

situation diagram

vector diagram

change in velocity directed in towards centre of circle

$$v_A + \text{change} = v_B$$

The object is shown moving between two points A and B on a horizontal circle. Its velocity has changed from v_A to v_B. The magnitude of velocity is always the same, but the direction has changed. Since velocities are vector quantities we need to use vector mathematics to work out the average change in velocity. This vector diagram is also shown above.

In this example, the direction of the average change in velocity is towards the centre of the circle. This is always the case and thus true for the instantaneous acceleration. For a mass m moving at a speed v in uniform circular motion of radius r,

Centripetal acceleration $a_{\text{centripetal}} = \dfrac{v^2}{r}$ [in towards the centre of the circle]

A force must have caused this acceleration. The value of the force is worked out using Newton's second law:

Centripetal force (CPF) $F_{\text{centripetal}} = m\, a_{\text{centripetal}}$

$$= \frac{m\, v^2}{r} \text{ [in towards the centre of the circle]}$$

For example, if a car of mass 1500 kg is travelling at a constant speed of 20 m s^{-1} around a circular track of radius 50 m, the resultant force that must be acting on it works out to be

$$F = \frac{1500(20)^2}{50} = 12\ 000 \text{ N}$$

It is really important to understand that centripetal force is NOT a new force that starts acting on something when it goes in a circle. It is a way of working out what the total force must have been. This total force must result from all the other forces on the object. See the examples below for more details.

One final point to note is that the centripetal force does NOT do any work. (Work done = force × distance **in the direction of the force**.)

EXAMPLES

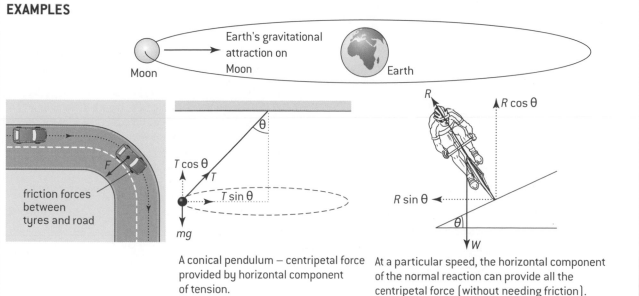

A conical pendulum – centripetal force provided by horizontal component of tension.

At a particular speed, the horizontal component of the normal reaction can provide all the centripetal force (without needing friction).

Angular velocity and vertical circular motion

RADIANS

Angles measure the fraction of a complete circle that has been achieved. They can, of course, be measured in degrees (symbol: °) but in studying circular motion, the radian (symbol: rad) is a more useful measure.

radius r

θ

s distance along circular arc

The fraction of the circle that has been achieved is the ratio of arc length s to the circumference:

$$\text{fraction of circle} = \frac{s}{2\pi r}$$

In degrees, the whole circle is divided up into 360° which defines the angle θ as:

$$\theta(\text{in degrees}) = \frac{s}{2\pi r} \times 360$$

In radians, the whole circle is divided up into 2π radians which defines the angle θ as:

$$\theta(\text{in radians}) = \frac{s}{2\pi r} \times 2\pi = \frac{s}{r}$$

For small angles (less than about 0.1 rad or 5°), the arc and the two radii form a shape that approximates to a triangle. Since radians are just a ratio, the following relationship applies if working in radians:

$$\sin\theta \approx \tan\theta \approx \theta$$

Angle/°	Angle/radian
0	0.00
5	0.09
45	$0.74 = \frac{\pi}{4}$
60	1.05
90	$1.57 = \frac{\pi}{2}$
180	$3.14 = \pi$
270	$4.71 = \frac{3\pi}{2}$
360	$6.28 = 2\pi$

ANGULAR VELOCITY, ω, AND TIME PERIOD, T

An object travelling in circular motion must be constantly changing direction. As a result its velocity is constantly changing even if its speed is constant (uniform circular motion). We define the average **angular velocity**, symbol ω (omega) as:

$$\omega_{\text{average}} = \frac{\text{angle turned}}{\text{time taken}} = \frac{\Delta\theta}{\Delta t}$$

The units of angular velocity are radians per second (rad s^{-1}).

The instantaneous angular velocity is the rate of change of angle:

$$\omega = \text{rate of change of angle} = \frac{d\theta}{dt}$$

1. Link between ω and v

In a time Δt, the object rotates an angle $\Delta\theta$

$$\theta = \frac{s}{r} \therefore s = r\Delta\theta$$

$$v = \frac{s}{\Delta t} = \frac{r\Delta\theta}{\Delta t} = r\omega$$

$$v = r\omega$$

2. Link between ω and time period T

The time period T is the time taken to complete one full circle. In this time, the total angle turned is 2π radians, so:

$$\omega = \frac{2\pi}{T} \text{ or } T = \frac{2\pi}{\omega}$$

3. Circular motion equations

Substitution of the above equations into the formulae for centripetal force and centripetal acceleration (page 65) provide versions that are sometime more useful:

$$\text{centripetal acceleration, } a = \frac{v^2}{r} = r\omega^2 = \frac{4\pi^2 r}{T^2}$$

$$\text{centripetal force, } F = \frac{mv^2}{r} = mr\omega^2 = \frac{4\pi^2 mr}{T^2}$$

CIRCULAR MOTION IN A VERTICAL PLANE

Uniform circular motion of a mass on the end of a string in a horizontal plane requires a constant centripetal force to act and the magnitude of the tension in the string will not change. Circular motion in the vertical plane is more complicated as the weight of the object always acts in the same vertical direction. The object will speed up and slow down during its motion due to the component of its weight that acts along the tangent to the circle. The maximum speed will be when the object is at the bottom and the minimum speed will occur at the top. The tension in the string will also change during one revolution.

In a vertical circle, the tension of the string will always act at 90° to the object's velocity so this force does no work in speeding it up or slowing it down. The conservation of energy means that:

$$mgy + \frac{1}{2}mv^2 = \text{constant}$$

a) SITUATION DIAGRAM

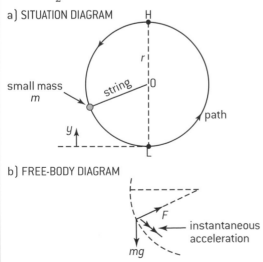

b) FREE-BODY DIAGRAM

instantaneous acceleration

1. At the top of the circle:,

The tension in the string, T, and the weight, mg, are in the same direction and add together to provide the CPF:

$$T_{\text{top}} + mg = \frac{mv_{\text{top}}^2}{r}$$

To remain in the vertical circle, the object must be moving with a certain minimum speed. At this minimum top speed, $v_{\text{top min}}$, the tension is zero and the centripetal force is provided by the object's weight:

$$mg = \frac{m(v_{\text{top min}})^2}{r}$$

$$v_{\text{top min}} = \sqrt{rg}$$

2. At the bottom of the circle:,

The tension in the string, T, and the weight, mg, are in opposite directions and the resultant force provides the CPF:

$$T_{\text{bottom}} - mg = \frac{mv_{\text{bottom}}^2}{r}$$

In order to complete the vertical circle, the KE at the bottom of the circle must be large enough for the object to arrive at the top of the circle with sufficient speed ($v_{\text{top min}} = \sqrt{rg}$) to complete the circle. Energetically the object gains PE ($= mg \times 2r$) so it must lose the same amount of KE:

$$\frac{1}{2}m(v_{\text{bottom min}})^2 - mg2r = \frac{1}{2}m(v_{\text{top min}})^2 = \frac{1}{2}mrg$$

$$\therefore (v_{\text{bottom min}})^2 - 4gr = rg$$

$$\therefore v_{\text{bottom min}} = \sqrt{5rg}$$

The mathematics in the above example (a mass on the end of a string) can also apply for any vehicle that is 'looping the loop'. In place of T, the tension in the rope, there is N, the normal reaction from the surface.

Newton's law of gravitation

NEWTON'S LAW OF UNIVERSAL GRAVITATION

If you trip over, you will fall down towards the ground.

Newton's theory of **universal gravitation** explains what is going on. It is called 'universal' gravitation because at the core of this theory is the statement that every mass in the Universe attracts all the other masses in the Universe. The value of the attraction between two **point** masses is given by an equation.

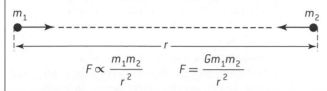

$$F \propto \frac{m_1 m_2}{r^2} \qquad F = \frac{G m_1 m_2}{r^2}$$

Universal gravitational constant G = 6.67×10^{-11} N m² kg⁻²

The following points should be noticed:

- The law only deals with point masses.
- There is a force acting on each of the masses. These forces are EQUAL and OPPOSITE (even if the masses are not equal).
- The forces are always attractive.
- Gravitation forces act between ALL objects in the Universe. The forces only become significant if one (or both) of the objects involved are massive, but they are there nonetheless.

The interaction between two spherical masses turns out to be the same as if the masses were concentrated at the centres of the spheres.

GRAVITATIONAL FIELD STRENGTH

The table below should be compared with the one on page 61.

	Gravitational field strength
Symbol	g
Caused by...	Masses
Affects...	Masses
One type of...	Mass
Simple force rule:	All masses attract

The gravitational field is therefore defined as the force per unit mass. $g = \frac{F}{m}$ m = small point test mass

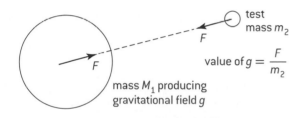

value of $g = \frac{F}{m_2}$

mass M_1 producing gravitational field g

The SI units for g are N kg⁻¹. These are the same as m s⁻². Field strength is a vector quantity and can be represented by the use of field lines.

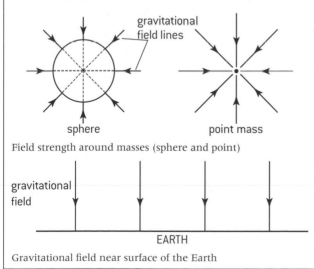

Field strength around masses (sphere and point)

Gravitational field near surface of the Earth

In the example on the left the numerical value for the gravitational field can be calculated using Newton's law:

$$F = \frac{GMm}{r^2} \qquad g = \frac{GM}{r^2}$$

The gravitational field strength at the surface of a planet must be the same as the acceleration due to gravity on the surface.

Field strength is defined to be $\frac{\text{force}}{\text{mass}}$

Acceleration = $\frac{\text{force}}{\text{mass}}$ (from $F = ma$)

For the Earth

$M = 6.0 \times 10^{24}$ kg

$r = 6.4 \times 10^6$ m

$g = \frac{6.67 \times 10^{-11} \times 6.0 \times 10^{24}}{(6.4 \times 10^6)^2} = 9.8$ m s⁻²

EXAMPLE

In order to calculate the overall gravitational field strength at any point we must use vector addition. The overall gravitational field strength at any point between the Earth and the Moon must be a result of both pulls.

There will be a single point somewhere between the Earth and the Moon where the total gravitational field due to these two masses is zero. Up to this point the overall pull is back to the Earth, after this point the overall pull is towards the Moon.

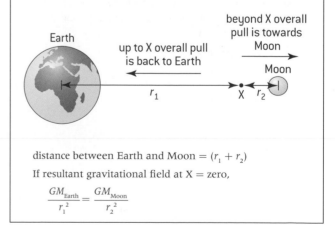

distance between Earth and Moon = $(r_1 + r_2)$

If resultant gravitational field at X = zero,

$$\frac{GM_{\text{Earth}}}{r_1^2} = \frac{GM_{\text{Moon}}}{r_2^2}$$

IB Questions – circular motion and gravitation

1. A ball is tied to a string and rotated at a uniform speed in a vertical plane. The diagram shows the ball at its lowest position. Which arrow shows the direction of the net force acting on the ball? [1]

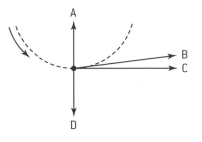

2. A particle of mass m is moving with constant speed v in uniform circular motion. What is the total work done by the centripetal force during one revolution? [1]

   A. Zero

   B. $\frac{mv^2}{2}$

   C. mv^2

   D. $2\pi mv^2$

3. A particle P is moving anti-clockwise with constant speed in a horizontal circle.

   Which diagram correctly shows the direction of the velocity v and acceleration a of the particle P in the position shown? [1]

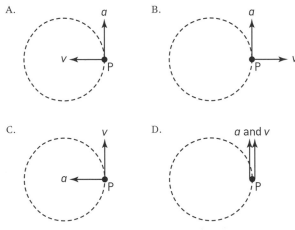

4. This question is about circular motion.

   A ball of mass 0.25 kg is attached to a string and is made to rotate with constant speed v along a horizontal circle of radius $r = 0.33$ m. The string is attached to the ceiling and makes an angle of 30° with the vertical.

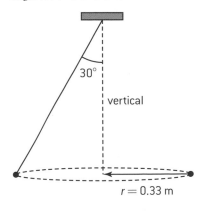

   a) (i) On the diagram above, draw and label arrows to represent the forces on the ball in the position shown. [2]

   (ii) State and explain whether the ball is in equilibrium. [2]

   b) Determine the speed of rotation of the ball. [3]

5. This question is about gravitational fields.

   a) Define *gravitational field strength*. [2]

   b) The gravitational field strength at the surface of Jupiter is 25 N kg^{-1} and the radius of Jupiter is 7.1×10^7 m.

   (i) Derive an expression for the gravitational field strength at the surface of a planet in terms of its mass M, its radius R and the gravitational constant G. [2]

   (ii) Use your expression in (b)(i) above to estimate the mass of Jupiter. [2]

6. Gravitational fields and potential

   a) Derive an expression for the gravitational field strength as a function of distance away from a point mass M. [3]

   b) The radius of the Earth is 6400 km and the gravitational field strength at its surface is 9.8 N kg^{-1}. Calculate a value for the mass of the Earth. [2]

   c) On the diagram below draw lines to represent the gravitational field outside the Earth. [2]

   d) A satellite that orbits the Earth is in the gravitational field of the Earth. Discuss why an astronaut inside the satellite feels weightless. [3]

Emission and absorption spectra

EMISSION SPECTRA AND ABSORPTION SPECTRA

When an element is given enough energy it emits light. This light can be analysed by splitting it into its various colours (or frequencies) using a prism or a diffraction grating. If all possible frequencies of light were present, this would be called a **continuous spectrum**. The light an element emits, its **emission spectrum**, is not continuous, but contains only a few characteristic colours. The frequencies emitted are particular to the element in question. For example, the yellow-orange light from a street lamp is often a sign that the element sodium is present in the lamp. Exactly the same particular frequencies are **absent** if a continuous spectrum of light is shone through an element when it is in gaseous form. This is called an **absorption** spectrum.

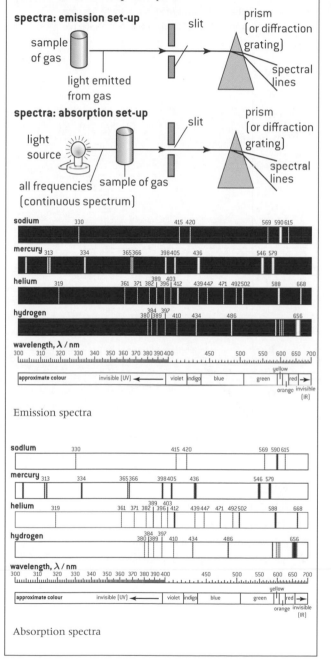

Emission spectra

Absorption spectra

EXPLANATION OF ATOMIC SPECTRA

In an atom, electrons are bound to the nucleus. See page 77, the atomic model. This means that they cannot 'escape' without the input of energy. If enough energy is put in, an electron can leave the atom. If this happens, the atom is now positive overall and is said to be ionized. Electrons can only occupy given energy levels – the energy of the electron is said to be **quantized**. These energy levels are fixed for particular elements and correspond to 'allowed' obitals. The reason why only these energies are 'allowed' forms a significant part of quantum theory (see HL topic 12).

When an electron moves between energy levels it must emit or absorb energy. The energy emitted or absorbed corresponds to the difference between the two allowed energy levels. This energy is emitted or absorbed as 'packets' of light called photons. A higher energy photon corresponds to a higher frequency (shorter wavelength) of light.

The energy of a photon is given by

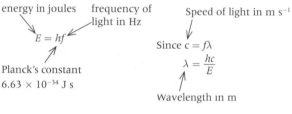

energy in joules   frequency of light in Hz

$$E = hf$$

Planck's constant
6.63×10^{-34} J s

Speed of light in m s^{-1}

Since $c = f\lambda$

$$\lambda = \frac{hc}{E}$$

Wavelength in m

Thus the frequency of the light, emitted or absorbed, is fixed by the energy difference between the levels. Since the energy levels are unique to a given element, this means that the emission (and the absorption) spectrum will also be unique.

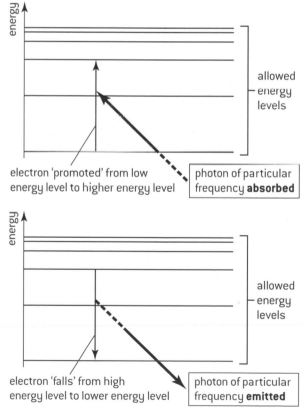

electron 'promoted' from low energy level to higher energy level

photon of particular frequency **absorbed**

electron 'falls' from high energy level to lower energy level

photon of particular frequency **emitted**

Nuclear stability

ISOTOPES

When a chemical reaction takes place, it involves the outer electrons of the atoms concerned. Different elements have different chemical properties because the arrangement of outer electrons varies from element to element. The chemical properties of a particular element are fixed by the amount of positive charge that exists in the nucleus – in other words, the number of protons. In general, different nuclear structures will imply different chemical properties. A **nuclide** is the name given to a particular species of atom (one whose nucleus contains a specified number of protons and a specified number of neutrons). Some nuclides are the same element – they have the same chemical properties and contain the same number of protons. These nuclides are called **isotopes** – they contain the same number of protons but different numbers of neutrons.

NOTATION

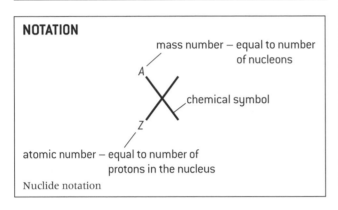

Nuclide notation

EXAMPLES

	Notation	Description	Comment
1	$_{6}^{12}\text{C}$	carbon-12	isotope of **2**
2	$_{6}^{13}\text{C}$	carbon-13	isotope of **1**
3	$_{92}^{238}\text{U}$	uranium-238	
4	$_{78}^{198}\text{Pt}$	platinum-198	same mass number as **5**
5	$_{80}^{198}\text{Hg}$	mercury-198	same mass number as **4**

Each element has a unique chemical symbol and its own atomic number. *No.1* and *No.2* are examples of two isotopes, whereas *No.4* and *No.5* are not.

In general, when physicists use this notation they are concerned with the nucleus rather than the whole atom. Chemists use the same notation but tend to include the overall charge on the atom. Thus $_{6}^{12}\text{C}$ can represent the carbon nucleus to a physicist or the carbon atom to a chemist depending on the context. If the charge is present the situation becomes unambiguous. $_{17}^{35}\text{Cl}^{-}$ must refer to a chlorine ion – an atom that has gained one extra electron.

NUCLEAR STABILITY

Many atomic nuclei are unstable. The process by which they decay is called radioactive decay (see page 72). It involves emission of alpha (α), beta (β) or gamma (γ) radiation. The stability of a particular nuclide depends greatly on the numbers of neutrons present. The graph below shows the stable nuclides that exist.

- For small nuclei, the number of neutrons tends to equal the number of protons.

- For large nuclei there are more neutrons than protons.

- Nuclides above the band of stability have 'too many neutrons' and will tend to decay with either alpha or beta decay (see page 72).

- Nuclides below the band of stability have 'too few neutrons' and will tend to emit positrons (see page 73).

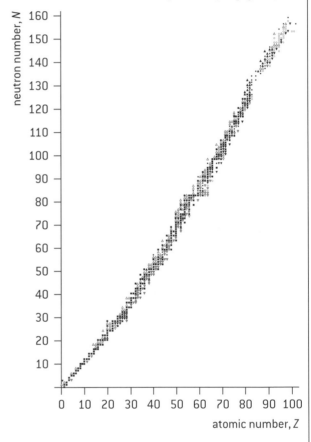

Key

N number of neutrons

Z number of protons

■ naturally occurring stable nuclide

● naturally occurring α-emitting nuclide

○ artificially produced α-emitting nuclide

▲ naturally occurring β^{-}-emitting nuclide

△ artificially produced β^{+}-emitting nuclide

▽ artificially produced β^{-}-emitting nuclide

▼ artificially produced electron-capturing nuclide

▼ artificial nuclide decaying by spontaneous fission

Fundamental forces

STRONG NUCLEAR FORCE

The protons in a nucleus are all positive. Since like charges repel, they must be repelling one another all the time. This means there must be another force keeping the nucleus together. Without it the nucleus would 'fly apart'. We know a few things about this force.

- It must be strong. If the proton repulsions are calculated it is clear that the gravitational attraction between the nucleons is far too small to be able to keep the nucleus together.

- It must be very short-ranged as we do not observe this force anywhere other than inside the nucleus.

- It is likely to involve the neutrons as well. Small nuclei tend to have equal numbers of protons and neutrons. Large nuclei need proportionately more neutrons in order to keep the nucleus together.

The name given to this force is the **strong nuclear force**.

WEAK NUCLEAR FORCE

The strong nuclear force (see box left) explains why nuclei do not fly apart and thus why they are stable. Most nuclei, however, are unstable. Mechanisms to explain alpha and gamma emission (see page 72) can be identified but another nuclear force must be involved if we wish to be able to explain all aspects of the nucleus including beta emission. We know a few things about this force:

- It must be weak. Many nuclei are stable and beta emission does not always occur.

- It must be very short-ranged as we do not observe this force anywhere other than inside the nucleus.

- Unlike the strong nuclear force, it involves the lighter particles (e.g. electrons, positrons and neutrinos) as well as the heavier ones (e.g. protons and neutrons).

The name given to this force is the **weak nuclear force**.

OTHER FUNDAMENTAL FORCES/INTERACTIONS

The standard model of particle physics is based around the forces that we observe on a daily basis along with the two 'new' forces that have been identified as being involved in nuclear stability (above). As a result in the standard model, there are only four fundament forces (or **interactions**) that are known to exist. These are Gravity, Electromagnetic, Strong and Weak. More detail about all these forces is discussed on page 78. Outline information about two 'everyday' interactions is listed below:

Gravity

- Gravity is the force of attraction between all objects that have mass.

- Gravity is always attractive – masses are pulled together.

- The range of the gravity force is infinite.

- Despite the above, the gravity force is relatively quite weak. At least one of the masses involved needs to be large for the effects to be noticeable. For example, the gravitational force of attraction between you and this book is negligible, but the force between this book and the Earth is easily demonstrable (drop it).

- Newton's law of gravitation describes the mathematics governing this force.

Electromagnetic

- This single force includes all the forces that we normally categorize as either electrostatic or magnetic.

- Electromagnetic forces involve charged matter.

- Electromagnetic forces can be attractive or repulsive.

- The range of the electromagnetic force is infinite.

- The electromagnetic force is relatively strong – tiny imbalances of charges on an atomic level give rise to significant forces on the laboratory scale.

- At the end of the 19th century, Maxwell showed that the electrostatic force and the magnetic force were just two different aspects of the more fundamental electromagnetic force.

- The mathematics of the electromagnetic force is described by Maxwell's equations.

- Friction (and many other 'everyday' forces) is simply the result of the force between atoms and this is governed by the electromagnetic interaction.

The electromagnetic force and the weak nuclear force are now considered to be aspects of the single electroweak force.

PARTICLES THAT EXPERIENCE AND MEDIATE THE FUNDAMENTAL FORCES.

See page 78 onwards for more details about the standard model for the fundamental structure of matter. The following table summarizes which particles experience these forces and how they are mediated.

	Gravitational	Weak	Electromagnetic	Strong
Particles experience	All	Quark, Gluon	Charged	Quark, Gluon
Particles mediate	Graviton	W^+, W^-, Z^0	γ	Gluon

Radioactivity 1

IONIZING PROPERTIES

Many atomic nuclei are unstable. The process by which they decay is called **radioactive decay**. Every decay involves the emission of one of three different possible radiations from the nucleus: alpha (α), beta (β) or gamma (γ).

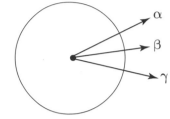

Alpha, beta and gamma all come from the nucleus

All three radiations are ionizing. This means that as they go through a substance, collisions occur which cause electrons to be removed from atoms. Atoms that have lost or gained electrons are called ions. This ionizing property allows the radiations to be detected. It also explains their dangerous nature. When ionizations occur in biologically important molecules, such as DNA, function can be affected.

EFFECTS OF RADIATION

At the molecular level, an ionization could cause damage directly to a biologically important molecule such as DNA or RNA. This could cause it to cease functioning. Alternatively, an ionization in the surrounding medium is enough to interfere with the complex chemical reactions (called **metabolic pathways**) taking place.

Molecular damage can result in a disruption to the functions that are taking place within the cells that make up the organism. As well as potentially causing the cell to die, this could just prevent cells from dividing and multiplying. On top of this, it could be the cause of the transformation of the cell into a malignant form.

As all body tissues are built up of cells, damage to these can result in damage to the body systems that have been affected. The non-functioning of these systems can result in death for the animal. If malignant cells continue to grow then this is called **cancer**.

RADIATION SAFETY

There is no such thing as a safe dose of ionizing radiation. Any hospital procedures that result in a patient receiving an extra dose (for example having an X-ray scan) should be justifiable in terms of the information received or the benefit it gives.

There are three main ways of protecting oneself from too large a dose. These can be summarized as follows:

- **Run away!**
  The simplest method of reducing the dose received is to increase the distance between you and the source. Only electromagnetic radiation can travel large distances and this follows an inverse square relationship with distance.

- **Don't waste time!**
  If you have to receive a dose, then it is important to keep the time of this exposure to a minimum.

- **If you can't run away, hide behind something!**
  Shielding can always be used to reduce the dose received. Lead-lined aprons can also be used to limit the exposure for both patient and operator.

PROPERTIES OF ALPHA, BETA AND GAMMA RADIATIONS

Property	Alpha, α	Beta, β	Gamma, γ
Effect on photographic film	Yes	Yes	Yes
Approximate number of ion pairs produced in air	10^4 per mm travelled	10^2 per mm travelled	1 per mm travelled
Typical material needed to absorb it	10^{-2} mm aluminium; piece of paper	A few mm aluminium	10 cm lead
Penetration ability	Low	Medium	High
Typical path length in air	A few cm	Less than one m	Effectively infinite
Deflection by E and B fields	Behaves like a positive charge	Behaves like a negative charge	Not deflected
Speed	About 10^7 m s^{-1}	About 10^8 m s^{-1}, very variable	3×10^8 m s^{-1}

NATURE OF ALPHA, BETA AND GAMMA DECAY

When a nucleus decays the mass numbers and the atomic numbers must balance on each side of the nuclear equation.

- Alpha particles are helium nuclei, $^4_2\alpha$ or $^4_2\text{He}^{2+}$. In alpha decay, a 'chunk' of the nucleus is emitted. The portion that remains will be a different nuclide.

$$^A_Z X \rightarrow {}^{(A-4)}_{(Z-2)} Y + {}^4_2\alpha$$

e.g. $^{241}_{95}\text{Am} \rightarrow {}^{237}_{93}\text{Np} + {}^4_2\alpha$

The atomic numbers and the mass numbers balance on each side of the equation.

$(95 = 93 + 2 \text{ and } 241 = 237 + 4)$

- Beta particles are electrons, $^0_{-1}\beta$ or $^0_{-1}e^-$, emitted **from the nucleus**. The explanation is that the electron is formed when a neutron decays. At the same time, another particle is emitted called an antineutrino.

$$^1_0 n \rightarrow {}^1_1 p + {}^0_{-1}\beta + \bar{\nu}$$

Since an antineutrino has no charge and virtually no mass it does not affect the equation.

$$^A_Z X \rightarrow {}^A_{(Z+1)} Y + {}^0_{-1}\beta + \bar{\nu}$$

e.g. $^{90}_{38}\text{Sr} \rightarrow {}^{90}_{39}\text{Y} + {}^0_{-1}\beta + \bar{\nu}$

- Gamma rays are unlike the other two radiations in that they are part of the electromagnetic spectrum. After their emission, the nucleus has less energy but its mass number and its atomic number have not changed. It is said to have changed from an **excited state** to a lower energy state.

$$^A_Z X^* \rightarrow {}^A_Z X + {}^0_0\gamma$$
Excited   Lower energy
state   state

Radioactivity 2

ANTIMATTER

The nuclear model given on page 77 is somewhat simplified. One important thing that is not mentioned there is the existence of antimatter. Every form of matter has its equivalent form of antimatter. If matter and antimatter came together they would annihilate each other. Not surprisingly, antimatter is rare but it does exist. For example, another form of radioactive decay that can take place is beta plus or positron decay. In this decay a proton decays into a neutron,

and the antimatter version of an electron, a positron, is emitted.

$$_{1}^{1}p \rightarrow \, _{0}^{1}n + \, _{+1}^{0}\beta^{+} + \nu$$
$$_{10}^{19}Ne \rightarrow \, _{9}^{19}F + \, _{+1}^{0}\beta^{+} + \nu$$

The positron, β^{+}, emission is accompanied by a neutrino.

The antineutrino is the antimatter form of the neutrino.

For more details see page 78.

BACKGROUND RADIATION

Radioactive decay is a natural phenomenon and is going on around you all the time. The activity of any given source is measured in terms of the number of individual nuclear decays that take place in a unit of time. This information is quoted in **becquerels** (Bq) with 1 Bq = 1 nuclear decay per second.

Experimentally this would be measured using a **Geiger counter**, which detects and counts the number of ionizations taking place inside the **GM tube**. A working Geiger counter will always detect some radioactive ionizations taking place even when there is no identified radioactive source: there is a **background count** as a result of the **background radiation**. A reading of 30 counts per minute, which corresponds to the detector registering 30 ionizing events, would not be unusual.

To analyse the activity of a given radioactive source, it is necessary to correct for the background radiation taking place. It would be necessary to record the background count without the radioactive source present and this value can then be subtracted from all readings with the source present.

Some cosmic gamma rays will be responsible, but there will also be α, β and γ radiation received as a result of radioactive decays that are taking place in the surrounding materials. The pie chart below identifies typical sources of background radiation, but the actual value varies from country to country and from place to place.

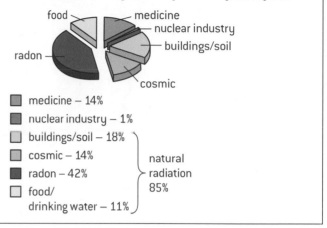

- medicine – 14%
- nuclear industry – 1%
- buildings/soil – 18%
- cosmic – 14%
- radon – 42%
- food/ drinking water – 11%

} natural radiation 85%

RANDOM DECAY

Radioactive decay is a **random** process and is not affected by external conditions. For example, increasing the temperature of a sample of radioactive material does not affect the rate of decay. This means that is there no way of knowing whether or not a particular nucleus is going to decay within a certain period of time. All we know is the *chances* of a decay happening in that time.

Although the process is random, the large numbers of atoms involved allows us to make some accurate predictions. If we

start with a given number of atoms then we can expect a certain number to decay in the next minute. If there were more atoms in the sample, we would expect the number decaying to be larger. On average the rate of decay of a sample is proportional to the number of atoms in the sample. This proportionality means that radioactive decay is an **exponential** process. The number of atoms of a certain element, N, decreases exponentially over time. Mathematically this is expressed as:

$$\frac{dN}{dt} \propto -N$$

Half-life

HALF-LIFE

There is a temptation to think that every quantity that decreases with time is an exponential decrease, but exponential curves have a particular mathematical property. In the graph shown below, the time taken for half the number of nuclides to decay is always the same, whatever starting value we choose. This allows us to express the chances of decay happening in a property called the **half-life**, $T_{\frac{1}{2}}$. The half-life of a nuclide is the time taken for half the number of nuclides present in a sample to decay. An equivalent statement is that the half-life is the time taken for the rate of decay (or activity) of a particular sample of nuclides to halve. A substance with a large half-life takes a long time to decay. A substance with a short half-life will decay quickly. Half-lives can vary from fractions of a second to millions of years.

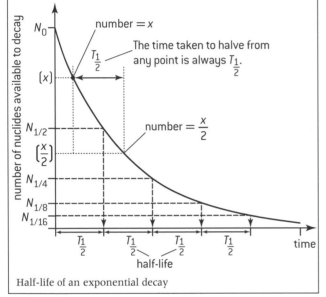

number $= x$

$T_{\frac{1}{2}}$

The time taken to halve from any point is always $T_{\frac{1}{2}}$.

(x)

number $= \dfrac{x}{2}$

$\left(\dfrac{x}{2}\right)$

half-life

Half-life of an exponential decay

INVESTIGATING HALF-LIFE EXPERIMENTALLY

When measuring the activity of a source, the background rate should be subtracted.

- If the half-life is short, then readings can be taken of activity against time.

  → A simple graph of activity against time would produce the normal exponential shape. Several values of half-life could be read from the graph and then averaged. This method is simple and quick but not the most accurate.

  → A graph of ln (activity) against time could be produced. This should give a straight line and the decay constant can be calculated from the gradient. See page 217.

- If the half-life is long, then the activity will effectively be constant over a period of time. In this case one needs to find a way to calculate the number of nuclei present, N, and then use

$$\frac{\mathrm{d}N}{\mathrm{d}t} = -\lambda N.$$

EXAMPLE

In simple situations, working out how much radioactive material remains is a matter of applying the half-life property several times. A common mistake is to think that if the half-life of a radioactive material is 3 days then it will all decay in six days. In reality, after six days (two half-lives) a 'half of a half' will remain, i.e. a quarter.

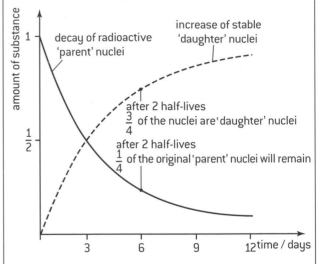

increase of stable 'daughter' nuclei

decay of radioactive 'parent' nuclei

after 2 half-lives $\dfrac{3}{4}$ of the nuclei are 'daughter' nuclei

after 2 half-lives $\dfrac{1}{4}$ of the original 'parent' nuclei will remain

The decay of parent into daughter

e.g. The half-life of $^{14}_{6}C$ is 5570 years.

Approximately how long is needed before less than 1% of a sample of $^{14}_{6}C$ remains?

Time	Fraction left
$T_{\frac{1}{2}}$	50%
$2T_{\frac{1}{2}}$	25%
$3T_{\frac{1}{2}}$	12.5%
$4T_{\frac{1}{2}}$	~ 6.3%
$5T_{\frac{1}{2}}$	~ 3.1%
$6T_{\frac{1}{2}}$	~ 1.6%
$7T_{\frac{1}{2}}$	~ 0.8%
6 half lives	= 33420 years
7 half lives	= 38990 years

∴ approximately 37000 years needed

SIMULATION

The result of the throw of a die is a random process and can be used to simulate radioactive decay. The dice represent nuclei available to decay. Each throw represents a unit of time. Every six represents a nucleus decaying meaning this die is no longer available.

Nuclear reactions

ARTIFICIAL TRANSMUTATIONS

There is nothing that we can do to change the likelihood of a certain radioactive decay happening, but under certain conditions we can make nuclear reactions happen. This can be done by bombarding a nucleus with a nucleon, an alpha particle or another small nucleus. Such reactions are called **artificial transmutations**. In general, the target nucleus first 'captures' the incoming object and then an emission takes place. The first ever artificial transmutation was carried out by Rutherford in 1919. Nitrogen was bombarded by alpha particles and the presence of oxygen was detected spectroscopically.

$$_2^4\text{He}^{2+} + \,_7^{14}\text{N} \rightarrow \,_8^{17}\text{O} + \,_1^1\text{p}$$

The mass numbers $(4 + 14 = 17 + 1)$ and the atomic numbers $(2 + 7 = 8 + 1)$ on both sides of the equation must balance.

UNIFIED MASS UNITS

The individual masses involved in nuclear reactions are tiny. In order to compare atomic masses physicists often use unified mass units, u. These are defined in terms of the most common isotope of carbon, carbon-12. There are 12 nucleons in the carbon-12 atom (6 protons and 6 neutrons) and one unified mass unit is defined as exactly one twelfth the mass of a carbon-12 atom. Essentially, the mass of a proton and the mass of a neutron are both 1 u as shown in the table below.

$1\text{ u} = \frac{1}{12}$ mass of a (carbon-12) atom $= 1.66 \times 10^{-27}$ kg

mass* of 1 proton $= 1.007\,276$ u

mass* of 1 neutron $= 1.008\,665$ u

mass* of 1 electron $= 0.000\,549$ u

= Technically these are all 'rest masses' – see option A

MASS DEFECT AND BINDING ENERGY

The table above shows the masses of neutrons and protons. It should be obvious that if we add together the masses of 6 protons, 6 neutrons and 6 electrons we will get a number bigger than 12 u, the mass of a carbon-12 atom. What has gone wrong? The answer becomes clear when we investigate what keeps the nucleus bound together.

The difference between the mass of a nucleus and the masses of its component nucleons is called the **mass defect**. If one imagined assembling a nucleus, the protons and neutrons would initially need to be brought together. Doing this takes work because the protons repel one another. Creating the bonds between the protons and neutrons releases a greater amount of energy than the work done in bringing them together. This energy released must come from somewhere. The answer lies in Einstein's famous mass–energy equivalence relationship.

$$\Delta E = \Delta mc^2$$

energy in joules   mass in kg   speed of light in m s^{-1}

In Einstein's equation, mass is another form of energy and it is possible to convert mass directly into energy and vice versa. The **binding energy** is the amount of energy that is released when a nucleus is assembled from its component nucleons. It comes from a decrease in mass. The binding energy would also be the energy that needs to be added in order to separate a nucleus into its individual nucleons. The mass defect is thus a measure of the binding energy.

UNITS

Using Einstein's equation, 1 kg of mass is equivalent to 9×10^{16} J of energy. This is a huge amount of energy. At the atomic scale other units of energy tend to be more useful. The electronvolt (see page 53), or more usually, the megaelectronvolt are often used.

$1\text{ eV} = 1.6 \times 10^{-19}$ J

$1\text{ MeV} = 1.6 \times 10^{-13}$ J

1 u of mass converts into 931.5 MeV

Since mass and energy are equivalent it is sometimes useful to work in units that avoid having to do repeated multiplications by the (speed of light)^2. A new possible unit for mass is thus MeV c^{-2}. It works like this:

If 1 MeV c^{-2} worth of mass is converted you get 1 MeV worth of energy.

WORKED EXAMPLES

Question:

How much energy would be released if 14 g of carbon-14 decayed as shown in the equation below?

$$_6^{14}\text{C} \rightarrow \,_7^{14}\text{N} + \,_{-1}^0\beta + \bar{\nu}$$

Answer:

Information given

atomic mass of carbon-14 $= 14.003242$ u;

atomic mass of nitrogen-14 $= 14.003074$ u;

mass of electron $= 0.000549$ u

mass of left-hand side = nuclear mass of $_6^{14}\text{C}$
$$= 14.003242 - 6(0.000549)\text{ u}$$
$$= 13.999948\text{ u}$$

nuclear mass of $_7^{14}\text{N} = 14.003074 - 7(0.000549)$ u
$$= 13.999231\text{ u}$$

mass of right-hand side $= 13.999231 + 0.000549$ u
$$= 13.999780\text{ u}$$

mass difference = LHS − RHS
$$= 0.000168\text{ u}$$

energy released per decay $= 0.000168 \times 931.5$ MeV
$$= 0.156492\text{ MeV}$$

14g of C-14 is 1 mol

∴ Total number of decays $= N_A = 6.022 \times 10^{23}$

∴ Total energy release $= 6.022 \times 10^{23} \times 0.156492$ MeV
$$= 9.424 \times 10^{22}\text{ MeV}$$
$$= 9.424 \times 10^{22} \times 1.6 \times 10^{-13}\text{ J}$$
$$= 1.51 \times 10^{10}\text{ J}$$
$$\approx 15\text{ GJ}$$

NB Many examination calculations avoid the need to consider the masses of the electrons by providing you with the *nuclear mass* as opposed to the *atomic mass*.

Fission and fusion

FISSION

Fission is the name given to the nuclear reaction whereby large nuclei are induced to break up into smaller nuclei and release energy in the process. It is the reaction that is used in nuclear reactors and atomic bombs. A typical single reaction might involve bombarding a uranium nucleus with a neutron. This can cause the uranium nucleus to break up into two smaller nuclei. A typical reaction might be:

$$^{1}_{0}n + \,^{235}_{92}U \rightarrow \,^{141}_{56}Ba + \,^{92}_{36}Kr + 3\,^{1}_{0}n + energy$$

Since the one original neutron causing the reaction has resulted in the production of three neutrons, there is the possibility of a **chain reaction** occurring. It is technically quite difficult to get the neutrons to lose enough energy to go on and initiate further reactions, but it is achievable.

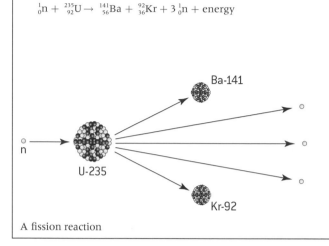

A fission reaction

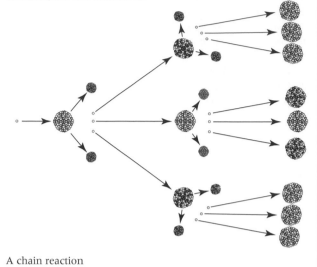

A chain reaction

FUSION

Fusion is the name given to the nuclear reaction whereby small nuclei are induced to join together into larger nuclei and release energy in the process. It is the reaction that 'fuels' all stars including the Sun. A typical reaction that is taking place in the Sun is the fusion of two different isotopes of hydrogen to produce helium.

$$^{2}_{1}H + \,^{3}_{1}H \rightarrow \,^{4}_{2}He + \,^{1}_{0}n + energy$$

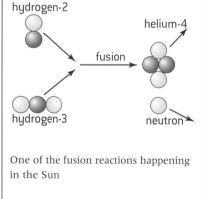

One of the fusion reactions happening in the Sun

BINDING ENERGY PER NUCLEON

Whenever a nuclear reaction (fission or fusion) releases energy, the products of the reaction are in a lower energy state than the reactants. Mass loss must be the source of this energy. In order to compare the energy states of different nuclei, physicists calculate the binding energy per nucleon. This is the total binding energy for the nucleus divided by the total number of nucleons. One of the nuclei with the largest binding energy per nucleon is iron-56, $^{56}_{26}Fe$.

A reaction is energetically feasible if the products of the reaction have a greater binding energy per nucleon when compared with the reactants.

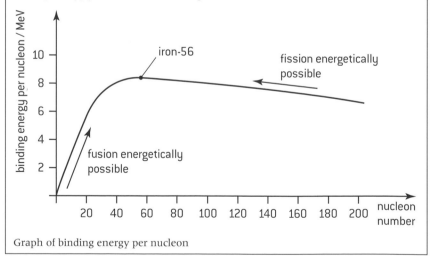

Graph of binding energy per nucleon

Structure of matter

INTRODUCTION

All matter that surrounds us, living or otherwise, is made up of different combinations of atoms. There are only a hundred, or so, different types of atoms present in nature. Atoms of a single type form an element. Each of these elements has a name and a chemical symbol; e.g. hydrogen, the simplest of all the elements, has the chemical symbol H. Oxygen has the chemical symbol O. The combination of two hydrogen atoms with one oxygen atom is called a water molecule – H_2O. The full list of elements is shown in a periodic table. Atoms consist of a combination of three things: protons, neutrons and electrons.

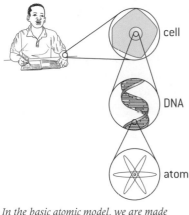

In the basic atomic model, we are made up of protons, neutrons, and electrons – nothing more.

ATOMIC MODEL

The basic atomic model, known as the nuclear model, was developed during the last century and describes a very small central nucleus surrounded by electrons arranged in different energy levels. The nucleus itself contains protons and neutrons (collectively called **nucleons**). All of the positive charge and almost all the mass of the atom is in the nucleus. The electrons provide only a tiny bit of the mass but all of the negative charge. Overall an atom is neutral. The vast majority of the volume is nothing at all – a vacuum. The nuclear model of the atom seems so strange that there must be good evidence to support it.

	Protons	Neutrons	Electrons
Relative mass	1	1	Negligible
Charge	+1	Neutral	−1

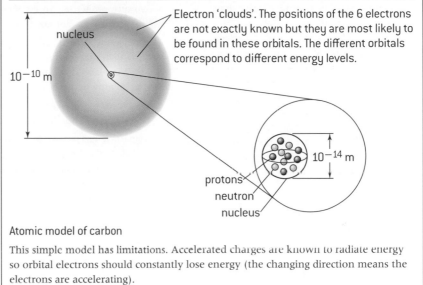

Electron 'clouds'. The positions of the 6 electrons are not exactly known but they are most likely to be found in these orbitals. The different orbitals correspond to different energy levels.

Atomic model of carbon

This simple model has limitations. Accelerated charges are known to radiate energy so orbital electrons should constantly lose energy (the changing direction means the electrons are accelerating).

EVIDENCE

One of the most convincing pieces of evidence for the nuclear model of the atom comes from the Rutherford–Geiger–Marsden experiment. Positive alpha particles were 'fired' at a thin gold leaf. The relative size and velocity of the alpha particles meant that most of them were expected to travel straight through the gold leaf. The idea behind this experiment was to see if there was any detectable structure within the gold atoms. The amazing discovery was that some of the alpha particles were deflected through huge angles. The mathematics of the experiment showed that numbers being deflected at any given angle agreed with an inverse square law of repulsion from the nucleus. Evidence for electron energy levels comes from emission and absorption spectra. The existence of isotopes provides evidence for neutrons.

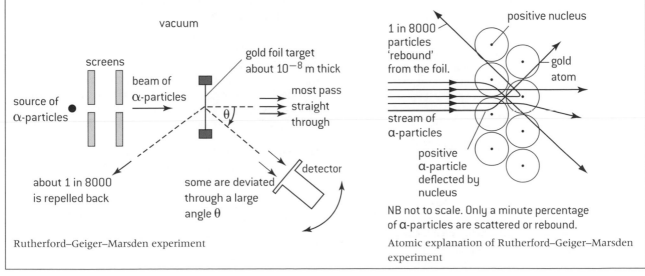

Rutherford–Geiger–Marsden experiment

Atomic explanation of Rutherford–Geiger–Marsden experiment

Description and classification of particles

CLASSIFICATION OF PARTICLES

Particle accelerator experiments identify many, many 'new' particles. Two original classes of particles were identified – the **leptons** (= 'light') and the **hadrons** (= 'heavy'). Protons and neutrons are hadrons whereas electrons are leptons. The hadrons were subdivided into **mesons** and **baryons**. Protons and neutrons are baryons. Another class of particles is involved in the mediation of the interactions between the particles. These were called **gauge bosons** or 'exchange bosons'.

Particles are called **elementary** if they have no internal structure, that is, they are not made out of smaller constituents. The classes of elementary particles are **quarks**, **leptons** and the **exchange particles**. Another particle, the Higgs boson, is also an elementry particle. Combinations of elementary particles are called **composite** particles. All hadrons are composed of combinations of quarks. Inside all baryons there are three quarks (or three antiquarks); inside all mesons there is one quark and one antiquark.

CONSERVATION LAWS

Not all reactions between particles are possible. The study of the reactions that did take place gave rise to some experimental conservation laws that applied to particle physics. Some of these laws were simply confirmation of conservation laws that were already known to physicists – charge, momentum (linear and angular) and mass-energy. On top of these fundamental laws there appeared to be other rules that were never broken e.g. the law of conservation of baryon number. If all baryons were assigned a 'baryon number' of 1 (and all antibaryons were assigned a baryon number of −1) then the total number of baryons before and after a collision was always the same. A similar law of conservation of lepton number applies.

Other reactions suggested new and different particle properties that were often, but not always, conserved in reactions. 'Strangeness' and 'charm' are examples of two such properties. Strangeness is conserved in all electromagnetic and strong interactions, but not always in weak interactions.

All particles, whether they are elementary or composite, can be specified in terms of their mass and the various quantum numbers that are related to the conservation laws that have been discovered. The quantum numbers that are used to identify particles include:

- electric charge, strangeness, charm, lepton number, baryon number and colour (this property is not the same as an object's actual colour – see page 79).

Every particle has its own **antiparticle**. An antiparticle has the same mass as its particle but all its quantum numbers (including charge, etc.) are opposite. There are some particles (e.g. the photon) that are their own antiparticle.

THE STANDARD MODEL – LEPTONS

There are six different leptons and six different antileptons. The six leptons are considered to be in three different generations or families in exactly the same way that there are considered to be three different generations of quarks (see page 79).

The electron and the electron neutrino have a lepton (electron family) number of +1. The antielectron and the antielectron neutrino have a lepton (electron family) number of −1.

Similar principles are used to assign lepton numbers of +1 or −1 to the muon and the tau family members.

Lepton family number is also conserved in all reactions. For example, whenever a muon is created, an antimuon or an antimuon neutrino must also be created so that the total number of leptons in the muon family is always conserved.

	Electric charge	**'Generation'** 1	2	3
Lepton	0	ν_e (electron-neutrino) $M = 0$ or almost 0	ν_μ (muon-neutrino) $M = 0$ or almost 0	ν_τ (tau-neutrino) $M = 0$ or almost 0
	−1	e (electron) $M = 0.511$ MeV c^{-2}	μ (muon) $M = 105$ MeV c^{-2}	τ (tau) $M = 1784$ MeV MeV c^{-2}

EXCHANGE PARTICLES

There are only four fundamental interactions that exist: Gravity, Electromagnetic, Strong and Weak.

All four interactions can be thought of as being mediated by an exchange of particles. Each interaction has its own exchange particle or particles. The bigger the mass of the exchange boson, the smaller the range of the force concerned.

The exchange results in repulsion between the two particles

From the point of view of quantum mechanics, the energy needed to create these virtual particles, ΔE is available so long as the energy of the particle does not exist for a longer time Δt than is proscribed by the uncertainty principle (see page 126).

The greater the mass of the exchange particle, the smaller the time for which it can exist. The range of the weak interaction is small as the masses of its exchange particles (W^+, W^- and Z^0) are large.

In particle physics, all real particles can be thought of as being surrounded by a cloud of virtual particles that appear and disappear out of the surrounding vacuum. The lifetime of these particles is inversely proportional to their mass. The interaction between two particles takes place when one or more of the virtual particles in one cloud is absorbed by the other particle.

Interaction	**Relative strength**	**Range (m)**	**Exchange particle**	**Particles experience**
Strong	1	$\sim 10^{-15}$	8 different gluons	Quarks, gluons
Electromagnetic	10^{-2}	infinite	photon	Charged
Weak	10^{-13}	$\sim 10^{-18}$	W^+, W^-, Z^0	Quarks, lepton
Gravity	10^{-39}	infinite	graviton	All

Leptons and bosons are unaffected by the strong force.

Quarks

STANDARD MODEL – QUARKS

The **standard model** of particle physics is the theory that says that all matter is considered to be composed of combinations of six types of quark and six types of lepton. This is the currently accepted theory. Each of these particles is considered to be fundamental, which means they do not have any deeper structure. Gravity is not explained by the standard model.

All hadrons are made up from different combinations of fundamental particles called **quarks**. There are six different types of quark and six types of antiquark. This very neatly matches the six leptons that are also known to exist. Quarks are affected by the strong force (see below), whereas leptons are not. **The weak interaction can change one type of quark into another.**

<table>
<tr><td rowspan="2" style="writing-mode: vertical">Quarks</td><td rowspan="2">Electric charge</td><td colspan="3">'Generation'</td></tr>
<tr><td>1</td><td>2</td><td>3</td></tr>
<tr><td>$+\frac{2}{3}e$</td><td>u (up) $M = 5$ MeV c^{-2}</td><td>c (charm) $M = 1500$ MeV c^{-2}</td><td>t (top) $M = 174$ MeV c^{-2}</td></tr>
<tr><td>$-\frac{1}{3}e$</td><td>d (down) $M = 10$ MeV c^{-2}</td><td>s (strange) $M = 200$ MeV c^{-2}</td><td>b (bottom) $M = 4700$ MeV c^{-2}</td></tr>
</table>

All quarks have a baryon number of $+\frac{1}{3}$,

All antiquarks have a baryon number of $-\frac{1}{3}$

All quarks have a strangeness number of 0 except the s quark that has a strangeness number of -1.

The c quark is the only quark with a charm number $= +1$, all other quarks have charm number of 0.

Isolated quarks cannot exist. They can exist only in twos or threes. Mesons are made from two quarks (a quark and an antiquark) whereas baryons are made up of a combination of three quarks (either all quarks or all antiquarks).

	Name of particle	Quark structure
Baryons	proton (p)	u u d
	neutron (n)	u d d
	lambda Λ	u d s
	antiproton ($\bar{\text{p}}$)	$\bar{\text{u}}\ \bar{\text{u}}\ \bar{\text{d}}$
Mesons	π^- (pi-minus)	d $\bar{\text{u}}$
	π^+ (pi-plus)	u $\bar{\text{d}}$
	K^0 (K$_{\text{zero}}$)	d $\bar{\text{s}}$

The force between quarks is still the strong interaction but the full description of this interaction is termed QCD theory – quantum chromodynamics. The quantum difference between the quarks is a property called colour. All quarks can be red (r), green (g) or blue (b). Antiquarks can be antired ($\bar{\text{r}}$), antigreen ($\bar{\text{g}}$) or antiblue ($\bar{\text{b}}$). The two up quarks in a proton are not identical because they have different colours.

Only white (**colour neutral**) combinations are possible. Baryons must contain r, g and b quarks (or $\bar{\text{r}}$, $\bar{\text{g}}$, $\bar{\text{b}}$) whereas mesons contain a colour and the anticolour (e.g. r and $\bar{\text{r}}$ or b and $\bar{\text{b}}$, etc.) The force between quarks is sometimes called the colour force. Eight different types of gluon mediate it.

The details of QCD do not need to be recalled.

QUANTUM CHROMODYNAMICS (QCD)

The interaction between objects with colour is called the colour interaction and is explained by a theory called quantum chromodynamics. The force-carrying particle is called the gluon. There are eight different types of gluon each with zero mass. Each gluon carries a combination of colour and anticolour and their emission and absorption by different quarks causes the colour force.

As the gluons themselves are coloured, there will be a colour interaction between gluons themselves as well as between quarks. The overall effect is that they bind quarks together. The force between quarks increases as the separation between quarks

increases. **Isolated quarks and gluons cannot be observed.** If sufficient energy is supplied to a hadron in order to attempt to isolate a quark, then more hadrons or mesons will be produced rather than isolated quarks. This is known as **quark confinement**.

The six colour-changing gluons are: $G_{r\bar{b}}$, $G_{r\bar{g}}$, $G_{b\bar{g}}$, $G_{b\bar{r}}$, $G_{g\bar{b}}$, $G_{g\bar{r}}$.

For example when a blue up quark emits the gluon $G_{b\bar{r}}$ it loses its blue colour and becomes a red up quark (the gluon contains antired, so red colour must be left behind). A red down quark absorbing this gluon will become a blue down quark.

There are two additional colour-neutral gluons: G_0 and G'_0, making a total of eight gluons.

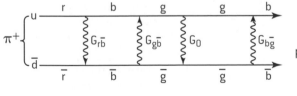

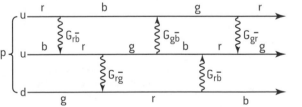

STRONG INTERACTION

The colour interaction and the strong interaction are essentially the same thing. Properly, the colour interaction is the fundamental force that binds quarks together into baryons and mesons. It is mediated by gluons. The **residual strong interaction** is the force that binds colour-neutral particles (such as the proton and neutron) together in a nucleus. The overall effect of the interactions between all the quarks in the nucleons is a short-range interaction between colour-neutral nucleons.

The particles mediating the strong interaction can be considered to involve the exchange of composite particles (π mesons: π^+, π^- or π^0) whereas the fundamental colour interaction is always seen as the exchange of gluons.

HIGGS BOSON

In addition to the three generations of leptons and quarks in the standard model there are the four classes of gauge boson and an additional highly massive boson, the Higgs boson. This was proposed in 1964 to explain the process by which particles can acquire mass. In 2013 scientists working with the Large Hadron Collider announced the experimental detection of a particle that that matched the standard model's predictions for the Higgs boson.

Feynman diagrams

RULES FOR DRAWING FEYNMAN DIAGRAMS

Feynman diagrams can be used to represent possible particle interactions. The diagrams are used to calculate the overall probability of an interaction taking place. In quantum mechanics, in order to find out the overall probability of an interaction, it is necessary to add together all the possible ways in which an interaction can take place. Used properly they are a mathematical tool for calculations but, at this level, they can be seen as a simple pictorial model of possible interactions.

In the Feynman diagrams below the x-axis represents time going from left to right and the y-axis represents space (some books reverse these two axes). To view them in the alternative way, turn the page anti-clockwise by 90°.

Some simple rules help in the construction of correct diagrams:

- Each junction in the diagram (vertex) has an arrow going in and one going out. These will represent a lepton–lepton transition or a quark–quark transition.
- Quarks or leptons are solid straight lines.
- Exchange particles are either wavy or broken (photons, $W^{\pm}$ or Z°) or curly (gluons).
- Time flows from left to right. Arrows from left to right represent particles travelling forward in time. Arrows from right to left represent antiparticles travelling forward in time.
- The labels for the different particles are shown at the end of the line.
- The junctions will be linked by a line representing the exchange particle involved.

EXAMPLES

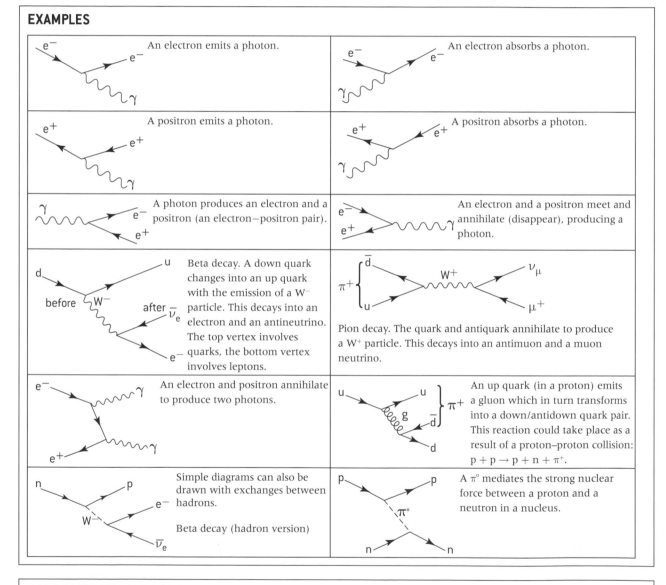

An electron emits a photon.

An electron absorbs a photon.

A positron emits a photon.

A positron absorbs a photon.

A photon produces an electron and a positron (an electron–positron pair).

An electron and a positron meet and annihilate (disappear), producing a photon.

Beta decay. A down quark changes into an up quark with the emission of a W^- particle. This decays into an electron and an antineutrino. The top vertex involves quarks, the bottom vertex involves leptons.

Pion decay. The quark and antiquark annihilate to produce a W^+ particle. This decays into an antimuon and a muon neutrino.

An electron and positron annihilate to produce two photons.

An up quark (in a proton) emits a gluon which in turn transforms into a down/antidown quark pair. This reaction could take place as a result of a proton–proton collision: $p + p \rightarrow p + n + \pi^+$.

Simple diagrams can also be drawn with exchanges between hadrons.

Beta decay (hadron version)

A π° mediates the strong nuclear force between a proton and a neutron in a nucleus.

USES OF FEYNMAN DIAGRAMS

Once a possible interaction has been identified with a Feynman diagram, it is possible to use it to calculate the probabilities for certain fundamental processes to take place. Each line and vertex corresponds to a mathematical term. By adding together all the terms, the probability of the interaction can be calculated using the diagram.

More complicated diagrams with the same overall outcome need to be considered in order to calculate the overall probability of a chosen outcome. The more diagrams that are included in the calculation, the more accurate the answer.

In a Feynman diagram, lines entering or leaving the diagram represent real particles and must obey mass, energy and momentum relationships. Lines in intermediate stages in the diagram represent virtual particles and do not have to obey energy conservation providing they exist for a short enough time for the uncertainty relationship to apply. Such virtual particles cannot be detected.

IB Questions – atomic, nuclear and particle physics

1. A sample of radioactive material contains the element Ra 226. The half-life of Ra 226 can be defined as the time it takes for

   A. the mass of the sample to fall to half its original value.

   B. half the number of atoms of Ra 226 in the sample to decay.

   C. half the number of atoms in the sample to decay.

   D. the volume of the sample to fall to half its original value.

2. Oxygen-15 decays to nitrogen-15 with a half-life of approximately 2 minutes. A pure sample of oxygen-15, with a mass of 100 g, is placed in an airtight container. After 4 minutes, the masses of oxygen and nitrogen in the container will be

Mass of oxygen	Mass of nitrogen
A. 0 g	100 g
B. 25 g	25 g
C. 50 g	50 g
D. 25 g	75 g

3. A radioactive nuclide $_z$X undergoes a sequence of radioactive decays to form a new nuclide $_{z+2}$Y. The sequence of emitted radiations could be

   A. β, β   B. α, β, β

   C. α, α   D. α, β, γ

4. In the Rutherford scattering experiment, a stream of α particles is fired at a thin gold foil. Most of the α particles

   A. are scattered randomly.

   B. rebound.

   C. are scattered uniformly.

   D. go through the foil.

5. A piece of radioactive material now has about 1/16 of its previous activity. If the half-life is 4 hours the difference in time between measurements is approximately

   A. 8 hours.

   B. 16 hours.

   C. 32 hours.

   D. 60 hours.

6. a) Use the standard model to describe, in terms of fundamental particles, the internal structure of:

   (i)   A proton

   (ii) An electron

   (iii) Baryons

   (iv) Mesons

   b) Draw Feynman diagram for β^+ decay.

7. A proton undergoes a strong interaction with a ϕ^- particle (quark content: $\overline{u}d$) to produce a neutron and another particle. Use conservation laws to deduce the structure of the particle produced in this reaction.

8. a) Two properties of the isotope of uranium, $^{238}_{92}$U are:

   (i) it decays radioactively (to $^{234}_{90}$Th)

   (ii) it reacts chemically (e.g. with fluorine to form UF_6).

   What features of the structure of uranium atoms are responsible for these two widely different properties? [2]

   b) A beam of deuterons (deuterium nuclei, $^2_1$H) are accelerated through a potential difference and are then incident on a magnesium target ($^{26}_{12}$Mg). A nuclear reaction occurs resulting in the production of a sodium nucleus and an alpha particle.

   (i) Write a balanced nuclear equation for this reaction. [2]

   (ii) Explain why it is necessary to give the deuterons a certain minimum kinetic energy before they can react with the magnesium nuclei. [2]

9. *Radioactive carbon dating*

   The carbon in trees is mostly carbon-12, which is stable, but there is also a small proportion of carbon-14, which is radioactive. When a tree is cut down, the carbon-14 present in the wood at that time decays with a half-life of 5,800 years.

   a) Carbon-14 decays by beta-minus emission to nitrogen-14. Write the equation for this decay. [2]

   b) For an old wooden bowl from an archaeological site, the average count-rate of beta particles detected per kg of carbon is 13 counts per minute. The corresponding count rate from newly cut wood is 52 counts per minute.

   (i) Explain why the beta activity from the bowl diminishes with time, even though the probability of decay of any individual carbon-14 nucleus is constant. [3]

   (ii) Calculate the approximate age of the wooden bowl. [3]

10. This question is about a nuclear fission reactor for providing electrical power.

   In a nuclear reactor, power is to be generated by the fission of uranium-235. The absorption of a neutron by ^{235}U results in the splitting of the nucleus into two smaller nuclei plus a number of neutrons and the release of energy. The splitting can occur in many ways; for example

   $$n + {}^{235}_{92}U \rightarrow {}^{90}_{38}Sr + {}^{143}_{54}Xe + neutrons + energy$$

   a) *The nuclear fission reaction*

   (i)   How many neutrons are produced in this reaction? [1]

   (ii) Explain why the release of several neutrons in each reaction is crucial for the operation of a fission reactor. [2]

   (iii) The sum of the rest masses of the uranium plus neutron before the reaction is 0.22 u greater than the sum of the rest masses of the fission products. What becomes of this 'missing mass'? [1]

   (iv) Show that the energy released in the above fission reaction is about 200 MeV. [2]

   b) *A nuclear fission power station*

   (i) Suppose a nuclear fission power station generates electrical power at 550 MW. Estimate the minimum number of fission reactions occurring each second in the reactor, stating any assumption you have made about efficiency. [4]

11. Which of the following is a correct list of particles upon which the strong nuclear force may act?

   A. protons and neutrons   B. protons and electrons

   C. neutrons and electrons   D. protons, neutrons and electrons

Energy and power generation – Sankey diagram

ENERGY CONVERSIONS

The production of electrical power around the world is achieved using a variety of different systems, often starting with the release of thermal energy from a fuel. In principle, thermal energy can be completely converted to work in a single process, but the continuous conversion of this energy into work implies the use of machines that are continuously repeating their actions in a fixed cycle. Any cyclical process must involve the transfer of some energy from the system to the surroundings that is no longer available to perform useful work. This unavailable energy is known as **degraded energy**, in accordance with the principle of the second law of thermodynamics (see page 162).

Energy conversions are represented using **Sankey diagrams**. An arrow (drawn from left to right) represents the energy changes taking place. The width of the arrow represents the power or energy involved at a given stage. Created or degraded energy is shown with an arrow up or down.

Note that Sankey diagrams are to scale. The width of the useful electrical output in the diagram on the right is 2.0 mm compared with 12.0 mm for the width of the total energy from the fuel. This represents an overall efficiency of 16.7%.

ELECTRICAL POWER PRODUCTION

In all electrical power stations the process is essentially the same. A fuel is used to release thermal energy. This thermal energy is used to boil water to make steam. The steam is used to turn turbines and the motion of the turbines is used to generate electrical energy. Transformers alter the potential difference (see page 114).

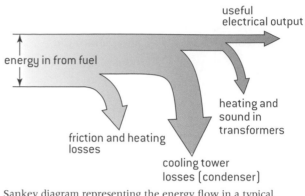

Sankey diagram representing the energy flow in a typical power station

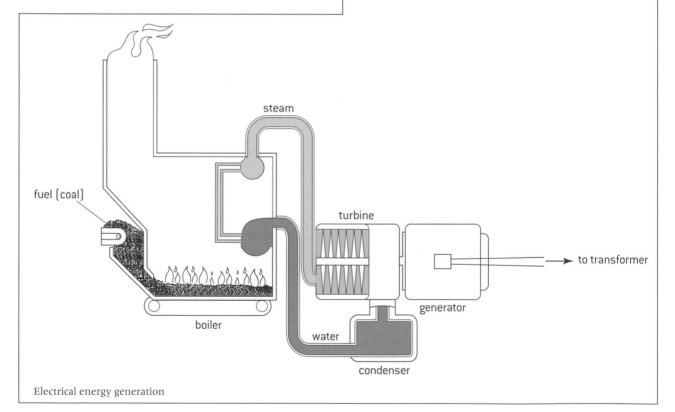

Electrical energy generation

POWER

Power is defined as the rate at which energy is converted. The units of power are J S^{-1} or W.

$$\text{Power} = \frac{\text{energy}}{\text{time}}$$

Primary energy sources

RENEWABLE / NON-RENEWABLE ENERGY SOURCES

The law of conservation of energy states that energy is neither created nor destroyed, it just changes form. As far as human societies are concerned, if we wish to use devices that require the input of energy, we need to identify sources of energy. **Renewable** sources of energy are those that cannot be used up, whereas **non-renewable** sources of energy can be used up and eventually run out.

Renewable sources	Non-renewable sources
hydroelectric	coal
photovoltaic cells	oil
active solar heaters	natural gas
wind	nuclear
biofuels	

Sometimes the sources are hard to classify so care needs to be taken when deciding whether a source is renewable or not. One point that sometimes worries students is that the Sun will eventually run out as a source of energy for the Earth, so no source is perfectly renewable! This is true, but all of these sources are considered from the point of view of life on Earth. When the Sun runs out, then so will life on Earth. Other things to keep in mind include:

- Nuclear sources (both fission and fusion) consume a material as their source so they must be non-renewable.

On the other hand, the supply available can make the source **effectively** renewable (fusion).

- It is possible for a fuel to be managed in a renewable or a non-renewable way. For example, if trees are cut down as a source of wood to burn then this is clearly non-renewable. It is, however, possible to replant trees at the same rate as they are cut down. If this is properly managed, it could be a renewable source of energy.

Of course these possible sources must have got their energy from somewhere in the first place. Most of the energy used by humans can be traced back to energy radiated from the Sun, but not quite all of it. Possible sources are:

- the Sun's radiated energy
- gravitational energy of the Sun and the Moon
- nuclear energy stored within atoms
- the Earth's internal heat energy.

Although you might think that there are other sources of energy, the above list is complete. Many everyday sources of energy (such as coal or oil) can be shown to have derived their energy from the Sun's radiated energy. On the industrial scale, electrical energy needs to be generated from another source. When you plug anything electrical into the mains electricity you have to pay the electricity-generating company for the energy you use. In order to provide you with this energy, the company must be using one (or more) of the original list of sources.

SPECIFIC ENERGY AND ENERGY DENSITY

Two quantities are useful to consider when making comparisons between different energy sources – the **specific energy** and the **energy density**.

Specific energy provides a useful comparison between fuels and is defined as the energy liberated per unit mass of fuel consumed. Specific energy is measured in $J\ kg^{-1}$

specific energy

$$= \frac{\text{energy released from fuel}}{\text{mass of fuel consumed}}$$

Fuel choice can be particularly influenced by specific energy when the fuel needs to be transported: the greater the mass of fuel that needs to be transported, the greater the cost.

Energy density is defined as the energy liberated per unit volume of fuel consumed. The unit is $J\ m^{-3}$

energy density

$$= \frac{\text{energy release from fuel}}{\text{volume of fuel consumed}}$$

COMPARISON OF ENERGY SOURCES

Fuel	Renewable?	CO_2 emission	Specific energy($MJ\ kg^{-1}$) (values vary depending on type)	Energy density ($MJ\ m^{-3}$)
Coal	No	Yes	22–33	23,000
Oil	No	Yes	42	36,500
Gas	No	Yes	54	37
Nuclear (uranium)	No	No	8.3×10^7	1.5×10^{12}
Waste	No	Yes	10	variable
Solar	Yes	No	n/a	n/a
Wind	Yes	No	n/a	n/a
Hydro – water stored in dams	Yes	No	n/a	n/a
Tidal	Yes	No	n/a	n/a
Pumped storage	n/a	No	n/a	n/a
Wave	Yes	No	n/a	n/a
Geothermal	Yes	No	n/a	n/a
Biofuels e.g. ethanol	Some types	Yes	30	21,000

Fossil fuel power production

ORIGIN OF FOSSIL FUEL

Coal, oil and natural gas are known as **fossil fuels**. These fuels have been produced over a timescale that involves tens or hundreds of millions of years from accumulations of dead matter. This matter has been converted into fossil fuels by exposure to the very high temperatures and pressure that exist beneath the Earth's surface.

Coal is formed from the dead plant matter that used to grow in swamps. Layer upon layer of decaying matter decomposed.

As it was buried by more plant matter and other substances, the material became more compressed. Over the geological timescale this turned into coal.

Oil is formed in a similar manner from the remains of microscopic marine life. The compression took place under the sea. Natural gas, as well as occurring in underground pockets, can be obtained as a by-product during the production of oil. It is also possible to manufacture gas from coal.

ENERGY TRANSFORMATIONS

Fossil fuel power stations release energy in fuel by burning it. The thermal energy is then used to convert water into steam that once again can be used to turn turbines. Since all fossil fuels were originally living matter, the original source of this energy was the Sun. For example, millions of years ago energy radiated from the Sun was converted (by photosynthesis) into living plant matter. Some of this matter has eventually been converted into coal.

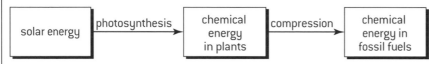

Energy storage in fossil fuels

EXAMPLE

Use the data on this page and the previous page to calculate the typical rate (in tonnes per hour) at which coal must be supplied to a 500 MW coal fired power station.

Answer

Electrical power supply $= 500 \text{ MW} = 5 \times 10^8 \text{ J s}^{-1}$

Power released from fuel $= 5 \times 10^8 \text{ / efficiency}$
$= 5 \times 10^8 / 0.35$
$= 1.43 \times 10^9 \text{ J s}^{-1}$

Rate of consumption of coal $= 1.43 \times 10^9 / 3.3 \times 10^7 \text{ kg s}^{-1}$
$= 43.3 \text{ kg s}^{-1}$
$= 43.3 \times 60 \times 60 \text{ kg hr}^{-1}$
$= 1.56 \times 10^5 \text{ kg hr}^{-1}$
$\approx 160 \text{ tonnes hr}^{-1}$

EFFICIENCY OF FOSSIL FUEL POWER STATIONS

The efficiency of different power stations depends on the design. At the time of publishing, the following figures apply.

Fossil fuel	Typical efficiency	Current maximum efficiency
Coal	35%	42%
Natural gas	45%	52%
Oil	38%	45%

Note that thermodynamic considerations limit the maximum achievable efficiency (see page 163).

ADVANTAGES AND DISADVANTAGES

Advantages

- Very high 'specific energy' and 'energy density' – a great deal of energy is released from a small mass of fossil fuel.
- Fossil fuels are relatively easy to transport.
- Still cheap when compared to other sources of energy.
- Power stations can be built anywhere with good transport links and water availability.
- Can be used directly in the home to provide heating.

Disadvantages

- Combustion products can produce pollution, notably acid rain.
- Combustion products contain 'greenhouse' gases.
- Extraction of fossil fuels can damage the environment.
- Non-renewable.
- Coal-fired power stations need large amounts of fuel.

Nuclear power – process

PRINCIPLES OF ENERGY PRODUCTION

Many nuclear power stations use uranium-235 as the 'fuel'. This fuel is not burned – the release of energy is achieved using a fission reaction. An overview of this process is described on page 76. In each individual reaction, an incoming neutron causes a uranium nucleus to split apart. The fragments are moving fast. In other words the temperature is very high. Among the fragments are more neutrons. If these neutrons go on to initiate further reactions then a chain reaction is created.

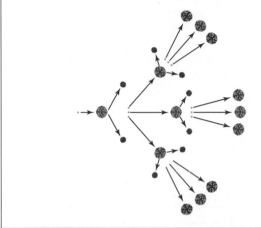

The design of a nuclear reactor needs to ensure that, on average, only one neutron from each reaction goes on to initiate a further reaction. If more reactions took place then the number of reactions would increase all the time and the chain reaction would run out of control. If fewer reactions took place, then the number of reactions would be decreasing and the fission process would soon stop.

The chance that a given neutron goes on to cause a fission reaction depends on several factors. Two important ones are:

- the number of potential nuclei 'in the way'
- the speed (or the energy) of the neutrons.

As a general trend, as the size of a block of fuel increases so do the chances of a neutron causing a further reaction (before it is lost from the surface of the block). As the fuel is assembled together a stage is reached when a chain reaction can occur. This happens when a so-called critical mass of fuel has been assembled. The exact value of the critical mass depends on the exact nature of the fuel being used and the shape of the assembly.

There are particular neutron energies that make them more likely to cause nuclear fission. In general, the neutrons created by the fission process are moving too fast to make reactions likely. Before they can cause further reactions the neutrons have to be slowed down.

MODERATOR, CONTROL RODS AND HEAT EXCHANGER

Three important components in the design of all nuclear reactors are the **moderator**, the **control rods** and the **heat exchanger**.

- Collisions between the neutrons and the nuclei of the moderator slow them down and allow further reactions to take place.

- The control rods are movable rods that readily absorb neutrons. They can be introduced or removed from the reaction chamber in order to control the chain reaction.

- The heat exchanger allows the nuclear reactions to occur in a place that is sealed off from the rest of the environment. The reactions increase the temperature in the core. This thermal energy is transferred to heat water and the steam that is produced turns the turbines.

A general design for one type of nuclear reactor (PWR or pressurized water reactor) is shown here. It uses water as the moderator and as a coolant.

ADVANTAGES AND DISADVANTAGES

Advantages
- Extremely high 'specific energy' – a great deal of energy is released from a very small mass of uranium.
- Reserves of uranium large compared to oil.

Disadvantages
- Process produces radioactive nuclear waste that is currently just stored.
- Larger possible risk if anything should go wrong.
- Non-renewable (but should last a long time).

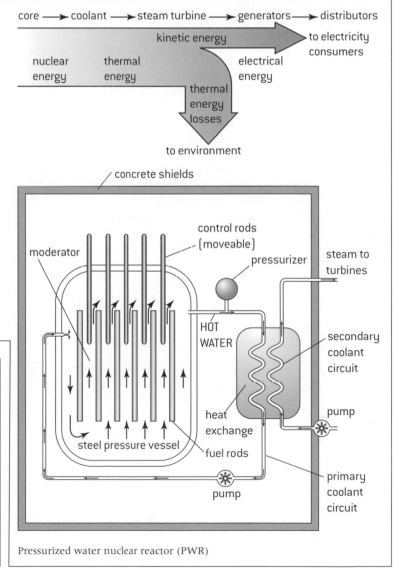

Pressurized water nuclear reactor (PWR)

Nuclear power – safety and risks

ENRICHMENT AND REPROCESSING

Naturally occurring uranium contains less than 1% of uranium-235. Enrichment is the process by which this percentage composition is increased to make nuclear fission more likely.

In addition to uranium-235, plutonium-239 is also capable of sustaining fission reactions. This nuclide is formed as a by-product of a conventional nuclear reactor. A uranium-238 nucleus can capture fast-moving neutrons to form uranium-239. This undergoes β-decay to neptunium-239 which undergoes further β-decay to plutonium-239:

$$^{238}_{92}U + ^{1}_{0}n \rightarrow ^{239}_{92}U$$

$$^{239}_{92}U \rightarrow ^{239}_{93}Np + ^{0}_{-1}\beta + \overline{\upsilon}$$

$$^{239}_{93}Np \rightarrow ^{239}_{94}Pu + ^{0}_{-1}\beta + \overline{\upsilon}$$

Reprocessing involves treating used fuel waste from nuclear reactors to recover uranium and plutonium and to deal with other waste products. A fast breeder reactor is one design that utilizes plutonium-239.

HEALTH, SAFETY AND RISK

Issues associated with the use of nuclear power stations for generation of electrical energy include:

- If the control rods were all removed, the reaction would rapidly increase its rate of production. Completely uncontrolled nuclear fission would cause an explosion and **thermal meltdown** of the core. The radioactive material in the reactor could be distributed around the surrounding area causing many fatalities. Some argue that the terrible scale of such a disaster means that the use of nuclear energy is a risk not worth taking. Nuclear power stations could be targets for terrorist attacks.

- The reaction produces radioactive nuclear waste. While much of this waste is of a low level risk and will radioactively decay within decades, a significant amount of material is produced which will remain dangerously radioactive for millions of years. The current solution is to bury this waste in geologically secure sites.

- The uranium fuel is mined from underground and any mining operation involves significant risk. The ore is also radioactive so extra precautions are necessary to protect the workers involved in uranium mines.

- The transportation of the uranium from the mine to a power station and of the waste from the nuclear power station to the reprocessing plant needs to be secure and safe.

- By-products of the civilian use of nuclear power can be used to produce nuclear weapons.

NUCLEAR WEAPONS

A nuclear power station involves controlled nuclear fission whereas an uncontrolled nuclear fission produces the huge amount of energy released in nuclear weapons. Weapons have been designed using both uranium and plutonium as the fuel. Issues associated with nuclear weapons include:

- Moral issues associated with any weapon of aggression that is associated with warfare. Nuclear weapons have such destructive capability that since the Second World War the threat of their deployment has been used as a deterrent to prevent non-nuclear aggressive acts against the possessors of nuclear capability.

- The unimaginable consequences of a nuclear war have forced many countries to agree to non-proliferation treaties, which attempt to limit nuclear power technologies to a small number of nations.

- A by-product of the peaceful use of uranium for energy production is the creation of plutonium-239 which could be used for the production of nuclear weapons. Is it right for the small number of countries that already have nuclear capability to prevent other countries from acquiring that knowledge?

FUSION REACTORS

Fusion reactors offer the theoretical potential of significant power generation without many of the problems associated with current nuclear fission reactors. The fuel used, hydrogen, is in plentiful supply and the reaction (if it could be sustained) would not produce significant amounts of radioactive waste.

The reaction is the same as takes place in the Sun (as outlined on page 76) and requires creating temperatures high enough to ionize atomic hydrogen into a plasma state (this is the 'fourth state of matter', in which electrons and protons are not bound in atoms but move independently). Currently the principal design challenges are associated with maintaining and confining the plasma at sufficiently high temperature and density for fusion to take place.

Solar power and hydroelectric power

SOLAR POWER (TWO TYPES)

There are two ways of harnessing the radiated energy that arrives at the Earth's surface from the Sun.

A **photovoltaic cell** (otherwise known as a solar cell or photocell) converts a portion of the radiated energy directly into a potential difference ('voltage'). It uses a piece of semiconductor to do this. Unfortunately, a typical photovoltaic cell produces a very small voltage and it is not able to provide much current. They are used to run electrical devices that do not require a great deal of energy. Using them in series would generate higher voltages and several in parallel can provide a higher current.

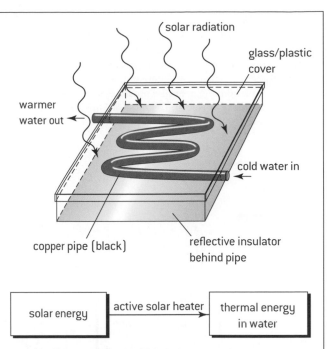

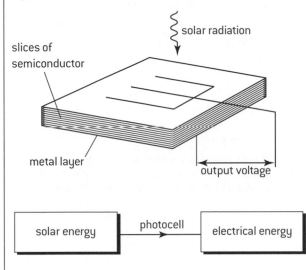

An **active solar heater** (otherwise known as a solar panel) is designed to capture as much thermal energy as possible. The hot water that it typically produces can be used domestically and would save on the use of electrical energy.

ADVANTAGES AND DISADVANTAGES

Advantages
- Very 'clean' production – no harmful chemical by-products.
- Renewable source of energy.
- Source of energy is free.

Disadvantages
- Can only be utilized during the day.
- Source of energy is unreliable – could be a cloudy day.
- A very large area would be needed for a significant amount of energy.

HYDROELECTRIC POWER

The source of energy in a hydroelectric power station is the gravitational potential energy of water. If water is allowed to move downhill, the flowing water can be used to generate electrical energy.

The water can gain its gravitational potential energy in several ways.

- As part of the 'water cycle', water can fall as rain. It can be stored in large reservoirs as high up as is feasible.

- Tidal power schemes trap water at high tides and release it during a low tide.

- Water can be pumped from a low reservoir to a high reservoir. Although the energy used to do this pumping must be more than the energy regained when the water flows back down hill, this '**pumped storage**' system provides one of the few large-scale methods of storing energy.

ADVANTAGES AND DISADVANTAGES

Advantages
- Very 'clean' production – no harmful chemical by-products.
- Renewable source of energy.
- Source of energy is free.

Disadvantages
- Can only be utilized in particular areas.
- Construction of dams will involve land being submerged under water.

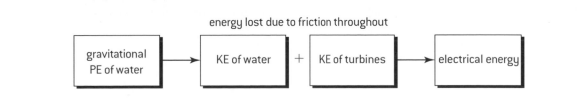

Wind power and other technologies

ENERGY TRANSFORMATIONS

There is a great deal of kinetic energy involved in the winds that blow around the Earth. The original source of this energy is, of course, the Sun. Different parts of the atmosphere are heated to different temperatures. The temperature differences cause pressure differences, due to hot air rising or cold air sinking, and thus air flows as a result.

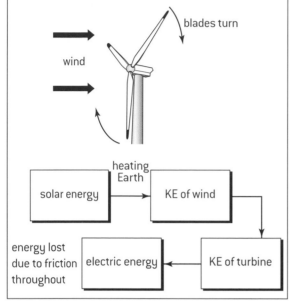

MATHEMATICS

The area 'swept out' by the blades of the turbine $= A = \pi r^2$

In one second the volume of air that passes the turbine $= vA$

So mass of air that passes the turbine in one second $= vA\rho$

Kinetic energy m available per second $= \frac{1}{2}mv^2$

$$= \frac{1}{2}(vA\rho)v^2$$

$$= \frac{1}{2}A\rho v^3$$

In other words, power available $= \frac{1}{2}A\rho v^3$

In practice, the kinetic energy of the incoming wind is easy to calculate, but it cannot all be harnessed as the air must continue to move – in other words the wind turbine cannot be one hundred per cent efficient. A doubling of the wind speed would mean that the available power would increase by a factor of eight.

ADVANTAGES AND DISADVANTAGES

Advantages

- Very 'clean' production – no harmful chemical by-products.
- Renewable source of energy.
- Source of energy is free.

Disadvantages

- Source of energy is unreliable – could be a day without wind.
- A very large area would need be covered for a significant amount of energy.
- Some consider large wind generators to spoil the countryside.
- Can be noisy.
- Best positions for wind generators are often far from centres of population.

SECONDARY ENERGY SOURCES

By far the most common primary energy sources in use worldwide are the three main fossil fuels: oil, coal and natural gas. With the inclusion of uranium, at the time of writing this guide, this accounts for 90% of the world's energy consumption. Other primary fuels include the renewables: solar, wind, tidal, biomass and geothermal.

With global energy demand expected to rise in the future, the hope is that developments with renewable energy can help to reduce the dependence on fossil fuels.

Primary energy sources are not convenient for individual users and typically a conversion process takes place that results in a

secondary energy source that can be widely used in society. The most common secondary sources are electrical energy (a very versatile secondary source) or refined fuels (e.g. petrol).

The storage of electrical energy is a challenge, with everyday devices (e.g. batteries or capacitors) having a very limited capability when compared with typical everyday demands. Power companies need to vary the generation of electrical energy to match consumer demand. Currently pumped storage hydroelectric systems are the only viable large-scale method of storing spare electrical energy capacity for future use. The efficiency of a typical system is approximately 75% meaning that one quarter of the energy supplied is wasted.

NEW AND DEVELOPING TECHNOLOGIES

It is impossible to predict technological developments that are going to take place over the coming years. Current models, however, predict a continuing dependence on the use of fossil fuels for many years to come. The hope is that we will be able to decrease this dependency over time. It is important to be

aware of the development of new technologies particularly those associated with:

- renewable energy sources
- improving the efficiency of our energy conversion process.

Thermal energy transfer

PROCESSES OF THERMAL ENERGY TRANSFER

There are several processes by which the transfer of thermal energy from a hot object to a cold object can be achieved. Three very important processes are called **conduction**, **convection** and **radiation**. Any given practical situation probably involves more than one of these processes happening at the same time. There is a fourth process called **evaporation**. This involves the faster moving molecules leaving the surface of a liquid that is below its boiling point. Evaporation causes cooling.

CONDUCTION

In thermal conduction, thermal energy is transferred along a substance without any bulk (overall) movement of the substance. For example, one end of a metal spoon soon feels hot if the other end is placed in a hot cup of tea.

Conduction is the process by which kinetic energy is passed from molecule to molecule.

macroscopic view

Thermal energy flows along the material as a result of the temperature difference across its ends.

microscopic view

The faster-moving molecules at the hot end pass on their kinetic energy to the slower-moving molecules as a result of intermolecular collisions.

Points to note:
- Poor conductors are called thermal **insulators**.
- Metals tend to be very good thermal conductors. This is because a different mechanism (involving the electrons) allows quick transfer of thermal energy.
- All gases (and most liquids) tend to be poor conductors.

Examples:
- Most clothes keep us warm by trapping layers of air – a poor conductor.
- If one walks around a house in bare feet, the floors that are better conductors (e.g. tiles) will feel colder than the floors that are good insulators (e.g. carpets) even if they are at the same temperature. (For the same reason, on a cold day a piece of metal feels colder than a piece of wood.)
- When used for cooking food, saucepans conduct thermal energy from the source of heat to the food.

EXAMPLE

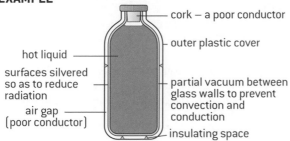

- cork – a poor conductor
- outer plastic cover
- hot liquid
- surfaces silvered so as to reduce radiation
- partial vacuum between glass walls to prevent convection and conduction
- air gap (poor conductor)
- insulating space

A thermos flask prevents heat loss

CONVECTION

In convection, thermal energy moves between two points because of a bulk movement of matter. This can only take place in a **fluid** (a liquid or a gas). When part of the fluid is heated it tends to expand and thus its density is reduced. The colder fluid sinks and the hotter fluid rises up. Central heating causes a room to warm up because a **convection current** is set up as shown below.

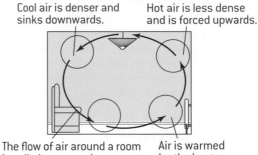

Cool air is denser and sinks downwards.

Hot air is less dense and is forced upwards.

The flow of air around a room is called a convection current.

Air is warmed by the heater.

Convection in a room

Points to note:
- Convection cannot take place in a solid.

Examples:
- The pilots of gliders (and many birds) use naturally occurring convection currents in order to stay above the ground.
- Sea breezes (winds) are often due to convection. During the day the land is hotter than the sea. This means hot air will rise from above the land and there will be a breeze onto the shore. During the night, the situation is reversed.
- Lighting a fire in a chimney will mean that a breeze flows in the room towards the fire.

RADIATION

Matter is not involved in the transfer of thermal energy by radiation. All objects (that have a temperature above zero kelvin) radiate **electromagnetic waves**. If you hold your hand up to a fire to 'feel the heat', your hands are receiving the radiation.

For most everyday objects this radiation is in the **infra-red** part of the **electromagnetic spectrum**. For more details of the electromagnetic spectrum, see page 37.

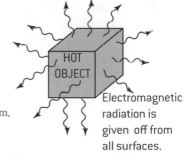

HOT OBJECT

Electromagnetic radiation is given off from all surfaces.

Points to note:
- An object at room temperature absorbs and radiates energy. If it is at constant temperature (and not changing state) then the rates are the same.
- A surface that is a good radiator is also a good absorber.
- Surfaces that are light in colour and smooth (shiny) are poor radiators (and poor absorbers).
- Surfaces that are dark and rough are good radiators (and good absorbers).
- If the temperature of an object is increased then the frequency of the radiation increases. The total rate at which energy is radiated will also increase.
- Radiation can travel through a vacuum (space).

Examples:
- The Sun warms the Earth's surface by radiation.
- Clothes in summer tend to be white – so as not to absorb the radiation from the Sun.

Radiation: Wien's law and the Stefan–Boltzmann law

BLACK-BODY RADIATION: STEFAN-BOLTZMANN LAW

In general, the radiation given out from a hot object depends on many things. It is possible to come up with a theoretical model for the 'perfect' emitter of radiation. The 'perfect' emitter will also be a perfect absorber of radiation – a black object absorbs all of the light energy falling on it. For this reason the radiation from a theoretical 'perfect' emitter is known as **black-body radiation**.

Black-body radiation does not depend on the nature of the emitting surface, but it does depend upon its temperature. At any given temperature there will be a range of different wavelengths (and hence frequencies) of radiation that are emitted. Some wavelengths will be more intense than others. This variation is shown in the graph below.

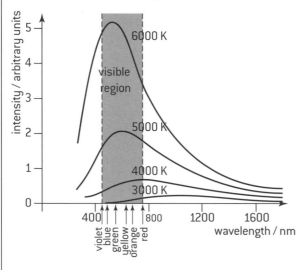

To be absolutely precise, it is not correct to label the y-axis on the above graph as the intensity, but this is often done. It is actually something that could be called the intensity function. This is defined so that the area under the graph (between two wavelengths) gives the intensity emitted in that wavelength range. The total area under the graph is thus a measure of the total power radiated. The power radiated by a Black-body (See page 195) is given by:

Surface area in m²   absolute temperature in kelvins

$$P = \sigma A T^4$$

Total power radiated in W   Stefan-Boltzmann constant

Although stars and planets are not perfect emitters, their radiation spectrum is approximately the same as black-body radiation.

WIEN'S LAW

Wien's displacement law relates the wavelength at which the intensity of the radiation is a maximum λ_{max} to the temperature of the black body T. This states that

$\lambda_{max} T = $ constant

The value of the constant can be found by experiment. It is 2.9×10^{-3} m K. It should be noted that in order to use this constant, the wavelength should be substituted into the equation in metres and the temperature in kelvin.

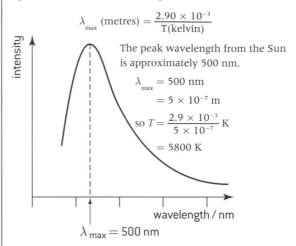

$$\lambda_{max} \text{ (metres)} = \frac{2.90 \times 10^{-3}}{T \text{(kelvin)}}$$

The peak wavelength from the Sun is approximately 500 nm.

$\lambda_{max} = 500$ nm

     $= 5 \times 10^{-7}$ m

so $T = \dfrac{2.9 \times 10^{-3}}{5 \times 10^{-7}}$ K

     $= 5800$ K

We can analyse light from a star and calculate a value for its surface temperature. This will be much less than the temperature in the core. Hot stars will give out all frequencies of visible light and so will tend to appear white in colour. Cooler stars might well only give out the higher wavelengths (lower frequencies) of visible light – they will appear red. Radiation emitted from planets will peak in the infra-red.

INTENSITY, I

The intensity of radiation is the power per unit area that is received by the object. The unit is W m^{-2}.

$$I = \frac{\text{Power}}{A}.$$

EQUILIBRIUM AND EMISSIVITY

If the temperature of a planet is constant, then the power being absorbed by the planet must equal the rate at which energy is being radiated into space. The planet is in **thermal equilibrium**. If it absorbs more energy than it radiates, then the temperature must go up and if the rate of loss of energy is greater than its rate of absorption then its temperature must go down.

In order to estimate the power absorbed or emitted, the following concepts are useful.

Emissivity

The Earth and its atmosphere are not a perfect black body. Emissivity, e, is defined as the ratio of power radiated per unit area by an object to the power radiated per unit area by a black body at the same temperature. It is a ratio and so has no units.

$$e = \frac{\text{power radiated by object per unit area}}{\text{power radiated per unit area by black body at same temperature}}$$

thus

$$p = e\sigma A T^4$$

ALBEDO

Some of the radiation received by a planet is reflected straight back into space. The fraction that is reflected back is called the **albedo**, α.

The Earth's albedo varies daily and is dependent on season (cloud formations) and latitude. Oceans have a low value but snow has a high value. The global annual mean albedo is 0.3 (30%) on Earth.

$$\text{albedo} = \frac{\text{total scattered power}}{\text{total incident power}}$$

Solar power

SOLAR CONSTANT

The amount of power that arrives from the Sun is measured by the solar constant. It is properly defined as the amount of solar energy that falls per second on an area of 1 m² above the Earth's atmosphere that is at right angles to the Sun's rays. Its average value is about 1400 W m⁻².

This is not the same as the power that arrives on 1 m² of the Earth's surface. Scattering and absorption in the atmosphere means that often less than half of this arrives at the Earth's surface. The amount that arrives depends greatly on the weather conditions.

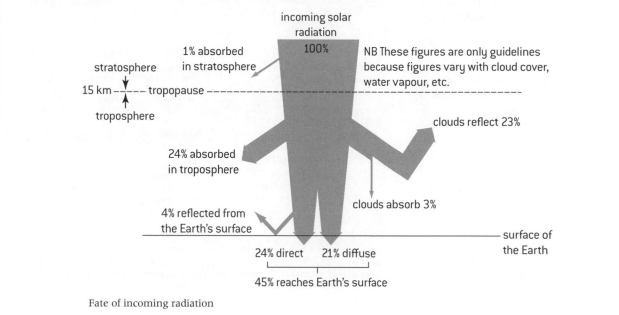

Fate of incoming radiation

Different parts of the Earth's surface (regions at different latitudes) will receive different amounts of solar radiation. The amount received will also vary with the seasons since this will affect how spread out the rays have become.

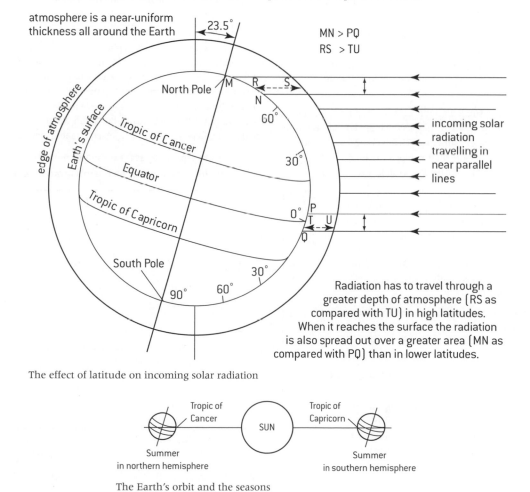

The effect of latitude on incoming solar radiation

The Earth's orbit and the seasons

The greenhouse effect

PHYSICAL PROCESSES

Short wavelength radiation is received from the Sun and causes the surface of the Earth to warm up. The Earth will emit infra-red radiation (longer wavelengths than the radiation coming from the Sun) because the Earth is cooler than the Sun. Some of this infra-red radiation is absorbed by gases in the atmosphere and re-radiated in all directions.

This is known as the **greenhouse effect** and the gases in the atmosphere that absorb infra-red radiation are called **greenhouse gases**. The net effect is that the upper atmosphere and the surface of the Earth are warmed. The name is potentially confusing, as real greenhouses are warm as a result of a different mechanism.

The temperature of the Earth's surface will be constant if the rate at which it radiates energy equals the rate at which it absorbs energy. The greenhouse effect is a natural process and without it the temperature of the Earth would be much lower; the average temperature of the Moon is more than 30 °C colder than the Earth.

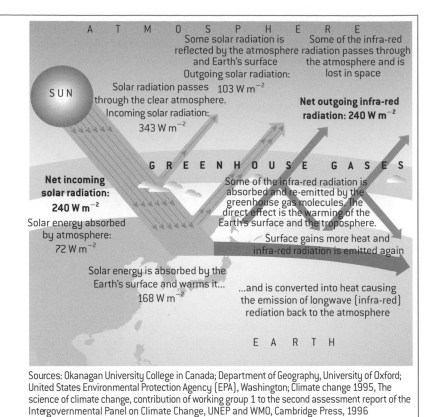

Sources: Okanagan University College in Canada; Department of Geography, University of Oxford; United States Environmental Protection Agency (EPA), Washington; Climate change 1995, The science of climate change, contribution of working group 1 to the second assessment report of the Intergovernmental Panel on Climate Change, UNEP and WMO, Cambridge Press, 1996

GREENHOUSE GASES

The main greenhouse gases are naturally occurring but the balance in the atmosphere can be altered as a result of their release due to industry and technology. They are:

*   **Methane**, CH_4. This is the principal component of natural gas and the product of decay, decomposition or fermentation. Livestock and plants produce significant amounts of methane.

*   **Water**, H_2O. The small amounts of water vapour in the upper atmosphere (as opposed to clouds which are condensed water vapour) have a significant effect. The average water vapour levels in the atmosphere do not appear to alter greatly as a result of industry, but local levels can vary.

*   **Carbon dioxide**, CO_2. Combustion releases carbon dioxide into the atmosphere which can significantly increase the greenhouse effect. Overall, plants (providing they are growing) remove carbon dioxide from the atmosphere during photosynthesis. This is known as **carbon fixation**.

*   **Nitrous oxide**, N_2O. Livestock and industries (e.g. the production of Nylon) are major sources of nitrous oxide. Its effect is significant as it can remain in the upper atmosphere for long periods.

In addition the following gases also contribute to the greenhouse effect:

*   **Ozone**, O_3. The **ozone layer** is an important region of the atmosphere that absorbs high energy UV photons which would otherwise be harmful to living organisms. Ozone also adds to the greenhouse effect.

*   **Chlorofluorocarbons** (CFCs). Used as refrigerants, propellants and cleaning solvents. They also have the effect of depleting the ozone layer.

Each of these gases absorbs infra-red radiation as a result of resonance (see page 168). The natural frequency of oscillation of the bonds within the molecules of the gas is in the infra-red region. If the driving frequency (from the radiation emitted from the Earth) is equal to the natural frequency of the molecule, resonance will occur. The amplitude of the molecules' vibrations increases and the temperature will increase. The absorption will take place at specific frequencies depending on the molecular energy levels.

Absorption spectra for major natural greenhouse gases in the Earth's atmosphere

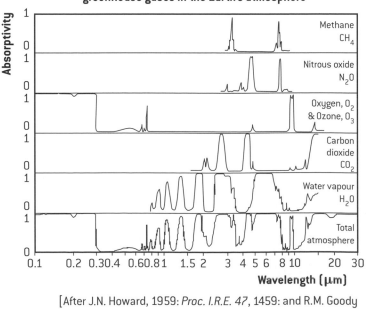

[After J.N. Howard, 1959: *Proc. I.R.E. 47*, 1459: and R.M. Goody and G.D. Robinson, 1951: *Quart. J. Roy Meteorol. Soc. 77*, 153]

Global warming

POSSIBLE CAUSES OF GLOBAL WARMING

Records show that the mean temperature of the Earth has been increasing in recent years.

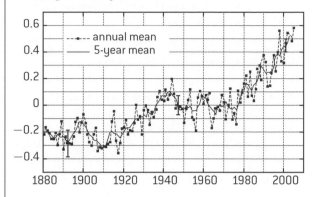

All atmospheric models are highly complicated. Some possible suggestions for this increase include.

- Changes in the composition of greenhouse gases in the atmosphere.

- Changes in the intensity of the radiation emitted by the Sun linked to, for example, increased solar flare activity.

- Cyclical changes in the Earth's orbit and volcanic activity.

The first suggestion could be caused by natural effects or could be caused by human activities (e.g. the increased burning of fossil fuels). An **enhanced greenhouse effect** is an increase in the greenhouse effect caused by human activities.

In 2013, the IPCC (Intergovernmental Panel on Climate Change) report stated that 'It is extremely likely that human influence has been the dominant cause of the observed warming since the mid–20th century'.

Although it is still being debated, the generally accepted view is that that the increased combustion of fossil fuels has released extra carbon dioxide into the atmosphere, which has enhanced the greenhouse effect.

EVIDENCE FOR GLOBAL WARMING

One piece of evidence that links global warming to increased levels of greenhouse gases comes from ice core data. The ice core has been drilled in the Russian Antarctic base at Vostok. Each year's new snow fall adds another layer to the ice.

Isotopic analysis allows the temperature to be estimated and air bubbles trapped in the ice cores can be used to measure the atmospheric concentrations of greenhouse gases. The record provides data from over 400,000 years ago to the present. The variations of temperature and carbon dioxide are very closely correlated.

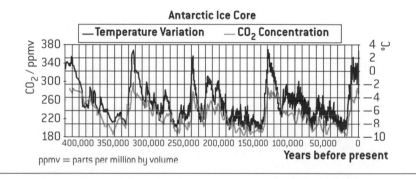

ppmv = parts per million by volume

MECHANISMS

Predicting the future effects of global warming involves a great deal of uncertainty, as the interactions between different systems in the Earth and its atmosphere are extremely complex.

There are many mechanisms that may increase the rate of global warming.

- Global warming reduces ice/snow cover, which in turn reduces the albedo. This will result in an increase in the overall rate of heat absorption.

- Temperature increase reduces the solubility of CO_2 in the sea and thus increases atmospheric concentrations.

- Continued global warming will increase both evaporation and the atmosphere's ability to hold water vapour. Water vapour is a greenhouse gas.

- Regions with frozen subsoil exist (called tundra) that support simple vegetation. An increase in temperature may cause a significant release of trapped CO_2.

- Not only does deforestation result in the release of further CO_2 into the atmosphere, the reduction in number of trees reduces carbon fixation.

The first four mechanisms are examples of processes whereby a small initial temperature increase has gone on to cause a further increase in temperature. This process is known as **positive feedback**. Some people have suggested that the current temperature increases may be 'corrected' by a process which involves negative feedback, and temperatures may fall in the future.

IB Questions – energy production

1. A wind generator converts wind energy into electric energy. The source of this wind energy can be traced back to solar energy arriving at the Earth's surface.

   a) Outline the energy transformations involved as solar energy converts into wind energy. [2]

   b) List **one** advantage and **one** disadvantage of the use of wind generators. [2]

   The expression for the maximum theoretical power, P, available from a wind generator is

   $$P = \frac{1}{2} A \rho v^3$$

   where A is the area swept out by the blades,

   ρ is the density of air and

   v is the wind speed.

   c) Calculate the maximum theoretical power, P, for a wind generator whose blades are 30 m long when a 20 m s^{-1} wind blows. The density of air is 1.3 kg m^{-3}. [2]

   d) In practice, under these conditions, the generator only provides 3 MW of electrical power.

      (i) Calculate the efficiency of this generator. [2]

      (ii) Give **two** reasons explaining why the actual power output is less than the maximum theoretical power output. [2]

2. This question is about energy sources.

   a) Give **one** example of a renewable energy source and **one** example of a non-renewable energy source and explain why they are classified as such. [4]

   b) A wind farm produces 35,000 MWh of energy in a year. If there are ten wind turbines on the farm show that the average power output of **one** turbine is about 400 kW. [3]

   c) State **two** disadvantages of using wind power to generate electrical power. [2]

3. This question is about energy transformations.

   Wind power can be used to generate electrical energy.

   Construct an energy flow diagram which shows the energy transformations, starting with solar energy and ending with electrical energy, generated by windmills. Your diagram should indicate where energy is degraded. [7]

4. This question is about a coal-fired power station which is water cooled.

   Data:

Electrical power output from the station	= 200 MW
Temperature at which water enters cooling tower	= 288 K
Temperature at which water leaves cooling tower	= 348 K
Rate of water flow through tower	= 4000 kg s^{-1}
Energy content of coal	= 2.8 × 10^7 J kg^{-1}
Specific heat of water	= 4200 J kg^{-1} K^{-1}

   Calculate

   a) the energy per second carried away by the water in the cooling tower; [2]

   b) the energy per second produced by burning the coal; [2]

   c) the overall efficiency of the power station; [2]

   d) the mass of coal burnt each second. [1]

5. This question is about tidal power systems.

   a) Describe the principle of operation of such a system. [2]

   b) Outline **one** advantage and **one** disadvantage of using such a system. [2]

   c) A small tidal power system is proposed. Use the data in the table below to calculate the total energy available and hence estimate the useful output power of this system.

Height between high tide and low tide	4 m
Trapped water would cover an area of	1.0 × 10^6 m^2
Density of water	1.0 × 10^3 kg m^{-3}
Number of tides per day	2

   [4]

6. Solar power and climate models.

   a) Distinguish, in terms of the energy changes involved, between a solar heating panel and a photovoltaic cell. [2]

   b) State an appropriate domestic use for a

      (i) solar heating panel. [1]

      (ii) photovoltaic cell. [1]

   c) The radiant power of the Sun is 3.90 × 10^{26} W. The average radius of the Earth's orbit about the Sun is 1.50 × 10^{11} m. The albedo of the atmosphere is 0.300 and it may be assumed that no energy is absorbed by the atmosphere.

   Show that the intensity incident on a solar heating panel at the Earth's surface when the Sun is directly overhead is 966 W m^{-2}. [3]

   d) Show, using your answer to (c), that the average intensity incident on the Earth's surface is 242 Wm^{-2}. [3]

   e) Assuming that the Earth's surface behaves as a black-body and that no energy is absorbed by the atmosphere, use your answer to (d) to show that the average temperature of the Earth's surface is predicted to be 256 K. [2]

   f) Outline, with reference to the greenhouse effect, why the average surface temperature of the Earth is higher than 256 K. [4]

(HL) Simple harmonic motion

SIMPLE HARMONIC MOTION (SHM) EQUATION

SHM occurs when the forces on an object are such that the resultant acceleration, a, is directed towards, and is proportional to, its displacement, x, from a fixed point.

$$a \propto -x \text{ or } a = -(\text{constant}) \times x$$

The mathematics of SHM is simplified if the constant of proportionality between a and x is identified as the square of another constant ω which is called the angular frequency. Thus the general form for the equation that defines SHM is:

$$a = -\omega^2 x$$

The solutions for this equation follow below. The angular frequency ω has the units of rad s^{-1} and is related to the time period, T, of the oscillation by the following equation.

$$\omega = \frac{2\pi}{T}$$

IDENTIFICATION OF SHM

In order to analyse a situation to decide if SHM is taking place, the following procedure should be followed.

- Identify all the forces acting on an object when it is displaced an arbitrary distance x from its rest position using a free-body diagram.

- Calculate the resultant force using Newton's second law. If this force is proportional to the displacement and always points back towards the mean position (i.e. $F \propto -x$) then the motion of the object must be SHM.

- Once SHM has been identified, the equation of motion must be in the following form:

$$a = -\left(\frac{\text{restoring force per unit displacement, } k}{\text{oscillating mass, } m} \right) \times x$$

- This identifies the angular frequency ω as $\omega^2 = \left(\frac{k}{m} \right)$ or $\omega = \sqrt{\left(\frac{k}{m} \right)}$. Identification of ω allows quantitative equations to be applied.

ACCELERATION, VELOCITY AND DISPLACEMENT DURING SHM

The variation with time of the acceleration, a, velocity, v, and displacement, x, of an object doing SHM depends on the angular frequency ω.

The precise format of the relationships depends on where the object is when the clock is started (time t = zero). The left hand set of equations correspond to an oscillation when the object is in the mean position when $t = 0$. The right hand set of equations correspond to an oscillation when the object is at maximum displacement when $t = 0$.

$x = x_0 \sin \omega t$	$x = x_0 \cos \omega t$
$v = \omega x_0 \cos \omega t$	$v = -\omega x_0 \sin \omega t$
$a = -\omega^2 x_0 \sin \omega t$	$a = -\omega^2 x_0 \cos \omega t$

The first two equations can be rearranged to produce the following relationship:

$$v = \pm \omega \sqrt{(x_0 - x^2)}$$

x_0 is the amplitude of the oscillation measured in m
t is the time taken measured in s
ω is the angular frequency measured in rad s^{-1}
ωt is an angle that increases with time measured in radians. A full oscillation is completed when $(\omega t) = 2\pi$ rad.

The angular frequency is related to the time period T by the following equation.

$$T = \frac{2\pi}{\omega} = 2\pi \sqrt{\frac{m}{k}}$$

TWO EXAMPLES OF SHM

Two common situations that approximate to SHM are:

1. Mass, m, on a vertical spring

   Provided that:

   - the mass of the spring is negligible compared to the mass of the load
   - friction (air friction) is negligible
   - the spring obeys Hooke's law with spring constant, k at all times (i.e. elastic limit is not exceeded)
   - the gravitational field strength g is constant
   - the fixed end of the spring cannot move.

   Then it can be shown that:

   $$\omega^2 = \frac{k}{m}$$

   Or $T = 2\pi \sqrt{\frac{m}{k}}$

2. The simple pendulum of length l and mass m

   Provided that:

   - the mass of the string is negligible compared with the mass of the load
   - friction (air friction) is negligible
   - the maximum angle of swing is small ($\leq$ 5° or 0.1 rad)
   - the gravitational field strength g is constant
   - the length of the pendulum is constant.

   Then it can be shown that:

   $$\omega^2 = \frac{g}{l}$$

   Or $T = 2\pi \sqrt{\frac{l}{g}}$

   Note that the mass of the pendulum bob, m, is not in this equation and thus does not affect the time period of the pendulum, T.

EXAMPLE

A 600 g mass is attached to a light spring with spring constant 30 N m^{-1}.

(a) Show that the mass does SHM.

(b) Calculate the frequency of its oscillation.

(a) Weight of mass = mg = 6.0 N

Additional displacement x down means that resultant force on mass = kx upwards. Since $F \propto -x$, the mass will oscillate with SHM.

(b) Since SHM, $T = 2\pi \sqrt{\left(\frac{m}{k} \right)} = 2\pi \sqrt{\left(\frac{0.6}{30} \right)} = 0.889$ s

$$f = \frac{1}{T} = \frac{1}{0.889} = 1.1 \text{ Hz}$$

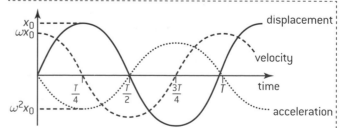

- acceleration leads velocity by 90°
- velocity leads displacement by 90°
- acceleration and displacement are 180° out of phase
- displacement lags velocity by 90°
- velocity lags acceleration by 90°

HL Energy changes during simple harmonic motion

During SHM, energy is interchanged between KE and PE. Providing there are no resistive forces which dissipate this energy, the total energy must remain constant. The oscillation is said to be **undamped**.

The kinetic energy can be calculated from

$$E_k = \frac{1}{2} mv^2 = \frac{1}{2} m \omega^2 (x_0 - x^2)$$

The potential energy can be calculated from

$$E_p = \frac{1}{2} m \omega^2 x^2$$

The total energy is

$$E = E_k + E_p = \frac{1}{2} m \omega^2 (x_0 - x^2) + \frac{1}{2} m\omega^2 x^2 = \frac{1}{2} m \omega^2 x_0$$

Energy in SHM is proportional to:

- the mass m
- the (amplitude)^2
- the (frequency)^2

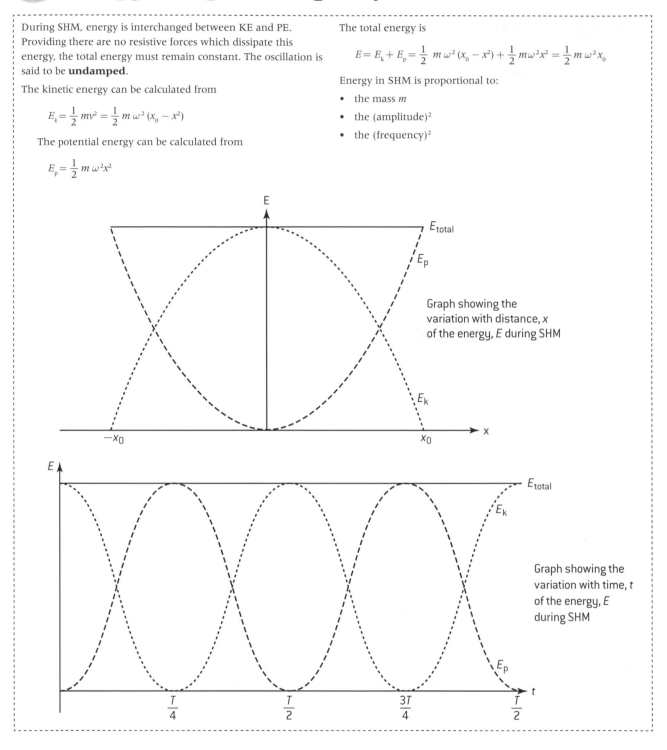

Graph showing the variation with distance, x of the energy, E during SHM

Graph showing the variation with time, t of the energy, E during SHM

 # Diffraction

BASIC OBSERVATIONS

Diffraction is a wave effect. The objects involved (slits, apertures, etc.) have a size that is of the same order of magnitude as the wavelength of visible light.

Nature of obstacle	Geometrical shadow	Diffraction pattern
(a) straight edge		
(b) single long slit $b \sim 3\lambda$		
(c) circular aperture		
(d) single long slit $b \sim 5\lambda$		

The intensity plot for a single slit is:

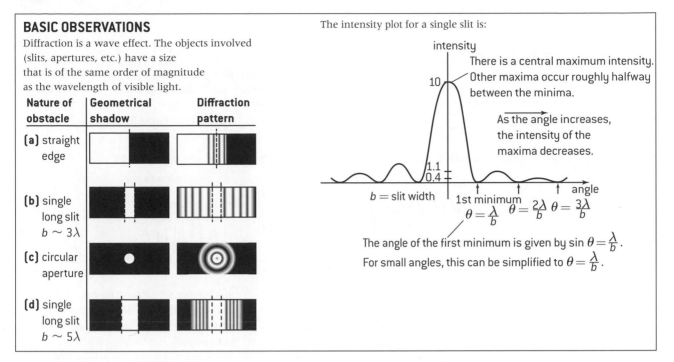

There is a central maximum intensity. Other maxima occur roughly halfway between the minima.

As the angle increases, the intensity of the maxima decreases.

$b = $ slit width

1st minimum
$\theta = \frac{\lambda}{b}$ $\theta = \frac{2\lambda}{b}$ $\theta = \frac{3\lambda}{b}$

The angle of the first minimum is given by $\sin \theta = \frac{\lambda}{b}$.
For small angles, this can be simplified to $\theta = \frac{\lambda}{b}$.

EXPLANATION

The shape of the relative intensity versus angle plot can be derived by applying an idea called **Huygens' principle**. We can treat the slit as a series of secondary wave sources. In the forward direction ($\theta = $ zero) these are all in phase so they add up to give a maximum intensity. At any other angle, there is a path difference between the rays that depends on the angle.

The overall result is the addition of all the sources. The condition for the first minimum is that the angle must make all of the sources across the slit cancel out.

The condition for the first maximum out from the centre is when the path difference across the whole slit is $\frac{3\lambda}{2}$. At this angle the slit can be analysed as being three equivalent sections each having a path difference of $\frac{\lambda}{2}$ across its length. Together, two of these sections will destructively interfere leaving the resulting amplitude to be $\frac{1}{3}$ of the maximum. Since intensity $\propto$ (amplitude)², the first maximum intensity out from the centre will be $\frac{1}{9}$ of the central maximum intensity. By a similar argument, the second maximum intensity out from the centre will have $\frac{1}{5}$ of the maximum amplitude and thus be $\frac{1}{25}$ of the central maximum intensity.

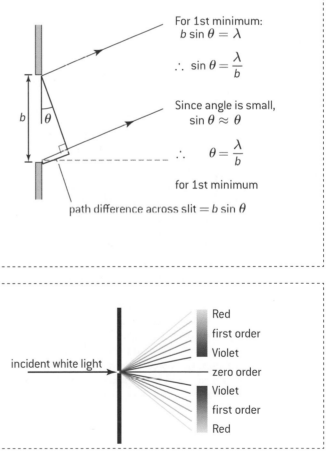

For 1st minimum:
$b \sin \theta = \lambda$

$\therefore$ $\sin \theta = \frac{\lambda}{b}$

Since angle is small,
$\sin \theta \approx \theta$

$\therefore$ $\theta = \frac{\lambda}{b}$

for 1st minimum

path difference across slit $= b \sin \theta$

SINGLE-SLIT DIFFRACTION WITH WHITE LIGHT

When a single slit is illuminated with white light, each component colour has a specific wavelength and so the associated maxima and minima for each wavelength will be located at a different angle. For a given slit width, colours with longer wavelengths (red, orange, etc.) will diffract more than colours with short wavelengths (blue, violet, etc.). The maxima for the resulting diffraction pattern will show all the colours of the rainbow with blue and violet nearer to the central position and red appearing at greater angles.

incident white light

Red
first order
Violet
zero order
Violet
first order
Red

HL Two-source interference of waves: Young's double-slit experiment

DOUBLE-SLIT INTERFERENCE

The double-slit interference pattern shown on page 47 was derived assuming that each slit was behaving like a perfect point source. This can only take place if the slits are infinitely small. In practice they have a finite width. The diffraction pattern of each slit needs to be taken into account when working out the overall double slit interference pattern as shown below.

Decreasing the slit width will mean that the observed pattern becomes more and more 'idealized'. Unfortunately, it will also mean that the total intensity of light will be decreased. The interference pattern will become harder to observe.

(a) Young's fringes for infinitely narrow slits

(b) diffraction pattern for a finite-width slit

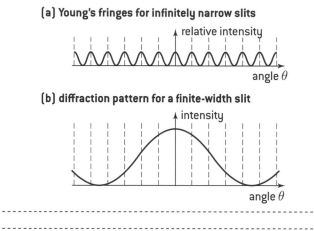

(c) Young's fringes for slits of finite width

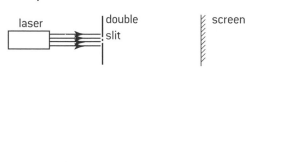

$s = \dfrac{\lambda D}{d}$ still applies but different fringes will have different intensities with it being possible for some fringes to be missing.

INVESTIGATING YOUNG'S DOUBLE-SLIT EXPERIMENTALLY

Possible set-ups for the double-slit experiment are shown on page 47.

Set-up 1

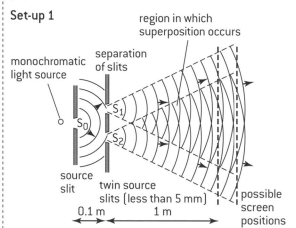

In the original set-up (set-up 1) light from the monochromatic source is diffracted at S_0 so as to ensure that S_1 and S_2 are receiving coherent light. Diffraction takes place providing S_1 and S_2 are narrow enough. The slit separations need to be approximately 1 mm (or less) thus the slit widths are of the order of 0.1 mm (or less). This would provide fringes that were separated by approximately 0.5mm on a screen (semi-transparent or translucent) situated 1m away. The laboratory will need to be darkened to allow the fringes to be visible and they can be viewed using a microscope.

The most accurate measurements for slit separation and fringe width are achieved using a **travelling microscope**. This is a microscope that is mounted on a frame so that it can be moved perpendicular to the direction in which it is pointing. The microscope is moved across ten or more fringes and the distance moved by the microscope can be read off from the scale. The precision of this measurement is often improved by utilizing a vernier scale.

In the simplified version (set-up 2) of the experiment, fringes can still be bright enough to be viewed several metres away from the slits and thus they can be projected onto an opaque screen (it is dangerous to look into a laser beam). Their separation can be then be directly measured with a ruler.

Set-up 2

Ⓗ Multiple-slit diffraction

THE DIFFRACTION GRATING

The diffraction that takes place at an individual slit affects the overall appearance of the fringes in Young's double-slit experiment (see page 98 for more details). This section considers the effect on the final interference pattern of adding further slits. A series of parallel slits (at a regular separation) is called a **diffraction grating**.

Additional slits at the same separation will not affect the condition for constructive interference. In other words, the angle at which the light from slits adds constructively will be unaffected by the number of slits. The situation is shown below.

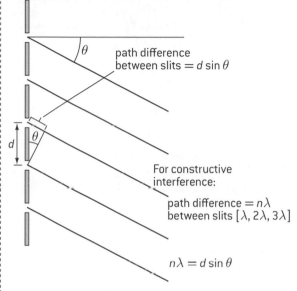

path difference between slits = $d \sin \theta$

For constructive interference:

path difference = $n\lambda$
between slits [$\lambda, 2\lambda, 3\lambda$]

$$n\lambda = d \sin \theta$$

This formula also applies to the Young's double-slit arrangement. The difference between the patterns is most noticeable at the angles where perfect constructive interference does not take place. If there are only two slits, the maxima will have a significant angular width. Two sources that are just out of phase interfere to give a resultant that is nearly the same amplitude as two sources that are exactly in phase.

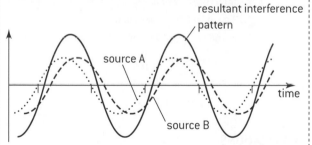

resultant interference pattern

source A

source B

time

The addition of more slits will mean that each new slit is just out of phase with its neighbour. The overall interference pattern will be totally destructive.

overall interference pattern is totally destructive

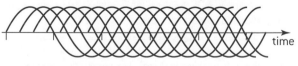

time

The addition of further slits at the same slit separation has the following effects:

* the principal maxima maintain the same separation
* the principal maxima become much sharper
* the overall amount of light being let through is increased, so the pattern increases in intensity.

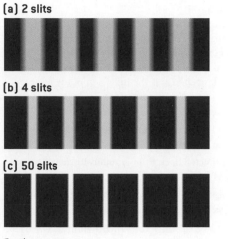

(a) 2 slits

(b) 4 slits

(c) 50 slits

Grating patterns

USES

One of the main uses of a diffraction grating is the accurate experimental measurement of the different wavelengths of light contained in a given spectrum. If white light is incident on a diffraction grating, the angle at which constructive interference takes place depends on wavelength. Different wavelengths can thus be observed at different angles. The accurate measurement of the angle provides the experimenter with an accurate measurement of the exact wavelength (and thus frequency) of the colour of light that is being considered. The apparatus that is used to achieve this accurate measurement is called a **spectrometer**.

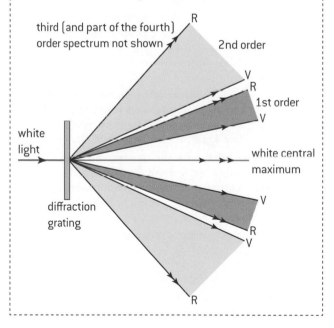

third (and part of the fourth) order spectrum not shown

2nd order

1st order

white light

white central maximum

diffraction grating

PHASE CHANGES

There are many situations when interference can take place that also involve the reflection of light. When analysing in detail the conditions for constructive or destructive interference, one needs to take any **phase changes** into consideration. A phase change is the inversion of the wave that can take place at a reflection interface, but it does not always happen. It depends on the two media involved.

The technical term for the inversion of a wave is that it has 'undergone a phase change of π'.

- When light is reflected back from an optically denser medium there is a phase change of π.
- When light is reflected back from an optically less dense medium there is no phase change.

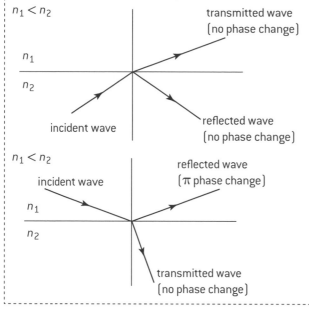

EXAMPLE

The equations in the box on the right work out the angles for which constructive and destructive interference take place for a given wavelength. If the source of light is an extended source, the eye receives rays leaving the film over a range of values for θ.

If white light is used then the situation becomes more complex. Provided the thickness of the film is small, then one or two colours may reinforce along a direction in which others cancel. The appearance of the film will be bright colours, such as can be seen when looking at

- an oil film on the surface of water or
- soap bubbles.

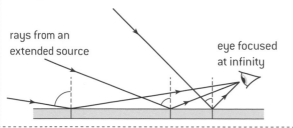

CONDITIONS FOR INTERFERENCE PATTERNS

A parallel-sided film can produce interference as a result of the reflections that are taking place at both surfaces of the film.

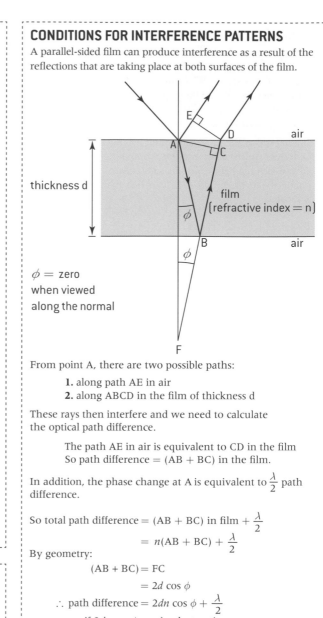

ϕ = zero
when viewed
along the normal

From point A, there are two possible paths:

    1. along path AE in air
    2. along ABCD in the film of thickness d

These rays then interfere and we need to calculate the optical path difference.

    The path AE in air is equivalent to CD in the film
    So path difference = (AB + BC) in the film.

In addition, the phase change at A is equivalent to $\frac{\lambda}{2}$ path difference.

So total path difference = (AB + BC) in film + $\frac{\lambda}{2}$
$$= n(AB + BC) + \frac{\lambda}{2}$$
By geometry:
$$(AB + BC) = FC$$
$$= 2d \cos \phi$$
$\therefore$ path difference = $2dn \cos \phi + \frac{\lambda}{2}$
    if $2dn \cos \phi = m\lambda$: destructive
    or when $\phi = 0$, $2dn = m\lambda$: destructive
    if $2dn \cos \phi = \left(m + \frac{\lambda}{2}\right)\lambda$: constructive
    or when $\phi = 0$, $2dn = m\lambda$: constructive
$$m = 0,1,2,3,4$$

APPLICATIONS

Applications of parallel thin films include:

- The design of non-reflecting radar coatings for military aircraft. If the thickness of the extra coating is designed so that radar signals destructively interfere when they reflect from both surfaces, then no signal will be reflected and an aircraft could go undetected.
- Measurements of thickness of oil slicks caused by spillage. Measurements of the wavelengths of electromagnetic signals that give constructive and destructive interference (at known angles) allow the thickness of the oil to be calculated.
- Design of non-reflecting surfaces for lenses (blooming), solar panels and solar cells. A strong reflection at any of these surfaces would reduce the amount of energy being usefully transmitted. A thin surface film can be added so that destructive interference takes place for a typical wavelength and thus maximum transmittance takes place at this wavelength.

 Resolution

DIFFRACTION AND RESOLUTION

If two sources of light are very close in angle to one another, then they are seen as one single source of light. If the eye can tell the two sources apart, then the sources are said to be **resolved**. The diffraction pattern that takes place at apertures affects the eye's ability to resolve sources. The examples to the right show how the appearance of two line sources will depend on the diffraction that takes place at a slit. The resulting appearance is the addition of the two overlapping diffraction patterns. The graph of the resultant relative intensity of light at different angles is also shown.

These examples look at the situation of a line source of light and the diffraction that takes place at a slit. A more common situation would be a point source of light, and the diffraction that takes place at a circular aperture. The situation is exactly the same, but diffraction takes place all the way around the aperture. As seen on page 97, the diffraction pattern of the point source is thus concentric circles around the central position. The geometry of the situation results in a slightly different value for the first minimum of the diffraction pattern.

For a **slit**, the first minimum was at the angle

$$\theta = \frac{\lambda}{b}$$

For a **circular aperture**, the first minimum is at the angle

$$\theta = \frac{1.22\,\lambda}{b}$$

If two sources are just resolved, then the first minimum of one diffraction pattern is located on top of the maximum of the other diffraction pattern. This is known as the **Rayleigh criterion**.

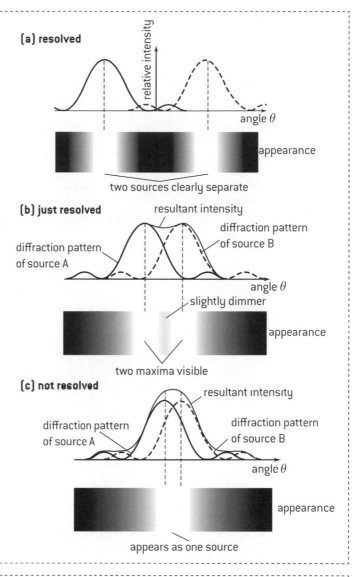

(a) resolved

two sources clearly separate

(b) just resolved

resultant intensity

diffraction pattern of source B

diffraction pattern of source A

slightly dimmer

appearance

two maxima visible

(c) not resolved

resultant intensity

diffraction pattern of source A

diffraction pattern of source B

appearance

appears as one source

EXAMPLE

Late one night, a student was observing a car approaching from a long distance away. She noticed that when she first observed the headlights of the car, they appeared to be one point of light. Later, when the car was closer, she became able to see two separate points of light. If the wavelength of the light can be taken as 500 nm and the diameter of her pupil is approximately 4 mm, calculate how far away the car was when she could first distinguish two points of light. Take the distance between the headlights to be 1.8 m.

When just resolved

$$\theta = \frac{1.22 \times \lambda}{b}$$

$$= \frac{1.22 \times 5 \times 10^{-7}}{0.004}$$

$$= 1.525 \times 10^{-4}$$

Since θ small

$$\theta = \frac{1.8}{x} \quad [x \text{ is distance to car}]$$

$$\Rightarrow x = \frac{1.8}{1.525 \times 10^{-4}}$$

$$= 11.803$$

$$\simeq 12 \text{ km}$$

RESOLVANCE OF DIFFRACTION GRATINGS

As a result of Rayleigh's criterion, there is a limit placed on a grating's ability to resolve different wavelengths. The resolvance, R, of a diffraction grating is defined as the ratio between a wavelength being investigated, λ, and the smallest possible resolvable wavelength difference, $\Delta\lambda$.

$$R = \frac{\lambda}{\Delta\lambda}$$

For any given grating, R is dependent on the diffraction order, m, being observed (first order: $m = 1$; second order: $m = 2$, etc.) and the total number of slits, N, on the grating that are being illuminated.

$$R = \frac{\lambda}{\Delta\lambda} = mN$$

Example:

In the sodium emission spectrum there are two wavelengths that are close to one another (the Na D-lines). These are 589.00 nm and 589.59 nm. In order for these to be resolved by a diffraction grating, the resolvance must be

$$R = \frac{\lambda}{\Delta\lambda} = \frac{589.00}{0.59} = 1000$$

In the first order spectrum, at least 1000 slits must be illuminated whereas in the second order spectrum, the requirement drops to only 500 slits.

HL The Doppler effect

DOPPLER EFFECT

The Doppler effect is the name given to the change of frequency of a wave as a result of the movement of the source or the movement of the observer.

When a source of sound is moving:

- Sound waves are emitted at a particular frequency from the source.
- The speed of the sound wave in air does not change, but the motion of the source means that the wave fronts are all 'bunched up' ahead of the source.
- This means that the stationary observer receives sound waves of reduced wavelength.
- Reduced wavelength corresponds to an increased frequency of sound.

The overall effect is that the observer will hear sound at a higher frequency than it was emitted by the source. This applies when the source is moving towards the observer. A similar analysis quickly shows that if the source is moving away from the observer, sound of a lower frequency will be received. A change of frequency can also be detected if the source is stationary, but the observer is moving.

- When a police car or ambulance passes you on the road, you can hear the pitch of the sound change from high to low frequency. It is high when it is approaching and low when it is going away.
- Radar detectors can be used to measure the speed of a moving object. They do this by measuring the change in the frequency of the reflected wave.
- For the Doppler effect to be noticeable with light waves, the source (or the observer) needs to be moving at high speed. If a source of light of a particular frequency is moving away from an observer, the observer will receive light of a lower frequency. Since the red part of the spectrum has lower frequency than all the other colours, this is called a **red shift**.
- If the source of light is moving towards the observer, there will be a **blue shift**.

MOVING SOURCE

Source moves from A to D with velocity, u_s, speed of waves is v.

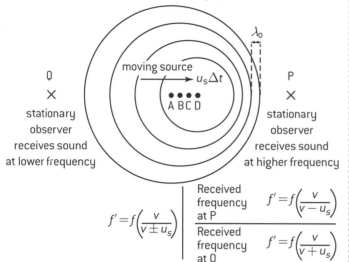

$$f' = f\left(\frac{v}{v \pm u_s}\right)$$

Received frequency at P	$f' = f\left(\dfrac{v}{v - u_s}\right)$
Received frequency at Q	$f' = f\left(\dfrac{v}{v + u_s}\right)$

MATHEMATICS OF THE DOPPLER EFFECT

Mathematical equations that apply to sound are stated on this page.

Unfortunately the same analysis does not apply to light – the velocities can not be worked out relative to the medium. It is, however, possible to derive an equation for light that turns out to be in exactly the same form as the equation for sound as long as two conditions are met:

- the relative velocity of source and detector is used in the equations.
- this relative velocity is a lot less than the speed of light.

Providing $v \ll c$

change in frequency — change wavelength due to relative motion — relative speed of source and observer

frequency of source

wavelength when no relative motion

speed of light

$$\frac{\Delta f}{f} = \frac{\Delta \lambda}{\lambda} \approx \frac{v}{c}$$

MOVING OBSERVER

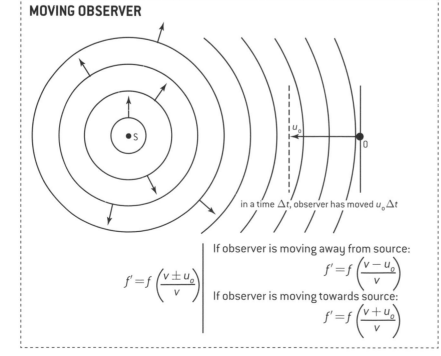

in a time Δt, observer has moved $u_o \Delta t$

$$f' = f\left(\frac{v \pm u_o}{v}\right)$$

If observer is moving away from source:
$$f' = f\left(\frac{v - u_o}{v}\right)$$

If observer is moving towards source:
$$f' = f\left(\frac{v + u_o}{v}\right)$$

EXAMPLE

The frequency of a car's horn is measured by a stationary observer as 200 Hz when the car is at rest. What frequency will be heard if the car is approaching the observer at 30 m s^{-1}? (Speed of sound in air is 330 m s^{-1}.)

$f = 200$ Hz

$f' = ?$

$u_s = 30$ m s^{-1}

$v = 330$ m s^{-1}

$f' = 200\left(\dfrac{300}{300 - 30}\right)$

$= 200 \times 1.1$

$= 220$ Hz

ⒽⓁ Examples and applications of the Doppler effect

1. **Train going through a station**

   The sound emitted by a moving train's whistle is of constant frequency, but the sound received by a passenger standing on the platform will change. At any instant of time, it is the resolved component of the train's velocity towards the passenger that is used to calculate the frequency received.

   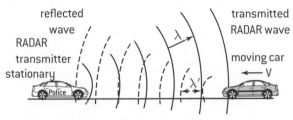

2. **Radars – speed measurement**

   In many countries the police use radar to measure speed of vehicles to see if they are breaking the speed limit.

   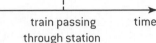

   - Pulse of microwave radiation of known frequency emitted.
   - Pulse is reflected off moving car and received back at source.
   - Difference in emitted and received frequencies is used to calculate speed of car.
   - Double Doppler effect taking place:
     ◊ Moving car receives a frequency that is higher than emitted as it is a moving observer.
     ◊ Moving car acts as a moving source when sending signal back.

3. **Medical physics – blood flow measurements**

   Doctors can use a pulse of ultrasound to measure the speed of red blood cells in an analogous way that a pulse of microwaves is used to measure the speed of a moving car (above).

   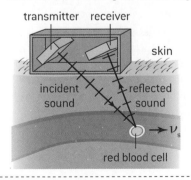

4. **Receding galaxies – red shift**

   - The relative intensities of the different wavelengths of light received from the stars in distant galaxies can be analysed.
   - The light shows a characteristic absorption spectrum.
   - The measured wavelengths are not the same as those associated with particular elements as measured in the laboratory.
   - For the vast majority of stars, all the received frequencies have been shifted towards the red end of the visible spectrum (i.e. to lower frequencies). The light shows a **red shift** (see page 202).
   - The magnitude of the red shift is used to calculate the recessional velocity and provides evidence for the Big Bang model of the creation of the Universe.

5. **Rotating object**

   The rotation of luminous objects (e.g. the Sun) can be measured by looking for a different Doppler shift on one side of the object compared with the other.

   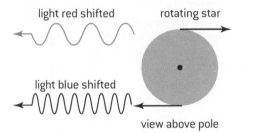

6. **Broadening of spectral lines**

   - Absorption and emission spectra provide evidence for discrete atomic energy levels (see page 69).
   - Precise measurements show that each individual level is actually equivalent to a small but defined wavelength range.
   - The gas molecules are moving so light from molecules will be subjected to Doppler shift.
   - Different molecules have a range of speeds so there will be a general **Doppler broadening** of the discrete wavelengths.
   - A higher temperature means a wider distribution of kinetic energies and hence more broadening to the spectral line.

HL IB Questions – wave phenomena

1. When a train travels towards you sounding its whistle, the pitch of the sound you hear is different from when the train is at rest. This is because

   A. the sound waves are travelling faster toward you.

   B. the wave fronts of the sound reaching you are spaced closer together.

   C. the wave fronts of the sound reaching you are spaced further apart.

   D. the sound frequency emitted by the whistle changes with the speed of the train.

2. A car is travelling at constant speed towards a stationary observer whilst its horn is sounded. The frequency of the note emitted by the horn is 660 Hz. The observer, however, hears a note of frequency 720 Hz.

   a) With the aid of a diagram, explain why a higher frequency is heard. [2]

   b) If the speed of sound is 330 m s⁻¹, calculate the speed of the car. [2]

3. This question is about using a diffraction grating to view the emission spectrum of sodium.

   Light from a sodium discharge tube is incident normally upon a diffraction grating having 8.00×10^5 lines per metre. The spectrum contains a double yellow line of wavelengths 589 nm and 590 nm.

   a) Determine the angular separation of the two lines when viewed in the second order spectrum. [4]

   b) State why it is more difficult to observe the double yellow line when viewed in the first order spectrum. [1]

4. This question is about thin film interference.

   A transparent thin film is sometimes used to coat spectacle lenses as shown in the diagram below.

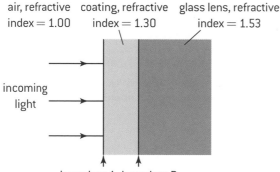

air, refractive index = 1.00   coating, refractive index = 1.30   glass lens, refractive index = 1.53

incoming light

boundary A   boundary B

   a) State the phase change which occurs to light that

   (i) is transmitted at boundary A into the film. [1]

   (ii) is reflected at boundary B. [1]

   (iii) is transmitted at boundary A from the film into the air. [1]

   b) Light of wavelength 570 nm in air is incident on the coating. Determine the smallest thickness of the coating required so that the reflection is minimized for normal incidence. [2]

5. Simple harmonic motion and the greenhouse effect

   a) A body is displaced from equilibrium. State the **two** conditions necessary for the body to execute simple harmonic motion. [2]

   b) In a simple model of a methane molecule, a hydrogen atom and the carbon atom can be regarded as two masses attached by a spring. A hydrogen atom is much less massive than the carbon atom such that any displacement of the carbon atom may be ignored.

   The graph below shows the variation with time t of the displacement x from its equilibrium position of a hydrogen atom in a molecule of methane.

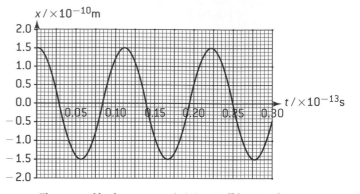

$x / \times 10^{-10}$m

$t / \times 10^{-13}$s

   The mass of hydrogen atom is 1.7×10^{-27} kg. Use data from the graph above

   (i) to determine its amplitude of oscillation. [1]

   (ii) to show that the frequency of its oscillation is 9.1×10^{13} Hz. [2]

   (iii) to show that the maximum kinetic energy of the hydrogen atom is 6.2×10^{-18} J. [2]

   c) Sketch a graph to show the variation with time t of the velocity v of the hydrogen atom for one period of oscillation starting at $t = 0$. (There is no need to add values to the velocity axis.) [3]

   d) Assuming that the motion of the hydrogen atom is simple harmonic, its frequency of oscillation f is given by the expression

   $$f = \frac{1}{2\pi}\sqrt{\frac{k}{m_p}},$$

   where k is the force per unit displacement between a hydrogen atom and the carbon atom and m_p is the mass of a proton.

   (i) Show that the value of k is approximately 560 N m⁻¹. [1]

   (ii) Estimate, using your answer to (d)(i), the maximum acceleration of the hydrogen atom. [2]

   e) Methane is classified as a greenhouse gas.

   (i) Describe what is meant by a greenhouse gas. [2]

   (ii) Electromagnetic radiation of frequency 9.1×10^{13} Hz is in the infrared region of the electromagnetic spectrum. Suggest, based on the information given in (b)(ii), why methane is classified as a greenhouse gas. [2]

Ⓗ Potential (gravitational and electric)

DESCRIBING FIELDS: g and E

The concept of field lines can be used to visually represent:

- the gravitational field, g, around a mass (or collection of masses)
- the electric field, E, around a charge (or collection of charges).

Magnetic fields can also be represented using field lines (see page 61). In all cases the field is the force per unit test point object placed at a particular point in the field with:

- gravitational field = force per unit test point mass (units: N kg^{-1})
- electric field = force per unit test point positive charge (units: N C^{-1})

Forces are vectors and field lines represent both the magnitude and the direction of the force that would be felt by a test object.

- The **magnitude** of the force is represented by how close the field lines are to one another (for an example of a more precise definition, see definition of magnetic flux on page 112).

- The **direction** of the force is represented by the direction of the field lines.

This means that, for both gravitational and electric fields, as a test object is moved:

- **along** a field line, work will be done (force and distance moved are in the same direction)
- **at right angles** to a field line, no work will be done (force and distance moved are perpendicular).

An alternative method of mapping the fields around an object is to consider the energy needed to move between points in the field. This defines the new concepts of electric potential and gravitational potential (see below).

POTENTIAL, V (GRAVITATIONAL OR ELECTRIC)

The *field* (gravitational or electric) is defined as the force per unit test point object placed at a particular point in the field. In an analogous definition, the *potential* (gravitational, V_g, or electric, V_e) is defined as the energy per unit test point object that the object has as a result of the field. The full mathematical relationships are shown on page 110.

Gravitational potential, $V_g = \dfrac{\text{energy}}{\text{mass}}$

Units of $V_g = $ J kg^{-1}

Electric potential, $V_e = \dfrac{\text{energy}}{\text{charge}}$

Units of $V_e = $ J C^{-1} (or volts)

POTENTIAL DIFFERENCE ΔV (ELECTRIC AND GRAVITATIONAL)

Potential is the energy per unit test object. In general, moving a mass between two points, A and B, in a gravitational field (or moving a charge between two points, A and B, in an electric field) means that work is done. When work is done, the potential at A and the potential at B will be different. Between the points A and B, there will be a potential difference, ΔV.

- If positive work is done *on* a test object as it moves between two points then the potential between the two points must increase.
- If work is done *by* the test object as it moves between the two points then the potential between the two points must decrease.

Gravitational potential difference between two points,

$$\Delta V_g = \frac{\text{work done moving a test mass}}{\text{test mass}}$$

Units of $\Delta V_g = $ J kg^{-1}

Electric potential difference between two points,

$$\Delta V_e = \frac{\text{work done moving a test charge}}{\text{test charge}}$$

Units of $\Delta V_e = $ J C^{-1} or V (volts)

Thus to calculate the work done, W, in moving a charge q or a mass m between two points in a field we have:

$$W = q\Delta V_e$$
$$W = m\Delta V_g$$

EQUIPOTENTIAL SURFACES

The best way of representing how the electric potential varies around a charged object is to identify the regions where the potential is the same. These are called **equipotential** surfaces. In two dimensions they would be represented as lines of equipotential. A good way of visualizing these lines is to start with the contour lines on a map.

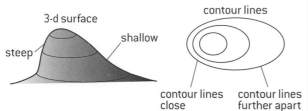

The contour diagram on the right represents the changing heights of the landscape on the left. Each line joins up points that are at the same height. Points that are high up represent a high value of gravitational potential and points that are low down represent a low gravitational potential. Contour lines are lines of equipotential in a gravitational field.

The same can be done with an electric field. Lines are drawn joining up points that have the same electric potential. The situation right shows the equipotentials for an isolated positive point charge.

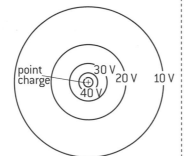

RELATIONSHIP TO FIELD LINES

There is a simple relationship between electric field lines and lines of equipotential – they are always at right angles to one another. Imagine the contour lines. If we move along a contour line, we stay at the same height in the gravitational field. This does not require work because we are moving at right angles to the gravitational force. Whenever we move along an electric equipotential line, we are moving between points that have the same electric potential – in other words, no work is being done. Moving at right angles to the electric field is the only way to avoid doing work in an electric field. Thus equipotential lines must be at right angles to field lines as shown below.

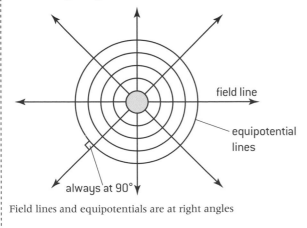

Field lines and equipotentials are at right angles

EXAMPLES OF EQUIPOTENTIALS

The diagrams below show equipotential lines for various situations.

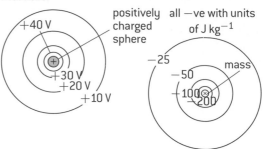

Equipotentials outside a charge-conducting sphere and a point mass.

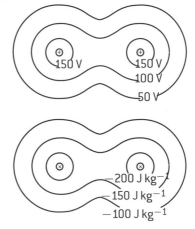

Equipotentials for two point charges (same charge) and two point masses.

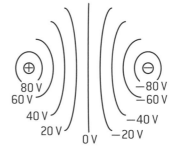

Equipotentials for two point charges (equal and opposite charges).

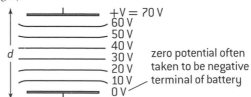

Equipotential lines between charged parallel plates.

It should be noted that although the correct definition of zero potential is at infinity, most of the time we are not really interested in the actual value of potential, we are only interested in the value of the difference in potential. This means that in some situations (such as the parallel conducting plates) it is easier to imagine the zero at a different point. This is just like setting sea level as the zero for gravitational contour lines rather than correctly using infinity for the zero.

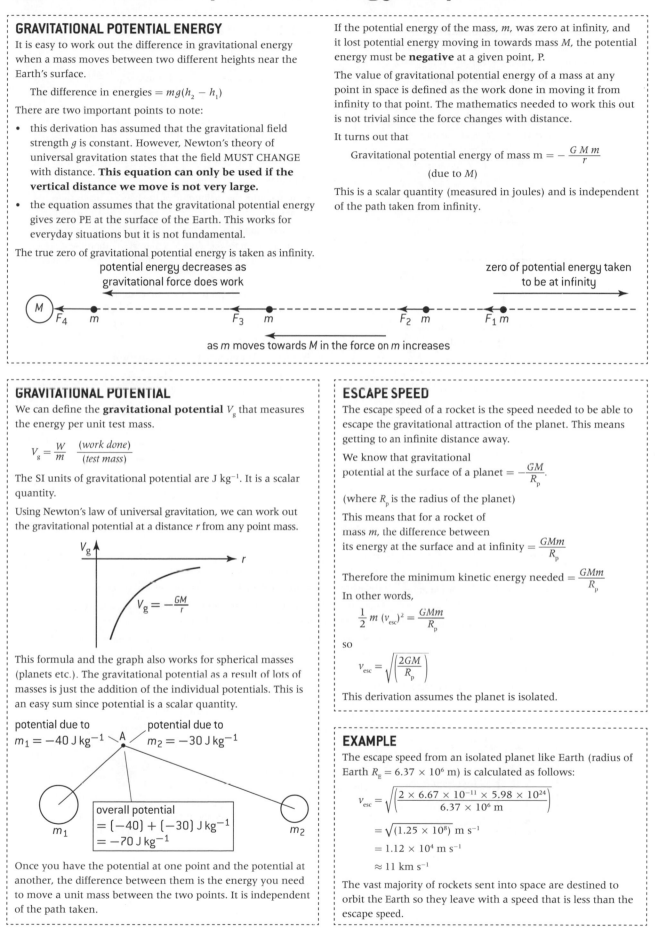 Gravitational potential energy and potential

GRAVITATIONAL POTENTIAL ENERGY

It is easy to work out the difference in gravitational energy when a mass moves between two different heights near the Earth's surface.

The difference in energies $= mg(h_2 - h_1)$

There are two important points to note:

- this derivation has assumed that the gravitational field strength g is constant. However, Newton's theory of universal gravitation states that the field MUST CHANGE with distance. **This equation can only be used if the vertical distance we move is not very large.**

- the equation assumes that the gravitational potential energy gives zero PE at the surface of the Earth. This works for everyday situations but it is not fundamental.

The true zero of gravitational potential energy is taken as infinity.

If the potential energy of the mass, m, was zero at infinity, and it lost potential energy moving in towards mass M, the potential energy must be **negative** at a given point, P.

The value of gravitational potential energy of a mass at any point in space is defined as the work done in moving it from infinity to that point. The mathematics needed to work this out is not trivial since the force changes with distance.

It turns out that

Gravitational potential energy of mass m $= -\dfrac{G M m}{r}$

(due to M)

This is a scalar quantity (measured in joules) and is independent of the path taken from infinity.

potential energy decreases as
gravitational force does work

zero of potential energy taken
to be at infinity

M F_4 m F_3 m F_2 m F_1 m

as m moves towards M in the force on m increases

GRAVITATIONAL POTENTIAL

We can define the **gravitational potential** V_g that measures the energy per unit test mass.

$$V_g = \frac{W}{m} \quad \frac{(work\ done)}{(test\ mass)}$$

The SI units of gravitational potential are J kg^{-1}. It is a scalar quantity.

Using Newton's law of universal gravitation, we can work out the gravitational potential at a distance r from any point mass.

$V_g = -\dfrac{GM}{r}$

This formula and the graph also works for spherical masses (planets etc.). The gravitational potential as a result of lots of masses is just the addition of the individual potentials. This is an easy sum since potential is a scalar quantity.

potential due to
$m_1 = -40$ J kg^{-1} A   potential due to
$m_2 = -30$ J kg^{-1}

overall potential
$= (-40) + (-30)$ J kg^{-1}
$= -70$ J kg^{-1}

m_1 m_2

Once you have the potential at one point and the potential at another, the difference between them is the energy you need to move a unit mass between the two points. It is independent of the path taken.

ESCAPE SPEED

The escape speed of a rocket is the speed needed to be able to escape the gravitational attraction of the planet. This means getting to an infinite distance away.

We know that gravitational potential at the surface of a planet $= -\dfrac{GM}{R_p}$.

(where R_p is the radius of the planet)

This means that for a rocket of mass m, the difference between its energy at the surface and at infinity $= \dfrac{GMm}{R_p}$

Therefore the minimum kinetic energy needed $= \dfrac{GMm}{R_p}$

In other words,

$$\frac{1}{2} m (v_{esc})^2 = \frac{GMm}{R_p}$$

so

$$v_{esc} = \sqrt{\left(\frac{2GM}{R_p}\right)}$$

This derivation assumes the planet is isolated.

EXAMPLE

The escape speed from an isolated planet like Earth (radius of Earth $R_E = 6.37 \times 10^6$ m) is calculated as follows:

$$v_{esc} = \sqrt{\left(\frac{2 \times 6.67 \times 10^{-11} \times 5.98 \times 10^{24}}{6.37 \times 10^6 \text{ m}}\right)}$$

$$= \sqrt{(1.25 \times 10^8)} \text{ m s}^{-1}$$

$$= 1.12 \times 10^4 \text{ m s}^{-1}$$

$$\approx 11 \text{ km s}^{-1}$$

The vast majority of rockets sent into space are destined to orbit the Earth so they leave with a speed that is less than the escape speed.

HL Orbital motion

GRAVITATIONAL POTENTIAL GRADIENT

In the diagram below, a point test mass m moves in a gravitational field from point A to point B.

The difference in gravitational potential, ΔV_g

$$= \frac{\text{average force} \times \text{distance moved}}{m} = -g \times \Delta r$$

The negative sign is because g is directed towards M, but the force doing the work is directed away from M and thus in the opposite direction from g. Since the gravitational force is attractive, work has to be done in going from A to B, so the potential at A < potential at B.

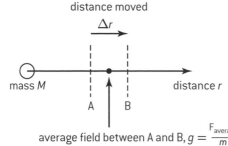

distance moved

Δr

mass M distance r

A   B

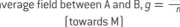

average field between A and B, $g = \dfrac{F_{average}}{m}$

(towards M)

$$\Delta V_g = -g \times \Delta r$$

$$g = -\frac{\Delta V}{\Delta r}$$

$\frac{\Delta V}{\Delta r}$ is called the potential gradient. It has units of J kg⁻¹ m⁻¹ (which are the same as N kg⁻¹ or m s⁻²).

The gravitational field strength is equal to minus the potential gradient. The equivalent relationship also applies for electric fields (see page 109).

ENERGY OF AN ORBITING SATELLITE

We already know that the gravitational energy $= -\dfrac{GMm}{r}$

The kinetic energy $= \dfrac{1}{2} m v^2$ but $v = \sqrt{\left(\dfrac{GM}{r}\right)}$ (Circular motion)

$\therefore$ kinetic energy $= \dfrac{1}{2} m \dfrac{GM}{r} = \dfrac{1}{2} \dfrac{GMm}{r}$

So total energy = KE + PE

$$= \frac{1}{2} \frac{GMm}{r} - \frac{GMm}{r} = -\frac{1}{2} \frac{GMm}{r}$$

Note that:

- In the orbit the magnitude of the KE $= \frac{1}{2}$ magnitude of the PE.

- The overall energy of the satellite is negative. (A satellite must have a total energy less than zero otherwise it would have enough energy to escape the Earth's gravitional field.)

- In order to move from a small radius orbit to a large radius orbit, the total energy must increase. To be precise, an increase in orbital radius makes the total energy go from a large negative number to a smaller negative number – this is an increase.

This can be summarized in graphical form.

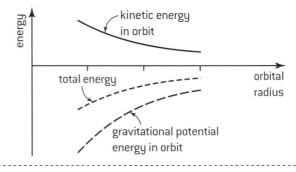

WEIGHTLESSNESS

One way of defining the weight of a person is to say that it is the value of the force recorded on a supporting scale.

If the scales were set up in a lift, they would record different values depending on the **acceleration** of the lift.

An extreme version of these situations occurs if the lift cable breaks and the lift (and passenger) accelerates down at 10 m s⁻².

accelerating down at 10 m s⁻²

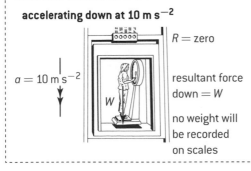

$a = 10$ m s⁻²

$R =$ zero

resultant force down $= W$

no weight will be recorded on scales

The person would appear to be weightless for the duration of the fall. Given the possible ambiguity of the term 'weight', it is better to call this situation the **apparent weightlessness** of objects in free-fall together.

An astronaut in an orbiting space station would also appear weightless. The space station and the astronaut are in free-fall together.

In the space station, the gravitational pull on the astronaut provides the centripetal force needed to stay in the orbit. This resultant force causes the centripetal acceleration. The same is true for the gravitational pull on the satellite and the satellite's acceleration. There is no contact force between the satellite and the astronaut so, once again, we have apparent weightlessness.

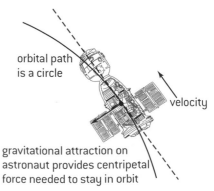

orbital path is a circle

velocity

gravitational attraction on astronaut provides centripetal force needed to stay in orbit

HL Electric potential energy and potential

POTENTIAL AND POTENTIAL DIFFERENCE

The concept of electrical potential difference between two points was introduced on page 105. As the name implies, potential difference is just the difference between the potential at one point and the potential at another. Potential is simply a measure of the total electrical energy per unit charge at a given point in space. The definition is very similar to that of gravitational potential.

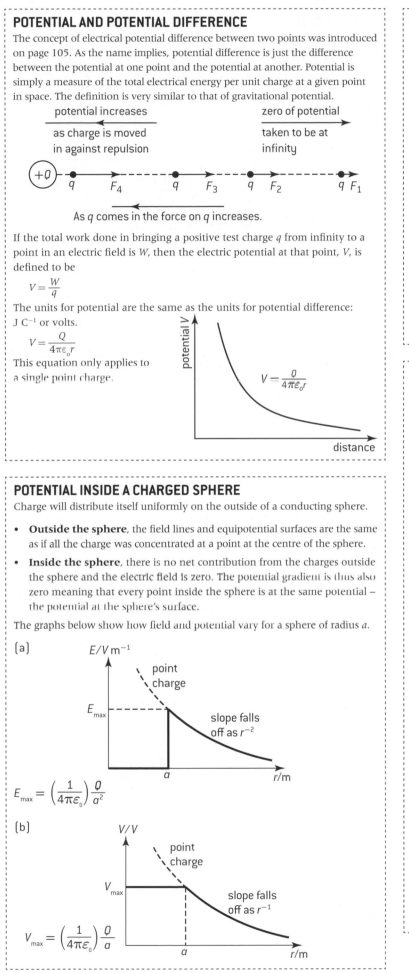

potential increases
as charge is moved
in against repulsion

zero of potential
taken to be at
infinity

As q comes in the force on q increases.

If the total work done in bringing a positive test charge q from infinity to a point in an electric field is W, then the electric potential at that point, V, is defined to be

$$V = \frac{W}{q}$$

The units for potential are the same as the units for potential difference: J C^{-1} or volts.

$$V = \frac{Q}{4\pi\varepsilon_0 r}$$

This equation only applies to a single point charge.

$V = \frac{Q}{4\pi\varepsilon_0 r}$

potential V

distance

POTENTIAL INSIDE A CHARGED SPHERE

Charge will distribute itself uniformly on the outside of a conducting sphere.

- **Outside the sphere**, the field lines and equipotential surfaces are the same as if all the charge was concentrated at a point at the centre of the sphere.

- **Inside the sphere**, there is no net contribution from the charges outside the sphere and the electric field is zero. The potential gradient is thus also zero meaning that every point inside the sphere is at the same potential – the potential at the sphere's surface.

The graphs below show how field and potential vary for a sphere of radius a.

(a)

$E/V\,m^{-1}$

point charge

E_{max}

slope falls off as r^{-2}

a r/m

$$E_{max} = \left(\frac{1}{4\pi\varepsilon_0}\right)\frac{Q}{a^2}$$

(b)

V/V

point charge

V_{max}

slope falls off as r^{-1}

a r/m

$$V_{max} = \left(\frac{1}{4\pi\varepsilon_0}\right)\frac{Q}{a}$$

POTENTIAL DUE TO MORE THAN ONE CHARGE

If several charges all contribute to the total potential at a point, it can be calculated by adding up the individual potentials due to the individual charges.

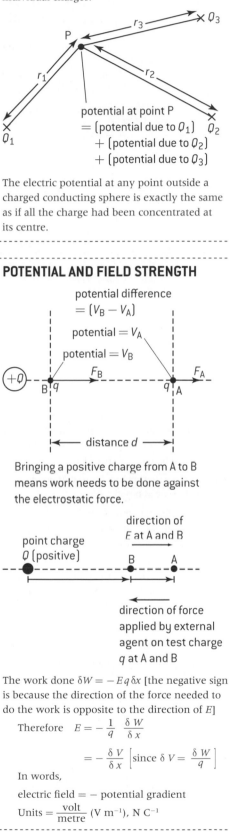

potential at point P
= (potential due to Q_1)
+ (potential due to Q_2)
+ (potential due to Q_3)

The electric potential at any point outside a charged conducting sphere is exactly the same as if all the charge had been concentrated at its centre.

POTENTIAL AND FIELD STRENGTH

potential difference
= $(V_B - V_A)$

potential = V_A

potential = V_B

distance d

Bringing a positive charge from A to B means work needs to be done against the electrostatic force.

direction of
F at A and B

point charge
Q (positive)

B   A

direction of force applied by external agent on test charge q at A and B

The work done $\delta W = -Eq\,\delta x$ [the negative sign is because the direction of the force needed to do the work is opposite to the direction of E]

Therefore $\quad E = -\frac{1}{q}\frac{\delta W}{\delta x}$

$\qquad = -\frac{\delta V}{\delta x}$ [since $\delta V = \frac{\delta W}{q}$]

In words,

electric field = − potential gradient

Units = $\frac{\text{volt}}{\text{metre}}$ (V m^{-1}), N C^{-1}

COMPARISON BETWEEN ELECTRIC & GRAVITATIONAL FIELD

Electrostatics	Gravitational
Force can be attractive or repulsive	Force always attractive
Coulomb's law – for point charges	Newton's law – for point masses
$$F_E = \frac{q_1 q_2}{4\pi\varepsilon_0 r^2} = k\frac{q_1 q_2}{r^2}$$	$$F = G\frac{m_1 m_2}{r^2}$$
Electric field	Gravitational field
electric field / charge producing field $$E = \frac{F}{q_2} = \frac{q_1}{4\pi\varepsilon_0 r^2} = k\frac{q_1}{r^2}$$ test charge	gravitational field / mass producing field $$g = \frac{F}{m_2} = \frac{Gm_1}{r^2}$$ test mass
Electric potential due to a point charge	Gravitational potential due to a point mass, m_1
$$V_e = \frac{q_1}{4\pi\varepsilon_0 r} = k\frac{q_1}{r}$$	$$V_g = -\frac{Gm_1}{r}$$
Electric potential gradient	Gravitational potential gradient
$$E = -\frac{\Delta V_e}{\Delta r}$$	$$g = -\frac{\Delta V_g}{\Delta r}$$
Electric potential energy	Gravitational potential energy
$$E_p = qV_e = \frac{q_1 q_2}{4\pi\varepsilon_0 r} = k\frac{q_1 q_2}{r}$$	$$E_p = mV_g = -\frac{GMm}{r}$$

UNIFORM FIELDS

Field strength is equal to minus the potential gradient.
A constant field thus means:

- A constant potential gradient i.e. a given increase in distance will equate to a fixed change in potential.

- In 3D this means that equipotential surfaces will be flat planes that are equally spaced apart. In 2D equipotential lines will be equally spaced.

- Field lines (perpendicular to equipotential surfaces) will be equally spaced parallel lines.

1. Constant gravitational field

   The gravitational field near the surface of a planet is effectively constant. At the surface of the Earth, the field lines will be perpendicular to the Earth's surface. Since $g = 9.81$ m s^{-2}, the potential gradient must also be 9.81 J kg^{-1} m^{-1}. Equipotential surfaces that are 1,000 m apart represent changes of potential approximately equal to 10 kJ kg^{-1}.

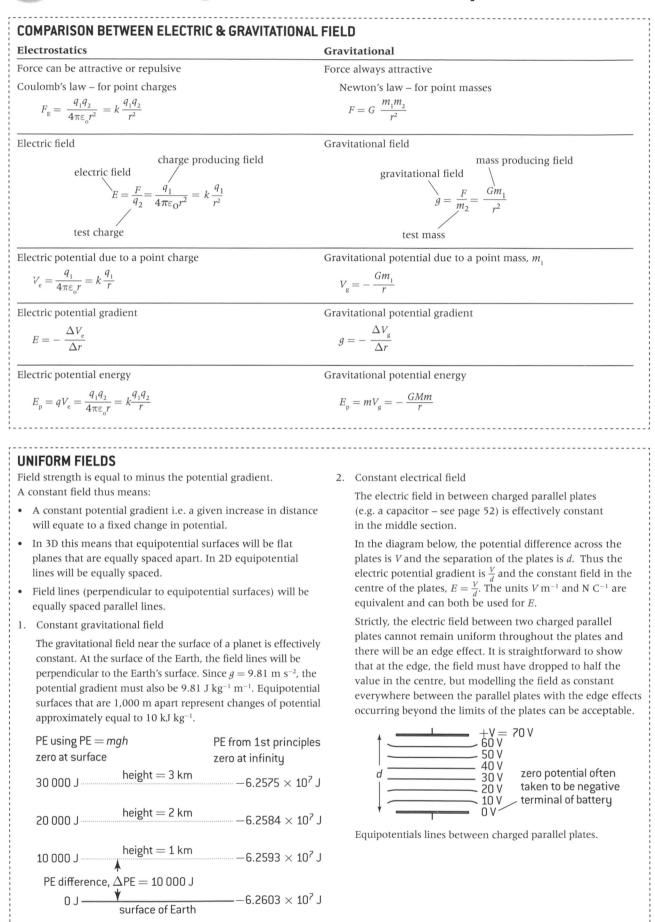

2. Constant electrical field

   The electric field in between charged parallel plates (e.g. a capacitor – see page 52) is effectively constant in the middle section.

   In the diagram below, the potential difference across the plates is V and the separation of the plates is d. Thus the electric potential gradient is $\frac{V}{d}$ and the constant field in the centre of the plates, $E = \frac{V}{d}$. The units V m^{-1} and N C^{-1} are equivalent and can both be used for E.

   Strictly, the electric field between two charged parallel plates cannot remain uniform throughout the plates and there will be an edge effect. It is straightforward to show that at the edge, the field must have dropped to half the value in the centre, but modelling the field as constant everywhere between the parallel plates with the edge effects occurring beyond the limits of the plates can be acceptable.

Equipotentials lines between charged parallel plates.

HL IB Questions – fields

1. Which **one** of the following graphs best represents the variation of the kinetic energy, KE, and of the gravitational potential energy, GPE, of an orbiting satellite at a distance r from the centre of the Earth?

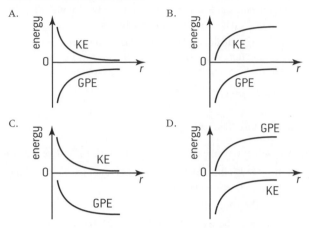

2. The diagram below illustrates some equipotential lines between two charged parallel metal plates.

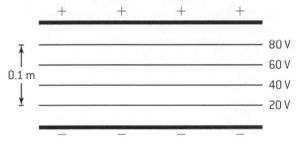

The electric field strength between the plates is

A. 6 NC^{-1} C. 600 NC^{-1}

B. 8 NC^{-1} D. 800 NC^{-1}

3. The diagram shows equipotential lines due to two objects

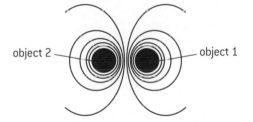

The two objects could be

A. electric charges of the same sign only.

B. masses only.

C. electric charges of opposite sign only.

D. masses or electric charges of any sign.   [1]

4. The Space Shuttle orbits about 300 km above the surface of the Earth. The shape of the orbit is circular, and the mass of the Space Shuttle is 6.8×10^4 kg. The mass of the Earth is 6.0×10^{24} kg, and radius of the Earth is 6.4×10^6 m.

a) (i) Calculate the change in the Space Shuttle's gravitational potential energy between its launch and its arrival in orbit.   [3]

(ii) Calculate the speed of the Space Shuttle whilst in orbit.   [2]

(iii) Calculate the energy needed to put the Space Shuttle into orbit.   [2]

b) (i) What forces, if any, act on the astronauts inside the Space Shuttle whilst in orbit?   [1]

(ii) Explain why astronauts aboard the Space Shuttle feel weightless.   [2]

c) Imagine an astronaut 2 m outside the exterior walls of the Space Shuttle, and 10 m from the centre of mass of the Space Shuttle. By making appropriate assumptions and approximations, calculate how long it would take for this astronaut to be pulled back to the Space Shuttle by the force of gravity alone.   [7]

5. a) The diagram below shows a planet of mass M and radius R_p:

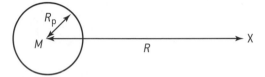

The gravitational potential V due to the planet at point X distance R from the centre of the planet is given by

$$V = -\frac{GM}{R}$$

where G is the universal gravitational constant.

Show that the gravitational potential V can be expressed as

$$V = -\frac{g_0 R_p^2}{R}$$

where g_0 is the acceleration of free-fall at the surface of the planet.   [3]

b) The graph below shows how the gravitational potential V due to the planet varies with distance R from the centre of the planet for values of R greater than R_p, where $R_p = 2.5 \times 10^6$ m.

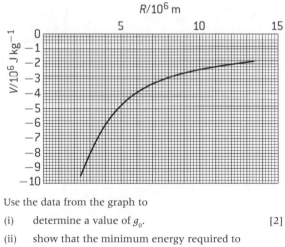

Use the data from the graph to

(i) determine a value of g_0.   [2]

(ii) show that the minimum energy required to raise a satellite of mass 3000 kg to a height 3.0×10^6 m above the **surface** of the planet is about 1.7×10^{10} J.   [3]

(HL) Induced electromotive force (emf)

INDUCED EMF

When a conductor moves through a magnetic field, an emf is induced. The emf induced depends on:

- The speed of the wire.
- The strength of the magnetic field.
- The length of the wire in the magnetic field.

We can calculate the magnitude of the induced emf by considering an electron at equilibrium in the middle of the wire. The induced electric force and the magnetic force are balanced.

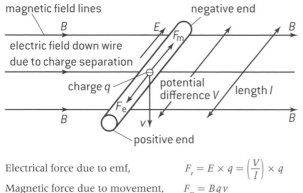

Electrical force due to emf, $\qquad F_e = E \times q = \left(\dfrac{V}{l}\right) \times q$

Magnetic force due to movement, $\qquad F_m = Bqv$

So $\qquad\qquad\qquad\qquad Bqv = \left(\dfrac{V}{l}\right)q$

$$V = Blv$$

As no current is flowing, the emf ε = potential difference

$$\varepsilon = Blv$$

If the wire was part of a complete circuit (outside the magnetic field), the emf induced would cause a current to flow.

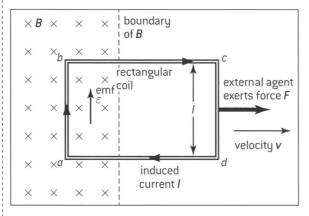

If this situation was repeated with a rectangular coil with N turns, each section ab would generate an emf equal to Bvl. The total emf generated will thus be

$$\varepsilon = BvlN$$

Note that in the situation above, a current only flows when one side of the coil (ab) is moving through the magnetic field and the other side (cd) is outside the field. If the whole coil was inside the magnetic field, each side would generate an emf. The two emfs would oppose one another and no current would flow.

PRODUCTION OF INDUCED EMF BY RELATIVE MOTION

An emf is induced in a conductor whenever lines of magnetic flux are cut. But flux is more than just a way of picturing the situation; it has a mathematical definition.

If the magnetic field is perpendicular to the surface, the magnetic flux $\Delta\phi$ passing through the area ΔA is defined in terms of the magnetic field strength B as follows.

$$\Delta\phi = B\,\Delta A, \text{ so } B = \frac{\Delta\phi}{\Delta A}$$

In a uniform field, $B = \dfrac{\phi}{A}$

An alternative name for 'magnetic field strength' is '**flux density**'.

If the area is not perpendicular, but at an angle θ to the field lines, the equation becomes

$$\phi = B\,A\cos\theta \text{ (units: T m}^2)$$

θ is the angle between **B** and the normal to the surface.

Flux can also be measured in webers (Wb), defined as follows.

$$1\text{ Wb} = 1\text{ T m}^2$$

These relationships allow us to calculate the induced emf ε in a moving wave is terms of flux.

in a time Δt:

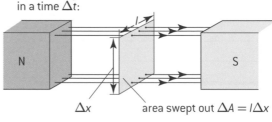

$\varepsilon = Blv$ since $v = \dfrac{\Delta x}{\Delta t}$ then $\varepsilon = \dfrac{B\,l\,\Delta x}{\Delta t}$

but $l\,\Delta x = \Delta A$, the area 'swept out' by the conductor in a time Δt so $\varepsilon = \dfrac{B\,\Delta A}{\Delta t}$

but $B\,\Delta A = \Delta\phi$ so $\varepsilon = \dfrac{\Delta\phi}{\Delta t}$

In words, 'the emf induced is equal to the rate of cutting of flux'. If the conductor is kept stationary and the magnets are moved, the same effect is produced.

EXAMPLE

An aeroplane flies at 200 m s^{-1}. Estimate the maximum pd that can be generated across its wings.

Vertical component
of Earth's magnetic field = 10^{-5} T (approximately)

Length across wings = 30 m (estimated)

emf = $10^{-5} \times 30 \times 200$

$= 6 \times 10^{-2}$ V

$= 0.06$ V

(HL) Lenz's law and Faraday's law

LENZ'S LAW

Lenz's law states that

'The direction of the induced emf is such that if an induced current were able to flow, it would oppose the change which caused it.'

(1)

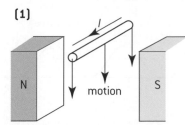

Current induced in this direction, the force would be upwards (left-hand rule)
∴ original motion would be opposed.

(2)

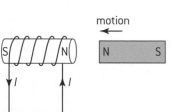

If current were induced this way, the induced field would repel the magnet — opposing motion.

Lenz's law can be explained in terms of the conservation of energy. The electrical energy generated within any system must result from work being done on the system. When a conductor is moved through a magnetic field and an induced current flows, an external force is needed to keep the conductor moving (the external force balances the opposing force that Lenz's law predicts). The external force does work and this provides the energy for the current to flow.

Put another way, if the direction of an induced current did not oppose the change that caused it, then it would be acting to support the change. If this was the case, then a force would be generated that further accelerated the moving object which would generate an even greater emf – electrical energy would be generated without work being done.

TRANSFORMER-INDUCED EMF

An emf is also produced in a wire if the magnetic field changes with time.

If the amount of flux passing through one turn of a coil is ϕ, then the total **flux linkage** with all N turns of the coil is given by

Flux linkage $= N\phi$

The universal rule that applies to all situations involving induced emf can now be stated as

'The magnitude of an induced emf is proportional to the rate of change of flux linkage.'

This is known as **Faraday's law** $\varepsilon = N\dfrac{\Delta\phi}{\Delta t}$

Faraday's law and Lenz's law can be combined together in the following mathematical statement for the emf, ε, generated in a coil of N turns with a rate of change of flux through the coil of $\frac{\Delta\phi}{\Delta t}$:

$$\varepsilon = -N\frac{\Delta\phi}{\Delta t}$$

The dependence on the rate of change of flux and the number of turns is Faraday's law and the negative sign (opposing the change) is Lenz's law.

APPLICATION OF FARADAY'S LAW TO MOVING AND ROTATING COILS

There are many situations involving magnetic fields with moving or rotating coils. To decide whether or not an emf is generated and, if it is, to calculate its value, the following procedure can be used:

- Choose the period of time, Δt, over which the motion of the coil is to be considered.

- At the beginning of the period, work out the flux passing through one turn of the coil, ϕ_{initial}. Note that the shape of the coil is not relevant just the cross-sectional area.

  $\phi = BA\cos\theta$.

- At the end of the period, work out the flux passing through one turn of the coil ϕ_{final} using the equation above. Note that the sense of the magnetic field is important. If the magnitude of the field is the same but it is passing through the coil in the opposite direction, then

  $\phi_{\text{final}} = -\phi_{\text{initial}}$

- Determine the change in flux, $\Delta\phi$:

  $\Delta\phi = \phi_{\text{final}} - \phi_{\text{initial}}$

- If there is no overall change of flux then, overall, no emf will be induced. If there is a change in flux then the emf induced in a coil of N turns will be:

  $\varepsilon = -N\dfrac{\Delta\phi}{\Delta t}$

Example:

A physicist holds her hand so that the magnetic field of the Earth (50 μT) passes through a ring on her hand.

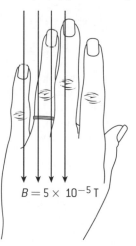

$B = 5 \times 10^{-5}$ T

In 0.1 s, she quickly turns her hand through 90° so that the magnetic field of the Earth no longer goes through the ring. Estimate the emf generated in the ring.

Answer:

Estimate of cross-sectional area of ring, A $\approx$ 1 cm² $= 10^{-4}$ m²

$\phi_{\text{initial}} = 5 \times 10^{-5} \times 10^{-4} \cos(0) = 5 \times 10^{-9}$ Wb

$\phi_{\text{final}} = 0$

∴ $\Delta\phi = 5 \times 10^{-9}$ Wb

magnitude of $\varepsilon = N\dfrac{\Delta\phi}{\Delta t} = \dfrac{5 \times 10^{-9}}{10^{-1}} = 5 \times 10^{-8}$ V

🄗 Alternating current (1)

COIL ROTATING IN A MAGNETIC FIELD – AC GENERATOR

The structure of a typical ac generator is shown below.

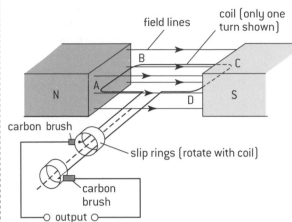

ac generator

The coil of wire rotates in the magnetic field due to an external force. As it rotates the flux linkage of the coil changes with time and induces an emf (Faraday's law) causing a current to flow. The sides AB and CD of the coil experience a force opposing the motion (Lenz's law). The work done rotating the coil generates electrical energy.

A coil rotating at constant speed will produce a sinusoidal induced emf. Increasing the speed of rotation will reduce the time period of the oscillation *and* increase the amplitude of the induced emf (as the rate of change of flux linkage is increased).

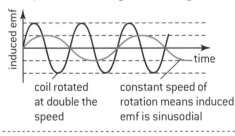

coil rotated at double the speed | constant speed of rotation means induced emf is sinusodial

RMS VALUES

If the output of an ac generator is connected to a resistor an alternating current will flow. A sinusoidal potential difference means a sinusoidal current.

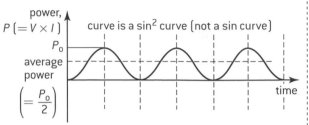

The graph shows that the average power dissipation is half the peak power dissipation for a sinusoidal current.

$$\text{Average power } \overline{P} = \frac{I_0^2 R}{2} = \left(\frac{I_0}{\sqrt{2}}\right)^2 R$$

Thus the effective current through the resistor is $\sqrt{\text{(mean value of } I^2)}$ and it is called the **root mean square** current or **rms** current, I_{rms}.

$$I_{rms} = \frac{I_0}{\sqrt{2}} \text{ (for sinusoidal currents)}$$

When ac values for voltage or current are quoted, it is the root mean square value that is being used. In Europe this value is 230 V, whereas in the USA it is 120 V.

$$V_{rms} = \frac{V_0}{\sqrt{2}}$$

$$\overline{P} = V_{rms} I_{rms} = \frac{1}{2} I_0 V_0$$

$$P_{max} = I_0 V_0$$

$$R = \frac{V}{I} = \frac{V_0}{I_0} = \frac{V_{rms}}{I_{rms}}$$

TRANSFORMER OPERATION

An **alternating** potential difference is put into the transformer, and an **alternating** potential difference is given out. The value of the output potential difference can be changed (increased or decreased) by changing the **turns ratio**. A **step-up** transformer increases the voltage, whereas a **step-down** transformer decreases the voltage.

The following sequence of calculations provides the correct method for calculating all the relevant values.

- The output voltage is fixed by the input voltage and the turns ratio.
- The value of the load that you connect fixes the output current (using $V = IR$).
- The value of the output power is fixed by the values above ($P = VI$).
- The value of the input power is equal to the output power for an ideal transformer.
- The value of the input current can now be calculated (using $P = VI$).

So how does the transformer manage to alter the voltages in this way?

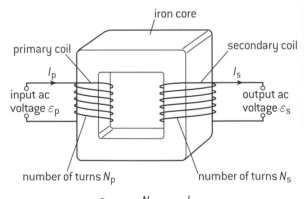

Transformer structure

- The alternating pd across the primary creates an ac within the coil and hence an alternating magnetic field in the iron core.
- This alternating magnetic field links with the secondary and induces an emf. The value of the induced emf depends on the rate of change of flux linkage, which increases with increased number of turns on the secondary. The input and output voltages are related by the turns ratio.

$$\frac{\varepsilon_p}{\varepsilon_s} = \frac{N_p}{N_s} = \frac{I_s}{I_p}$$

HL Alternating current (2)

TRANSMISSION OF ELECTRICAL POWER

Transformers play a very important role in the safe and efficient transmission of electrical power over large distances.

- If large amounts of power are being distributed, then the currents used will be high. (Power = $V\,I$)

- The wires cannot have zero resistance. This means they must dissipate some power

- Power dissipated is $P = I^2\,R$. If the current is large then the (current)^2 will be very large.

- Over large distances, the power wasted would be very significant.

- The solution is to choose to transmit the power at a very high potential difference.

- Only a small current needs to flow.

- A very high potential difference is much more efficient, but very dangerous to the user.

- Use step-up transformers to increase the voltage for the transmission stage and then use step-down transformers for the protection of the end user.

LOSSES IN THE TRANSMISSION OF POWER

In addition to power losses associated with the resistance of the power supply lines, which cause the power lines to warm up, there are also losses associated with non-ideal transformers:

- **Resistance of the windings (joule heating)** of a transformer result in the transformer warming up.

- **Eddy currents** are unwanted currents induced in the iron core. The currents are reduced by **laminating** the core into individually electrically insulated thin strips.

- **Hysteresis** losses cause the iron core to warm up as a result of the continued cycle of changes to its magnetism.

- **Flux** losses are caused by magnetic 'leakage'. A transformer is only 100% efficient if all of the magnetic flux that is produced by the primary links with the secondary.

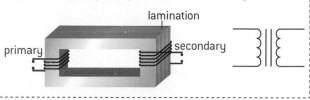

DIODE BRIDGES

The efficient transmission of electrical power is best achieved using alternating current (ac) and transformers can ensure the appropriate V_{rms} is supplied. Many electrical devices are, however, designed to operate using direct current (dc). The conversion from ac into dc is called **rectification** which relies on diodes.

A **diode** is a two-terminal electrical device that has different electrical characteristics depending on which way around it is connected. An ideal diode allows current to flow in the forward direction (negligible resistance with forward bias) but does not allow current to flow in the reverse direction (infinite resistance with reverse bias).

Symbol:

A ———|>|——— B

allowed current direction

Current is allowed to flow from A to B (A is positive and B is negative) but is prevented from following from B to A (A is negative and B is positive).

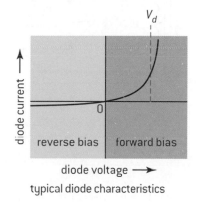

typical diode characteristics

HL Rectification and smoothing circuits

RECTIFICATION

1. Half-wave rectification

   A single diode will convert ac into a pulsating dc:

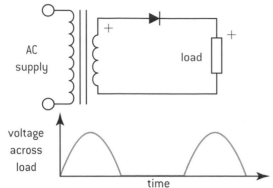

   In half-wave rectification, electrical energy that is available in the negative cycle of the ac is not utilized.

2. Full-wave rectification

   A diode bridge (using four diodes) can utilize all the electrical energy that is available during a complete cycle as shown below.

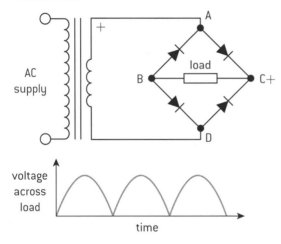

   In the positive half of the cycle, current flows through the diode bridge from A→C→B→D.

   In the negative half of the cycle, current flows through the diode bridge from D→C→B→A.

   Note that:

   - Current always flows through the load resistor in the same direction. (C→B)
   - Diodes on parallel sides point in the same directions.
   - The ac signal is fed to the points where opposite ends of two diodes join.
   - The positive output is taken from the junction of the negative side of two diodes.
   - The negative output is taken from the junction of the positive side of two diodes.
   - During each half-cycle one set of parallel-side diodes conducts.

SMOOTHING CIRCUITS

Diode-bridge circuits provide a current that flows in one direction (dc) but still pulsates. In order to achieve a steady pd, a **smoothing** device is required. One possibility is a **capacitor** (see page 117 for more details).

Note that:

- The output is still fluctuating slightly; this is known as the output **ripple**.
- The capacitor is acting as a short-term store of electrical energy.
- The capacitor is constantly charging and discharging.
- In order to ensure a slow discharge, the value of the capacitor C needs to be chosen to ensure that the time constant (see page 118) is sufficiently large.

INVESTIGATING A DIODE-BRIDGE RECTIFICATION CIRCUIT EXPERIMENTALLY

The display of the varying pd across the load is best achieved using a cathode ray oscilloscope (CRO).

The **y-input control**, allows the sensitivity of the CRO to appropriately display a changing pd on the y-axis. The **time-base** controls allows an appropriate calibration of the x-axis to match the time period of the oscillations.

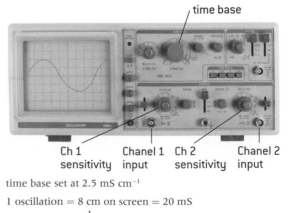

time base set at 2.5 mS cm⁻¹

1 oscillation = 8 cm on screen = 20 mS

∴ frequency = $\frac{1}{0.02}$ = 50 Hz

HL Capacitance

CAPACITANCE

Capacitors are devices that can store charge. The charge stored q is proportional to the pd across the capacitor V and the constant of proportionality is called the capacitance C.

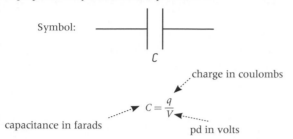

Symbol:

C

charge in coulombs

$C = \dfrac{q}{V}$

capacitance in farads

pd in volts

The farad (F) is a very large unit and practical capacitances are measured in μF, nF or pF.

$$1 \text{ F} = 1 \text{ C V}^{-1}$$

A measurement of the pd across a capacitance allows the charge stored to be calculated.

The capacitance of a parallel plate capacitor depends on three different factors:

- The area of each plate, A. Each plate is assumed to have the same area A and the plates overlap one another completely.

- The separation of the plates, d

- The material between the plates which is called the **dielectric material**. Different materials will have different values of a constant called its **permittivity**, ε. The permittivity of air is effectively the same as the permittivity of a vacuum (free space), $\varepsilon_0 = 8.85 \times 10^{-12}$ C^2 N^{-1} m^{-2}. The permittivity of all substances is greater than ε_0.

The relationship is:

$$C = \frac{\varepsilon A}{d}$$

when a dielectric material is introduced, change separation across the dielectric is induced. This increases the capacitance.

CAPACITORS IN SERIES AND PARALLEL

The effective total capacitance, C_{total}, of the combination of capacitors (C_1, C_2, C_3, etc.) in a circuit depends on whether the capacitors are joined together in series or in parallel. The capacitor equation can be used on individual capacitors or on the combination.

$$C_{\text{total}} = \frac{q_{\text{total}}}{V_{\text{total}}} \text{ and } C_1 = \frac{q_1}{V_1}, C_2 = \frac{q_2}{V_2}, \text{ etc.}$$

1. In series

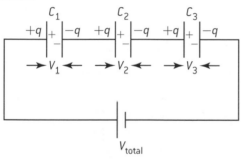

The charge stored in each capacitor is the same, q and the pds across the individual capacitors add together to give the total pd

$$q_{\text{total}} = q_1 = q_2 = q_3 = q$$

$$V_{\text{total}} = V_1 + V_2 + V_3$$

$$\therefore \frac{q_{\text{total}}}{C_{\text{total}}} = \frac{q_1}{C_1} + \frac{q_2}{C_2} + \frac{q_3}{C_3}$$

$$\therefore \frac{q}{C_{\text{series}}} = \frac{q}{C_1} + \frac{q}{C_2} + \frac{q}{C_3}$$

$$\frac{1}{C_{\text{series}}} = \frac{1}{C_1} + \frac{1}{C_2} + \cdots$$

e.g. if three capacitors 5 μF, 10 μF and 20 μF are added in series, the combined capacitance is:

$$\frac{1}{C_{\text{series}}} = \frac{1}{5} + \frac{1}{10} + \frac{1}{20} = \frac{7}{20} \ \mu\text{F}^{-1}$$

$$\therefore C_{\text{series}} = \frac{20}{7} = 2.86 \ \mu\text{F}$$

2. In parallel

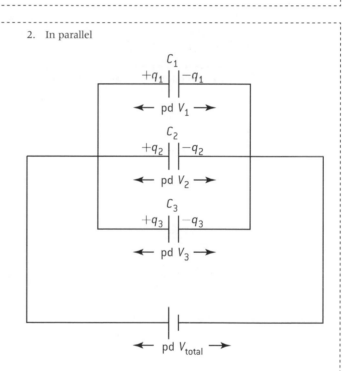

The pd across each capacitor is the same, V and the charges stored in each of the individual capacitors add together to give the total charge stored.

$$V_{\text{total}} = V_1 = V_2 = V_3 = V$$

$$q_{\text{total}} = q_1 + q_2 + q_3$$

$$\therefore C_{\text{total}} \ V_{\text{total}} = C_1 V_1 + C_2 V_2 + C_3 V_3$$

$$\therefore C_{\text{parallel}} \ V = C_1 V + C_2 V + C_3 V$$

$$\therefore C_{\text{parallel}} = C_1 + C_2 + \cdots$$

e.g. if three capacitors 5 μF, 10 μF and 20 μF are added in parallel, the combined capacitance is:

$$C_{\text{parallel}} = 5 + 10 + 20 = 35 \ \mu\text{F}$$

ⓗⓛ Capacitor discharge

CAPACITOR (RC) DISCHARGE CIRCUITS

If the two ends of a charged capacitor are joined together with a resistor, a current will flow until the capacitor is discharged.

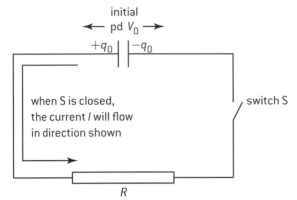

During the discharge process:

- the value of the discharge current, I, drops from an initial maximum I_0 down to zero
- the value of the stored charge, q, drops from an initial maximum q_0 down to zero
- the value of the pd across the capacitor (which is also the pd across the resistor), V, drops from an initial maximum V_0 down to zero.

Applying Kirchoff's law around the loop gives

$$0 = IR + \frac{q}{C}$$

Since I is the rate of flow of charge, $\frac{dq}{dt}$,

$$0 = R\frac{dq}{dt} + \frac{q}{C}$$

$$\frac{dq}{dt} = -\frac{q}{RC}$$

This has the rate of flow of charge proportional to the charge stored. The solution is an exponential decrease of charge stored given by:

$$q = q_0\, e^{-\frac{t}{RC}}$$

charge remaining — original charge — resistance (Ω) — capacitance (F) — time (s)

The product of RC is called the **time constant** for the circuit and is given the symbol τ (the Greek letter tau).

$$\tau = RC$$

The SI unit for τ will be seconds (NB: care needed with SI multipliers).

$$\therefore q = q_0\, e^{-\frac{t}{\tau}}$$

Since the current I and the pd V are both proportional to the charge, the following equations also apply:

$$I = I_0\, e^{-\frac{t}{\tau}}$$
$$V = V_0\, e^{-\frac{t}{\tau}}$$

Where

$$I_0 = \frac{q_0}{RC} = \frac{V_0}{R}$$

Example

A 10 μF capacitor is discharged through a 20 kΩ resistor. Calculate (a) the time constant τ for the circuit and (b) the fraction of charge remaining after one time constant

a) $\tau = RC = 10\ \mu\text{F} \times 20\ \text{k}\Omega = 200\ \text{ms}$

b) After one time constant,

$$q = q_0 e^{-1} = 0.37 q_0$$

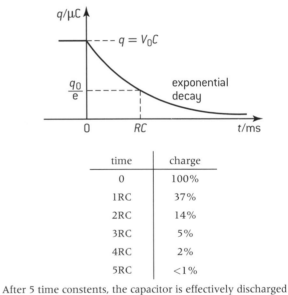

time	charge
0	100%
1RC	37%
2RC	14%
3RC	5%
4RC	2%
5RC	<1%

After 5 time constents, the capacitor is effectively discharged

ⓗ Capacitor charge

CAPACITOR CHARGING CIRCUITS

If the two ends of an uncharged capacitor are joined together with a resistor, a current will flow until the capacitor is charged.

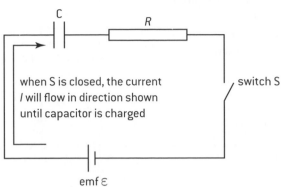

During the charging process:,

- the value of the charging current, I, drops from an initial maximum I_0 down to zero
- the value of the stored charge, q, increases from zero up to a final maximum value, q_0
- the value of the pd across the capacitor, V, increases from zero up to a final maximum value, ε
- the value of the pd across the resistor drops from an initial maximum ε down to zero.

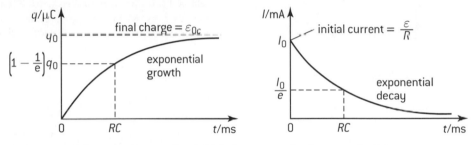

The equation for the increase of charge on the capacitor (which does not need to be memorized) is:

$$q = q_0 \left(1 - e^{-\frac{t}{\tau}}\right)$$

ENERGY STORED IN A CHARGED CAPACITOR

A charged capacitor can provide a temporary store of electrical energy when there is a potential difference V across the capacitor. The charge, q, that is stored is distributed with $+q$ on one plate and $-q$ on the other plate as shown below. There is an electric field between the plates.

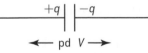

In the charging process, as more charge is added to the capacitor, the pd across it also increases proportionally. The graph (right) shows how the pd across the capacitor varies with charge stored in the capacitor during the charging process. The total energy stored, E, is represented by the area under the graph.

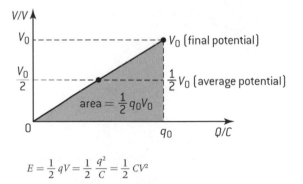

$$E = \frac{1}{2}qV = \frac{1}{2}\frac{q^2}{C} = \frac{1}{2}CV^2$$

Note that both charging and discharging are exponential processes. If a circuit is arranged in which a capacitor spends equal time charging and discharging through the same value resistor, then in one complete cycle, more charge will be added to the capacitor during the charging time than it loses during the discharging time. The result over several cycles will be for the capacitor to charge up to the same pd as the power supply.

HL IB Questions – electromagnetic induction

1. The **primary** of an ideal transformer has 1000 turns and the **secondary** 100 turns. The current in the primary is 2 A and the input power is 12 W.

   Which **one** of the following about the **secondary current** and the **secondary power output** is true?

	secondary current	secondary power output
A.	20 A	1.2 W
B.	0.2 A	12 W
C.	0.2 A	120 W
D.	20 A	12 W

2. This question is about electromagnetic induction.

   A small coil is placed with its plane parallel to a long straight current-carrying wire, as shown below.

   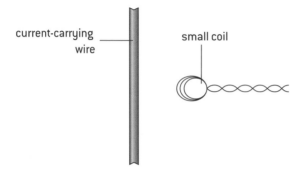

   a) (i) State Faraday's law of electromagnetic induction. [2]

       (ii) Use the law to explain why, when the current in the wire changes, an emf is induced in the coil. [1]

3. The diagram shows a simple generator with the coil rotating between magnetic poles. Electrical contact is maintained through two brushes, each touching a slip ring.

   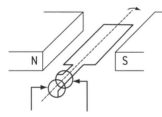

   At the instant when the rotating coil is oriented as shown, the voltage across the brushes

   A. is zero.

   B. has its maximum value.

   C. has the same constant value as in all other orientations.

   D. reverses direction.

4. The rms current rating of an electric heater is 4A. What direct current would produce the same power dissipation in the electric heater? [2]

   A. $\dfrac{4}{\sqrt{2}}$A   B. 4A

   C. $4\sqrt{2}$A   D. 8A

5. Two loops of wire are next to each other as shown here. There is a source of alternating emf connected to loop 1 and an ammeter in loop 2.

   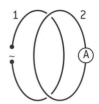

   The variation with time of the current in loop 1 is shown as line 1 in each of the graphs below. In which graph does line 2 best represent the current in loop 2?

   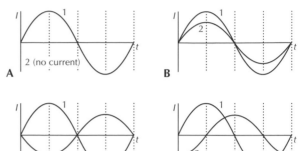

6. A loop of wire of negligible resistance is rotated in a magnetic field. A 4 Ω resistor is connected across its ends. A cathode ray oscilloscope measures the varying induced potential difference across the resistor as shown below.

   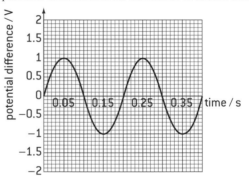

   a) If the coil is rotated at twice the speed, show on the axes above how potential difference would vary with time. [2]

   b) What is the rms value of the induced potential difference, V_{rms}, at the **original** speed of rotation? [1]

   c) Draw a graph showing how the power dissipated in the resistor varies with time, at the **original** speed of rotation. [3]

7. a) A 3μF capacitor is charged to 240 V. Calculate the charge stored. [1]

   b) Estimate the amount of time it would take for the charge you have calculated in (**a**) to flow through a 60 W light bulb connected to the 240 V mains electricity. [2]

   c) The charged capacitor in (**a**) is discharged through a 60 W 240 V light bulb.

       (i) Explain why the current during its discharge will not be constant. [2]

       (ii) Estimate the time taken for the capacitor to discharge through the light bulb. [2]

       (iii) Will the bulb light during discharge? Explain your answer. [2]

12 QUANTUM AND NUCLEAR PHYSICS

ⓗⓛ Photoelectric effect

PHOTOELECTRIC EFFECT

Under certain conditions, when light (ultra-violet) is shone onto a metal surface (such as zinc), electrons are emitted from the surface.

More detailed experiments (see below) showed that:

- Below a certain **threshold frequency**, f_0, no photoelectrons are emitted, no matter how long one waits.
- Above the threshold frequency, the maximum kinetic energy of these electrons depends on the frequency of the incident light.
- The number of electrons emitted depends on the intensity of the light and does not depend on the frequency.
- There is no noticeable delay between the arrival of the light and the emission of electrons.

These observations cannot be reconciled with the view that light is a wave. A wave of any frequency should eventually bring enough energy to the metal plate.

STOPPING POTENTIAL EXPERIMENT

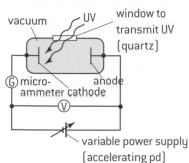

vacuum · UV · window to transmit UV (quartz)

Ⓖ micro-ammeter · anode · cathode · Ⓥ

variable power supply (accelerating pd)

In the apparatus above, photoelectrons are emitted by the cathode. They are then accelerated across to the anode by the potential difference.

The potential between cathode and anode can also be reversed.

In this situation, the electrons are decelerated. At a certain value of potential, the stopping potential, V_s, no more photocurrent is observed. The photoelectrons have been brought to rest before arriving at the anode.

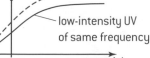

photocurrent · high-intensity UV · low-intensity UV of same frequency · V_s · potential

The stopping potential depends on the frequency of UV light in the linear way shown in the graph below.

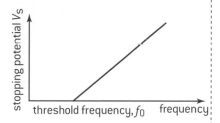

stopping potential V_s · threshold frequency, f_0 · frequency

The stopping potential is a measure of the maximum kinetic energy of the electrons.

Max KE of electrons $= V_s e$

[since pd $= \dfrac{\text{energy}}{\text{charge}}$

and e = charge on an electron]

$\therefore \frac{1}{2} mv^2 = V_s e \quad \therefore v = \sqrt{\dfrac{2 V_s e}{m}}$

EINSTEIN MODEL

Einstein introduced the idea of thinking of light as being made up of particles.

His explanation was:

- Electrons at the surface need a certain minimum energy in order to escape from the surface. This minimum energy is called the **work function** of the metal and given the symbol ϕ.
- The UV light energy arrives in lots of little packets of energy – the packets are called photons.
- The energy in each packet is fixed by the frequency of UV light that is being used, whereas the number of packets arriving per second is fixed by the intensity of the source.
- The energy carried by a photon is given by

$$E = hf$$

Planck's constant 6.63×10^{-34} J s

energy in joules · frequency of light in Hz

- Different electrons absorb different photons. If the energy of the photon is large enough, it gives the electron enough energy to leave the surface of the metal.
- Any 'extra' energy would be retained by the electron as kinetic energy.
- If the energy of the photon is too small, the electron will still gain this amount of energy but it will soon share it with other electrons.

Above the threshold frequency, incoming energy of photons = energy needed to leave the surface + kinetic energy.

In symbols,

$E_{max} = hf - \phi$

$hf = \phi + E_{max}$ or $hf = \phi + V_s e$

This means that a graph of frequency against stopping potential should be a straight line of gradient $\frac{e}{h}$.

EXAMPLE

What is the maximum velocity of electrons emitted from a zinc surface ($\phi = 4.2$ eV) when illuminated by EM radiation of wavelength 200 nm?

$\phi = 4.2$ eV $= 4.2 \times 1.6 \times 10^{-19}$ J $= 6.72 \times 10^{-19}$ J

Energy of photon $= h\dfrac{c}{\lambda} = \dfrac{6.63 \times 10^{-34} \times 3 \times 10^8}{2 \times 10^{-7}}$

$\quad = 9.945 \times 10^{-19}$ J

$\therefore$ KE of electron $= (9.945 - 6.72) \times 10^{-19}$ J

$\quad = 3.225 \times 10^{-19}$ J

$\therefore v = \sqrt{\dfrac{2\,KE}{m}}$

$\quad = \sqrt{\dfrac{2 \times 3.225 \times 10^{-19}}{9.1 \times 10^{-31}}}$

$\quad = 8.4 \times 10^5$ m s^{-1}

WAVE–PARTICLE DUALITY

The photoelectric effect of light waves clearly demonstrates that light can behave like particles, but its wave nature can also be demonstrated – it reflects, refracts, diffracts and interferes just like all waves. So what exactly is it? It seems reasonable to ask two questions.

1. *Is light a wave or is it a particle?*

The correct answer to this question is 'yes'! At the most fundamental and even philosophical level, light is just light. Physics tries to understand and explain what it is. We do this by imagining models of its behaviour. Sometimes it helps to think

of it as a wave and sometimes it helps to think of it as a particle, but neither model is complete. Light is just light. This dual nature of light is called **wave–particle duality**.

2. *If light waves can show particle properties, can particles such as electrons show wave properties?*

Again the correct answer is 'yes'. Most people imagine moving electrons as little particles having a definite size, shape, position and speed. This model does not explain why electrons can be diffracted through small gaps. In order to diffract they must have a wave nature. Once again they have a dual nature. See the experiment below.

DE BROGLIE HYPOTHESIS

If matter can have wave properties and waves can have matter properties, there should be a link between the two models. The de Broglie hypothesis is that all moving particles have a 'matter wave' associated with them. This matter wave can be thought of as a probability function associated with the moving particle. The (amplitude)² of the wave at any given point is a measure of the probability of finding the particle at that point. The wavelength of this matter wave is given by the de Broglie equation:

$$\lambda = \frac{hc}{pc} = \frac{hc}{E} \text{ for photons}$$

λ is the wavelength in m

h is Plank's constant $= 6.63 \times 10^{-34}$ J s

c is the speed of light $= 3.0 \times 10^8$ m s^{-1}

p is the momentum of the particle

The higher the energy, the lower the de Broglie wavelength. This equation was introduced on page 69 as the method of calculating a photon's wavelength from its energy, E. In order for the wave nature of particles to be observable in experiments, the particles often have very high velocities. In these situations the proper calculations are relativistic but simplifications are possible.

1. At very high energies: $pc = E$

In these situations, the rest energy of the particles can be negligible compared with their energy of motion.

For example, the rest energy of an electron (0.511 MeV) is negligible if it has been accelerated through an effective potential difference of 420 MV to have kinetic energy of 420 MeV. In these circumstances the total energy of an electron is effectively 420 MeV. The de Broglie wavelength of 420 MeV electrons is:

$$\lambda = \frac{6.6 \times 10^{-34} \times 3.0 \times 10^8}{420 \times 10^6 \times 1.6 \times 10^{-19}} = 2.9 \times 10^{-15} \text{ m}$$

2. At low energies

In these situations the relationship can be restated in terms of the momentum p of the particle measured in kg m s^{-1} (in non-relativistic mechanics, $P = $ mass $\times$ velocity):

$$\lambda = \frac{h}{p}$$

For example, electrons accelerated through 1 kV would gain a KE of 1.6×10^{-16} J. Since KE and non-relativistic momentum are related by $E_K = \frac{p^2}{2m}$, this gives $p = 1.7 \times 10^{-23}$ kg m s^{-1}

$$\lambda = \frac{6.6 \times 10^{-34}}{1.7 \times 10^{-23}} = 3.9 \times 10^{-11} \text{ m}$$

ELECTRON DIFFRACTION EXPERIMENT

In order to show diffraction, an electron 'wave' must travel through a gap of the same order as its wavelength. The atomic spacing in crystal atoms provides such gaps. If a beam of electrons impinges upon powdered carbon then the electrons will be diffracted according to the wavelength.

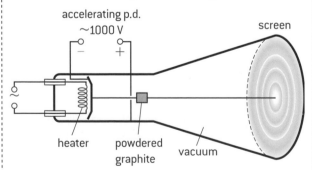

The circles correspond to the angles where constructive interference takes place. They are circles because the powdered carbon provides every possible orientation of gap. A higher accelerating potential for the electrons would result in a higher momentum for each electron. According to the de Broglie relationship, the wavelength of the electrons would thus decrease. This would mean that the size of the gaps is now proportionally bigger than the wavelength so there would be less diffraction. The circles would move in to smaller angles. The predicted angles of constructive interference are accurately verified experimentally.

DAVISSON AND GERMER EXPERIMENT (1927)

The diagram below shows the principle behind the Davisson and Germer electron diffraction experiment.

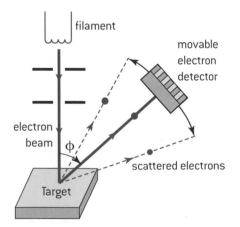

A beam of electrons strikes a target nickel crystal. The electrons are scattered from the surface. The intensity of these scattered electrons depends on the speed of the electrons (as determined by their accelerating potential difference) and the angle.

A maximum scattered intensity was recorded at an angle that quantitatively agrees with the constructive interference condition from adjacent atoms on the surface.

(HL) Atomic spectra and atomic energy states

INTRODUCTION

As we have already seen, atomic spectra (emission and absorption) provide evidence for the quantization of the electron energy levels. See page 69 for the laboratory set-up.

Different atomic models have attempted to explain these energy levels. The first quantum model of matter was the Bohr model for hydrogen: modern models describe the electrons by using wavefunctions (see page 125).

HYDROGEN SPECTRUM

The emission spectrum of atomic hydrogen consists of particular wavelengths. In 1885 a Swiss schoolteacher called Johann Jakob Balmer found that the visible wavelengths fitted a mathematical formula.

These wavelengths, known as the **Balmer series**, were later shown to be just one of several similar series of possible wavelengths that all had similar formulae. These can be expressed in one overall formula called the **Rydberg formula**.

$$\frac{1}{\lambda} = R_{\text{H}}\left(\frac{1}{n^2} - \frac{1}{m^2}\right)$$

λ – the wavelength

m – a whole number larger than 2 i.e. 3, 4, 5 etc

For the **Lyman series** of lines (in the ultra-violet range) $n = 1$. For the **Balmer series** $n = 2$. The other series are the **Paschen** ($n = 3$), **Brackett** ($n = 4$), and the **Pfund** ($n = 5$) series. In each case the constant R_{H}, called the **Rydberg constant**, has the one unique value, $1.097 \times 10^7\,\text{m}^{-1}$.

EXAMPLE

The diagram below represents some of the electron energy levels in the hydrogen atom. Calculate the wavelength of the photon emitted when an electron falls from $n = 3$ to $n = 2$.

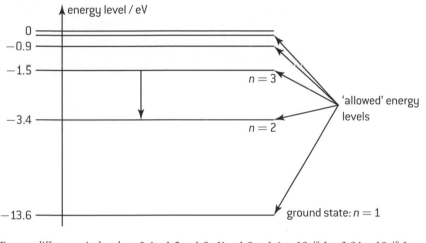

Energy difference in levels $= 3.4 - 1.5 - 1.9\,\text{eV} - 1.9 \times 1.6 \times 10^{-19}\,\text{J} - 3.04 \times 10^{-19}\,\text{J}$

Frequency of photon $f = \dfrac{E}{h} = \dfrac{3.04 \times 10^{-19}}{6.63 \times 10^{-34}} = 4.59 \times 10^{14}\,\text{Hz}$

Wavelength of photon $\lambda = \dfrac{c}{f} = \dfrac{3.00 \times 10^8}{4.59 \times 10^{14}} = 6.54 \times 10^{-7}\,\text{m} = 654\,\text{nm}$

This is in the visible part of the electromagnetic spectrum and one wavelength in the Balmer series.

PAIR PRODUCTION AND PAIR ANNIHILATION

Matter and radiation interactions are not restricted to the absorption or emission of radiation by matter (such as takes place in absorption or emission spectra, above). As introduced on page 73, for every 'normal' matter particle that exists, there will be a corresponding antimatter particle which has the same mass but every other property is opposite. For example:

- The antiparticle of an electron, e$^-$ (or β$^-$) is a positron, e$^+$ (or β$^+$)
- The antiparticle for a proton, p$^+$ is the antiproton, p$^-$
- The antiparticle for a neutrino, ν is an antineutrino, $\bar{\nu}$

When a particle and its corresponding antiparticle meet they **annihilate** one another and the mass is converted into radiation. As seen on page 78, these annihilations must obey certain conservations and in particular the conservation of energy, momentum and charge.

When an electron e$^-$ and a positron e$^+$ annihilate typically they create two photons. Each photon has a momentum and if combined momentum of the electron–positron pair was initially zero, then the two photons will be travelling in opposite directions. The reverse process is also possible – photons of sufficient energy can convert into a pair of particles (one matter and one antimatter). Much of the energy goes into the rest masses of the particles with any excess going into the kinetic energy of the particles that have been created. Typically for pair production to take place, the photon needs to interact with a nucleus. The nucleus is not changed in the interaction but is involved in the overall conservation of momentum and energy that must take place. Without its ability to 'absorb' some of the momentum, the interaction could not occur.

Bohr model of the atom

BOHR MODEL

Niels Bohr took the standard 'planetary' model of the hydrogen atom and filled in the mathematical details. Unlike planetary orbits, there are only a limited number of 'allowed' orbits for the electron. Bohr suggested that these orbits had fixed multiples of angular momentum. The orbits were quantized in terms of angular momentum. The energy levels predicted by this quantization were in exact agreement with the discrete wavelengths of the hydrogen spectrum. Although this agreement with experiment is impressive, the model has some problems associated with it.

Bohr postulated that:

- An electron does not radiate energy when in a stable orbit. The only stable orbits possible for the electron are ones where the **angular momentum** of the orbit is an integral multiple of $\frac{h}{2\pi}$ where h is a fixed number (6.6×10^{-34} J s) called **Planck's constant**. Mathematically

$$m_e v r = \frac{nh}{2\pi}$$

[angular momentum is equal to $m_e v r$]

- When electrons move between stable orbits they radiate (or absorb) energy.

$$F_{\text{electrostatic}} = \text{centripetal force}$$

$$\therefore \frac{e^2}{4\pi\varepsilon_0 r^2} = \frac{m_e v^2}{r}$$

but $v = \frac{nh}{2\pi m_e r}$ [from 1st postulate]

$$\therefore r = \frac{\varepsilon_0 n^2 h^2}{\pi m_e e^2} \text{ (by substitution)}$$

Total energy of electron $= KE + PE$

where $KE = \frac{1}{2}m_e v^2 = \frac{1}{2}\frac{e^2}{(4\pi\varepsilon_0 r)}$

and $PE = -\frac{e^2}{4\pi\varepsilon_0 r}$ [electrostatic PE]

so total energy $E_n = -\frac{1}{2}\frac{e^2}{4\pi\varepsilon_0 r} = -\frac{m_e e^4}{8\varepsilon_0^2 n^2 h^2}$

This final equation shows that:

- the electron is bound to (= 'trapped by') the proton because overall it has negative energy.
- the energy of an orbit is proportional to $-\frac{1}{n^2}$. In electronvolts

$$E_n = -\frac{13.6}{n^2}$$

The second postulate can be used (with the full equation) to predict the wavelength of radiation emitted when an electron makes a transition between stable orbits.

$$hf = E_2 - E_1$$

$$= \frac{m_e e^4}{8\varepsilon_0^2 h^2}\left(\frac{1}{n_1^2} - \frac{1}{n_2^2}\right)$$

but $f = \frac{c}{\lambda}$

$$\therefore \frac{1}{\lambda} = \frac{m_e e^4}{8\varepsilon_0^2 c h^3}\left(\frac{1}{n_1^2} - \frac{1}{n_2^2}\right)$$

It should be noted that:

- this equation is of the same form as the Rydberg formula.
- the values predicted by this equation are in very good agreement with experimental measurement.
- the Rydberg constant can be calculated from other (known) constants. Again the agreement with experimental data is good.

The limitations to this model are:

- if the same approach is used to predict the emission spectra of other elements, it fails to predict the correct values for atoms or ions with more than one electron.
- the first postulate (about angular momentum) has no theoretical justification.
- theory predicts that electrons should, in fact, not be stable in circular orbits around a nucleus. Any accelerated electron should radiate energy. An electron in a circular orbit is accelerating so it should radiate energy and thus spiral in to the nucleus.
- it is unable to account for relative intensity of the different lines.
- it is unable to account for the fine structure of the spectral lines.

NUCLEAR RADII AND NUCLEAR DENSITIES

Not surprisingly, more massive nuclei have larger radii. Detailed analysis of the data implies that the nuclei have a spherical distribution of positive charge with an essentially constant density. The results are consistent with a model in which the protons and neutrons can be imagined to be hard spheres that are bonded tightly together in a sphere of constant density. A nucleus that is twice the size of a smaller nucleus will have roughly 8 (=2^3) times the mass.

The nuclear radius R of element with atomic mass number A can be modelled by the relationship:

$$R = R_0 A^{\frac{1}{3}}$$

Where R_0 is a constant roughly equal to 10^{-15} m (or 1 fm).
$R_0 = 1.2 \times 10^{-15}$ m = 1.2 fm.

e.g. The radius of a uranium-238 nucleus is predicted to be

$$R = 1.2 \times 10^{-15} \times (238)^{\frac{1}{3}} \text{ m} = 7.4 \text{ fm}$$

The volume of a nucleus, V, of radius, R is given by:

$$V = \frac{4}{3}\pi R^3 = \frac{4}{3}\pi A R_0^3$$

Where the mass number A is equal to the number of nucleons

The number of nucleons per unit volume $= \frac{A}{V} = \frac{3A}{4\pi A R_0^3}$

$$= \frac{3}{4\pi R_0^3}$$

The mass of a nucleon is m ($\approx 1.7 \times 10^{-27}$ kg), so the nuclear density ρ is:

$$\rho = \frac{3m}{4\pi R_0^3} = \frac{3 \times 1.7 \times 10^{-27}}{4\pi(1.2 \times 10^{-15})^3} = 2 \times 10^{17} \text{ kg m}^{-3}$$

This is a vast density (a teaspoon of matter of this density has a mass $\approx 10^{12}$ kg). The only macroscopic objects with the same density as nuclei are neutron stars (see page 200).

ⓗ The Schrödinger model of the atom

SCHRÖDINGER MODEL

Erwin Schrödinger (1887–1961) built on the concept of matter waves and proposed an alternative model of the hydrogen atom using wave mechanics. The **Copenhagen interpretation** is a way to give a physical meaning to the mathematics of wave mechanics.

- The description of particles (matter and/or radiation) in quantum mechanics is in terms of a **wavefunction** ψ. This wavefunction has no physical meaning but the square of the wavefunction does.
- ψ is a **complex number**.
- At any instant of time, the wavefunction has different values at different points in space.
- The mathematics of how this wavefunction develops with time and interacts with other wavefunctions is like the mathematics of a travelling wave.
- The probability of finding the particle (electron or photon, etc.) at any point in space within the atom is given by the square of the amplitude of the wavefunction at that point.
- The square of the absolute value of ψ, $|\psi|^2$, is a real number corresponding to the probability density of finding the particle in a given place.
- When an observation is made the wavefunction is said to **collapse**, and the complete physical particle (electron or photon, etc.) will be observed to be at one location.

The standing waves on a string have a fixed wavelength but for energy reasons the same is not true for the electron wavefunctions. As an electron moves away from the nucleus it must lose kinetic energy because they have opposite charges. Lower kinetic energy means that it would be travelling with a lower momentum and the de Broglie relationship predicts a longer wavelength. This means that the possible wavefunctions that fit the boundary conditions have particular shapes.

The wavefunction provides a way of working out the probability of finding an electron at that particular radius. $|\psi|^2$ at any given point is a measure of the probability of finding the electron at that distance away from the nucleus – in any direction.

$$p(r) = |\psi|^2 \Delta V$$

    ↑

probability of detecting the electrons in a small volume of space, ΔV

The wavefunction exists in all three dimensions, which makes it hard to visualize. Often the electron orbital is pictured as a cloud. The exact position of the electron is not known but we know where it is more likely to be.

In Schrödinger's model there are different wavefunctions depending on the total energy of the electron. Only a few particular energies result in wavefunctions that fit the boundary conditions – electrons can only have these particular energies within an atom. An electron in the ground state has a total energy of –13.6 eV, but its position at any given time is undefined in this model. The wavefunction for an electron of this energy can be used to calculate the probability of finding it at a given distance away from the nucleus.

- The resulting **orbital** for the electron can be described in terms of the probability of finding the electron at a certain distance away. The probability of finding the electron at a given distance away is shown in the graph below.

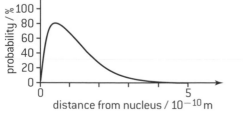

- The electron in this orbital can be visualized as a 'cloud' of varying electron density. It is more likely to be in some places than other places, but its actual position in space is undefined.

1s

Electron cloud for the 1s orbital in hydrogen

There are other fixed total energies for the electron that result in different possible orbitals. In general as the energy of the electron is increased it is more likely to be found at a further distance away from the nucleus.

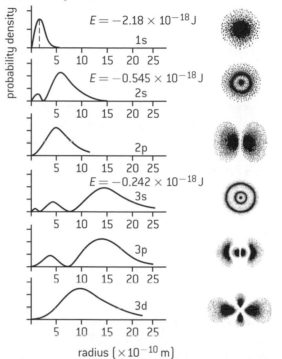

Probability density functions for some orbitals in the hydrogen atom. The scale on the vertical axis is different from graph to graph.

The wavefunction is central to quantum mechanics and, in principle, should be applied to all particles.

Example:

A particle is described as the following wavefunction:

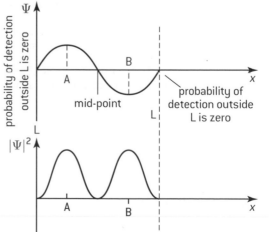

The particle will not be detected at the mid point and the probability of detection at A = probability of detection at B.

HL The Heisenberg uncertainty principle and the loss of determinism

HEISENBERG UNCERTAINTY PRINCIPLE

The **Heisenberg uncertainty principle** identifies a fundamental limit to the possible accuracy of any physical measurement. This limit arises because of the nature of quantum mechanics and not as a result of the ability (or otherwise) of any given experimenter. He showed that it was impossible to measure exactly the position **and** the momentum of a particle simultaneously. The more precisely the position is determined, the less precisely the momentum is known in this instant, and vice versa. They are linked variables and are called **conjugate quantities**.

There is a mathematical relationship linking these uncertainties.

$$\Delta x \Delta p \geq \frac{h}{4\pi}$$

Δx The uncertainty in the measurement of position

Δp The uncertainty in the measurement of momentum

Measurements of energy and time are also linked variables.

$$\Delta E \Delta t \geq \frac{h}{4\pi}$$

ΔE The uncertainty in the measurement of energy

Δt The uncertainty in the measurement of time

The implications of this lack of precision are profound. Before quantum theory was introduced, the physical world was best described by deterministic theories – e.g. Newton's laws. A deterministic theory allows us (in principle) to make absolute predictions about the future.

Quantum mechanics is not deterministic. It cannot ever predict exactly the results of a single experiment. It only gives us the probabilities of the various possible outcomes. The uncertainty principle takes this even further. Since we cannot know the precise position and momentum of a particle at any given time, its future can never be determined precisely. The best we can do is to work out a range of possibilities for its future.

It has been suggested that science would allow us to calculate the future so long as we know the present exactly. As Heisenberg himself said, it is not the conclusion of this suggestion that is wrong but the premise.

ESTIMATES FROM THE UNCERTAINTY PRINCIPLE

Example calculation: The position of a proton is measured with an accuracy of $\pm\,1.0 \times 10^{-11}$ m. What is the minimum uncertainty in the proton's position 1.0 s later?

$$\Delta x \Delta p \geq \frac{h}{4\pi} \qquad \therefore \quad \Delta x \times m\Delta v \geq \frac{h}{4\pi}$$

$$\Delta v \geq \frac{h}{4\pi m \Delta x} = \frac{6.63 \times 10^{-34}}{4\pi \times 1.67 \times 10^{-27} \times 1.0 \times 10^{-11}}$$

$$= 3200 \text{ m s}^{-1}$$

Thus uncertainty in position after 1.0 s = 3200 m = 3.2 km

The uncertainty principle can also be applied to illuminate some general principles but, to quote Richard Feynman (Feynman lectures on Physics, volume III, 1963), '[the application] *must not be taken too seriously; the idea is right but the analysis is not very accurate*'.

1. Estimate of the energy of an electron in an atom.

    When an electron is known to be confined within an atom, then the uncertainty in its position Δx must be less than the size of the atom, a. If we equate the two, this means the uncertainty for its momentum can be estimated as:

    $$\Delta p \approx \frac{h}{4\pi \Delta x} \approx \frac{h}{4\pi a}$$

    If we take this uncertainty in the momentum as a value for the momentum of the electron ($\Delta p \approx p$), the equations of classical mechanics can estimate the kinetic energy of the electron:

    $$E_K = \frac{p^2}{2m} \approx \frac{h^2}{32\pi^2 ma^2}$$

    The diameter of a hydrogen atom is approximately 10^{-10} m, so the estimation of the kinetic energy is:

    $$E_K \approx \frac{(6.6 \times 10^{-34})^2}{32\pi^2 \times 9.3 \times 10^{-31} \times (10^{-10})^2} = 1.5 \times 10^{-19} \text{ J}$$

    $$\approx 1 \text{ eV}$$

This calculation is a very rough estimate but correctly predicts the right order of magnitude for the electron's kinetic energy (ground state of electron in H atom is -13.6 eV).

2. Impossibility of an electron existing within a nucleus of an atom.

    The above calculation can be repeated imagining an electron being trapped inside the nucleus of size 10^{-14} m. If confined to a space this small, the electron's kinetic energy would be estimated to be a factor of 10^8 times bigger. An electron with an energy of the order of 100 MeV cannot be bound to a nucleus and thus it would have enough energy to escape.

3. Estimate of lifetime of an electron in an excited energy state.

    The **spectral linewidth** associated with an atom's emission spectrum is usually taken to be very small – only discrete wavelengths are observed. As a result of the uncertainty principle, the linewidth is, however, not zero. Practically, there will be a very limited range of wavelengths associated with any given transition and thus the uncertainty associated with the energy difference between the two levels involved is very small. An estimate of the lifetime of an electron in the excited state can be made using the uncertainty principle as the uncertainty in energy, ΔE, of a transition is inversely proportional to the average lifetime, Δt, in the excited state:

    $$\Delta E \Delta t \approx \frac{h}{4\pi}$$

    If $\Delta E = 5 \times 10^{-7}$ eV

    $$\Delta t \approx \frac{6.6 \times 10^{-34}}{4\pi \times 5 \times 10^{-7} \times 1.6 \times 10^{-19}}$$

    $$= 6.6 \times 10^{-10} \approx 1 \text{ ns}$$

HL Tunnelling, potential barrier and factors affecting tunnelling probability

Heisenberg's uncertainty relationship can be used to explain the quantum phenomenon of **tunnelling**. The situation being considered is a particle that is trapped because its energy E is less than the energy it needs to escape (U_0). In classical physics, if a particle does not have enough energy to escape from the **potential barrier** then it will always remain trapped inside the system. An example would be a 500 g tennis ball with a total energy of 4 J bouncing up and down between two walls that are 1.2 m high. In order to get over one of the walls, the tennis ball needs to have a potential energy of $mgh = 0.5 \times 10 \times 1.2 = 6$ J. Since it only has 4 J it must remain trapped by the walls.

In an equivalent microscopic situation (e.g. an electron trapped inside an atom with energy E which is less than the energy U_0 needed to escape), the rules of quantum physics mean that it is now possible for the particle to escape! The particle's wavefunction is continuous and does not drop immediately to zero when it meets the sides of the potential well but the amplitude decreases exponentially. This means that if the barrier has a finite width then the wavefunction does continue on the other side of the barrier (with reduced amplitude). Therefore there is a probability that particle will be able to escape despite not having enough energy to do so. Escaping the potential well does not use up any of the particle's total energy.

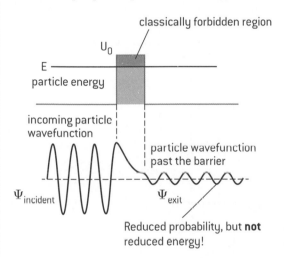

An explanation can be offered in terms of the uncertainty principle. In order for the particle to escape it would need a greater total energy ($E + \Delta E = U_0$). The particle can 'disobey' the law of conservation of energy by 'borrowing' an amount of energy ΔE provided it 'pays it back' in a time Δt such that the uncertainty principle applies:

$$\Delta E \Delta t \approx \frac{h}{4\pi}$$

The longer the barrier, the more time it takes the particle to tunnel. Increased tunnelling time will reduce the maximum possible uncertainty in the energy.

Example 1 – alpha decay
The protons and neutrons that form alpha particles already exist within nuclei and when emitted overall there is a release of energy. For example uranium-238 has a half-life of about 4.5 billion years. It decays by emitting an alpha particle:

$$^{238}_{92}\text{U} \rightarrow {}^{234}_{90}\text{Th} + {}^{4}_{2}\alpha$$

The energy of the emitted alpha particle is 4.25 MeV which is less than the total potential energy needed to escape the strong force within the nucleus. If we imagine an alpha particle being formed inside the uranium nucleus, it can only escape by tunnelling through the potential barrier. In this example, the very long half-life must mean that the probability of the tunnelling process taking place (given by $|\psi|^2$) must be very low.

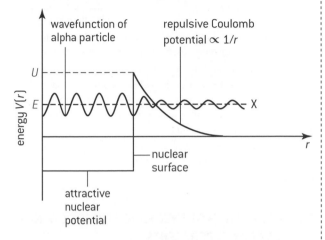

Example 2 – tunnelling electron microscope
In a **scanning tunnelling microscope**, a very fine metal tip is scanned close to, but not touching (separated by a few nm), a sample metal surface. There is a potential difference between the probe and the surface but the electrons in the surface do not have enough energy to escape the potential energy barrier as represented by the work function ϕ. Quantum tunnelling can, however, take place and a tunnelling current will flow as the wavefunction of an electron at the surface will extend beyond the metal surface. Some electrons will tunnel the gap and electrical current will be measurable. The value of the current depends on the separation of the tip and the surface and can be used to visualize atomic structure.

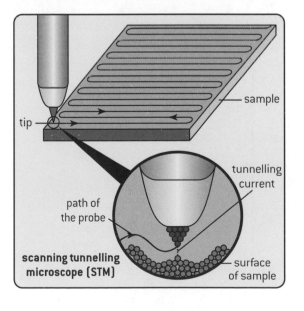

THE NUCLEUS – SIZE

In the example below, alpha particles are allowed to bombard gold atoms.

As they approach the gold nucleus, they feel a force of repulsion. If an alpha particle is heading directly for the nucleus, it will be reflected straight back along the same path. It will have got as close as it can. Note that none of the alpha particles actually collides with the nucleus – they do not have enough energy.

alpha particles

nucleus

closest approach, r

Alpha particles are emitted from their source with a known energy. As they come in they gain electrostatic potential energy and lose kinetic energy (they slow down). At the closest approach, the alpha particle is temporarily stationary and all its energy is potential.

Since electrostatic energy $= \frac{q_1 q_2}{4\pi\varepsilon_0 r}$, and we know q_1, the charge on an alpha particle and q_2, the charge on the gold nucleus we can calculate r.

DEVIATIONS FROM RUTHERFORD SCATTERING IN HIGH ENERGY EXPERIMENTS

Rutherford scattering is modelled in terms of the coulomb repulsion between the alpha particle and the target nucleus. At relatively low energies, detailed analysis of this model accurately predicts the relative intensity of scattered alpha particles at given angles of scattering. At high energies, however, the scattered intensity departs from predictions.

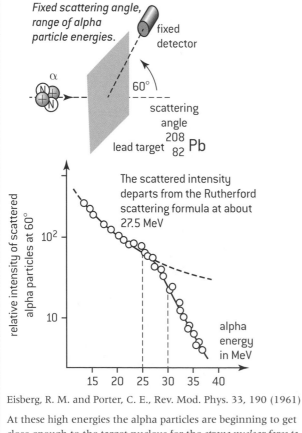

Fixed scattering angle, range of alpha particle energies.

fixed detector

α

60°

scattering angle

lead target $^{208}_{82}\text{Pb}$

The scattered intensity departs from the Rutherford scattering formula at about 27.5 MeV

relative intensity of scattered alpha particles at 60°

10^2

10

alpha energy in MeV

15 20 25 30 35 40

Eisberg, R. M. and Porter, C. E., Rev. Mod. Phys. 33, 190 (1961)

At these high energies the alpha particles are beginning to get close enough to the target nucleus for the *strong nuclear force* to begin to have an effect. In order to investigate the size of the nucleus in more detail, high energy electrons can be used (see box on the right).

EXAMPLE

If the α particles have an energy of 4.2 MeV, the closest approach to the gold nucleus ($Z = 79$) is given by

$$\frac{(2 \times 1.6 \times 10^{-19})(79 \times 1.6 \times 10^{-19})}{4 \times \pi \times 8.85 \times 10^{-12} \times r}$$

$$= 4.2 \times 10^6 \times 1.6 \times 10^{-19}$$

$$\therefore r = \frac{2 \times 1.6 \times 10^{-19} \times 79}{4 \times \pi \times 8.85 \times 10^{-12} \times 4.2 \times 10^6}$$

$$= 5.4 \times 10^{-14} \text{ m}$$

NUCLEAR SCATTERING EXPERIMENT INVOLVING ELECTRONS

Electrons, as leptons, do not feel the strong force. High-energy electrons have a very small de Broglie wavelength which can be of the right order to diffract around small objects such as nuclei. The diffraction pattern around a circular object of diameter D has its first minimum at an angle θ given by:

$$\sin \theta \approx \frac{\lambda}{D}$$

[Note that this small angle approximation is usually not appropriate to use to determine the location of the minimum intensity but this is being used to give an approximate answer around a spherical object. A more exact expression that is sometimes used for circular objects is $\sin \theta = 1.22 \frac{\lambda}{D}$]

High energy (400 MeV) electrons are directed at a target containing carbon-12 nuclei:

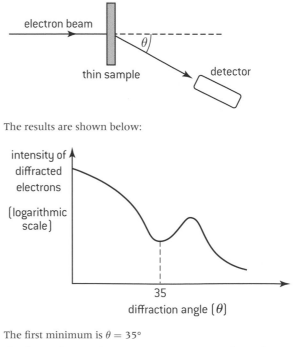

electron beam

θ

thin sample

detector

The results are shown below:

intensity of diffracted electrons

(logarithmic scale)

35

diffraction angle (θ)

The first minimum is $\theta = 35°$

The de Broglie wavelength for the electrons is effectively:

$$\lambda = \frac{hc}{E} = \frac{6.6 \times 10^{-34} \times 3.0 \times 10^8}{400 \times 10^6 \times 1.6 \times 10^{-19}} = 3.1 \times 10^{-15} \text{ m}$$

$$D \approx \frac{\lambda}{\sin \theta} = \frac{3.1 \times 10^{-15}}{\sin 35} = 5.4 \times 10^{-15} \text{ m}$$

So radius of nucleus $\approx 2.7 \times 10^{-15}$ m

⒣ Nuclear energy levels and radioactive decay

ENERGY LEVELS

The energy levels in a nucleus are higher than the energy levels of the electrons but the principle is the same. When an alpha particle or a gamma photon is emitted from the nucleus only discrete energies are observed. These energies correspond to the difference between two **nuclear energy levels** in the same way that the photon energies correspond to the difference between two **atomic energy levels**.

Beta particles are observed to have a continuous spectrum of energies. In this case there is another particle (the antineutrino in the case of beta minus decay) that shares the energy. Once again the amount of energy released in the decay is fixed by the difference between the nuclear energy levels involved. The beta particle and the antineutrino can take varying proportions of the energy available. The antineutrino, however, is very difficult to observe (see box below).

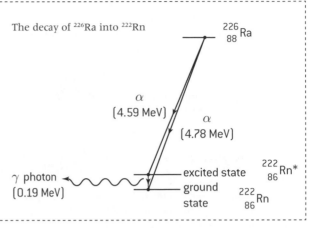

The decay of ^{226}Ra into ^{222}Rn

NEUTRINOS AND ANTINEUTRINOS

Understanding beta decay properly requires accepting the existence of a virtually undetectable particle, the neutrino. It is needed to account for the 'missing' energy and (angular) momentum when analysing the decay mathematically. Calculations involving mass difference mean that we know how much energy is available in beta decay. For example, an isotope of hydrogen, tritium, decays as follows:

$$^{3}_{1}\text{H} \rightarrow \, ^{3}_{2}\text{He} + \, ^{\ \ 0}_{-1}\beta$$

The mass difference for the decay is 19.5 keV c^{-2}. This means that the beta particles should have 19.5 keV of kinetic energy. In fact, a few beta particles are emitted with this energy, but all the others have less than this. The average energy is about half this value and there is no accompanying gamma photon. All beta decays seem to follow a similar pattern.

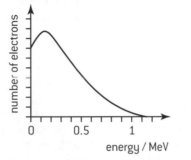

The energy distribution of the electrons emitted in the beta decay of bismuth-210. The kinetic energy of these electrons is between zero and 1.17 MeV.

The neutrino (and antineutrino) must be electrically neutral. Its mass would have to be very small, or even zero. It carries away the excess energy but it is very hard to detect. One of the triumphs of the particle physics of the last century was to be able to design experiments that

confirmed its existence. The full equation for the decay of tritium is:

$$^{3}_{1}\text{H} \rightarrow \, ^{3}_{2}\text{He} + \, ^{\ \ 0}_{-1}\beta + \bar{\nu}$$

where $\bar{\nu}$ is an antineutrino

As has been mentioned before, another form of radioactive decay can also take place, namely positron decay. In this decay, a proton within the nucleus decays into a neutron and the antimatter version of an electron, a positron, which is emitted.

$$^{1}_{1}\text{p} \rightarrow \, ^{1}_{0}\text{n} + \, ^{\ \ 0}_{+1}\beta^{+} + \nu$$

In this case, the positron, β^{+}, is accompanied by a neutrino.

The antineutrino is the antimatter form of the neutrino.

e.g. $^{19}_{10}\text{Ne} \rightarrow \, ^{19}_{9}\text{F} + \, ^{\ \ 0}_{+1}\beta^{+} + \nu$

$^{14}_{6}\text{C} \rightarrow \, ^{14}_{7}\text{N} + \, ^{\ \ 0}_{-1}\beta + \bar{\nu}$

MATHEMATICS OF EXPONENTIAL DECAY

The basic relationship that defines exponential decay as a random process is expressed as follows:

$$\frac{dN}{dt} \propto -N$$

The constant of proportionality between the rate of decay and the number of nuclei available to decay is called the decay constant and given the symbol λ. Its units are time^{-1} i.e. s^{-1} or yr^{-1} etc.

$$\frac{dN}{dt} = -\lambda N$$

The solution of this equation is:

$$N = N_0 e^{-\lambda t}$$

The activity of a source, A, $A = -\dfrac{dN}{dt}$

$$A = A_0 e^{-\lambda t} = \lambda N_0 e^{-\lambda t}$$

It is useful to take natural logarithms:

$$\ln (N) = \ln (N_0 e^{-\lambda t})$$
$$= \ln (N_0) + \ln (e^{-\lambda t})$$
$$= \ln (N_0) - \lambda t \ln (e)$$
$$\therefore \ln (N) = \ln (N_0) - \lambda t \quad (\text{since } \ln (e) = 1)$$

This is of the form $y = c + mx$ so a graph of $\ln N$ vs t will give a straight-line graph.

intercept = $\ln (N_0)$

gradient = $-\lambda$

$$N = N_0 e^{-\lambda t}$$

If $t = T_{\frac{1}{2}}$

$$N = \frac{N_0}{2}$$

So $\dfrac{N_0}{2} = N_0 e^{-\lambda T_{\frac{1}{2}}}$

$$\therefore \frac{1}{2} = e^{-\lambda T_{\frac{1}{2}}}$$

$$\therefore \ln \tfrac{1}{2} = -\lambda T_{\frac{1}{2}}$$

$$\therefore -\lambda T_{\frac{1}{2}} = -\ln \tfrac{1}{2}$$
$$= \ln 2$$

$$\therefore T_{\frac{1}{2}} = \frac{\ln 2}{\lambda}$$

EXAMPLE

The half-life of a radioactive isotope is 10 days. Calculate the fraction of a sample that remains after 25 days.

$$T_{\frac{1}{2}} = 10 \text{ days}$$

$$\lambda = \frac{\ln 2}{T_{\frac{1}{2}}}$$
$$= 6.93 \times 10^{-2} \text{ day}^{-1}$$

$$N = N_0 e^{-\lambda t}$$

Fraction remaining $= \dfrac{N}{N_0}$

$$= e^{-(6.93 \times 10^{-2} \times 25)}$$
$$= 0.187$$
$$= 18.7\%$$

(HL) IB Questions – quantum and nuclear physics

1. The diagrams show the variation with distance x of the wavefunction Ψ of four different electrons. The scale on the horizontal axis in all four diagrams is the same. For which electron is the uncertainty in the momentum the largest?

    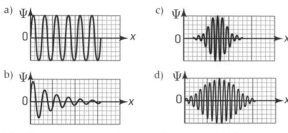

2. The diagram represents the available energy levels of an atom. How many emission lines could result from electron transitions between these energy levels?

    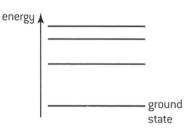

    A. 3   B. 6   C. 8   D. 12

3. A medical physicist wishes to investigate the decay of a radioactive isotope and determine its decay constant and half-life. A Geiger–Müller counter is used to detect radiation from a sample of the isotope, as shown.

    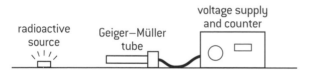

    a) Define the activity of a radioactive sample.   [1]

    Theory predicts that the activity A of the isotope in the sample should decrease exponentially with time t according to the equation $A = A_0 e^{-\lambda t}$, where A_0 is the activity at $t = 0$ and λ is the decay constant for the isotope.

    b) Manipulate this equation into a form which will give a straight line if a semi-log graph is plotted with appropriate variables on the axes. State what variables should be plotted.   [2]

    The Geiger counter detects a proportion of the particles emitted by the source. The physicist records the count-rate R of particles detected as a function of time t and plots the data as a graph of $\ln R$ versus t, as shown below.

    c) Does the plot show that the experimental data are consistent with an *exponential* law? Explain.   [1]

    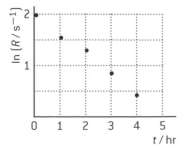

    d) The Geiger counter does not measure the total activity A of the sample, but rather the count-rate

    R of those particles that enter the Geiger tube. Explain why this will not matter in determining the decay constant of the sample.   [1]

    e) From the graph, determine a value for the decay constant λ.   [2]

    The physicist now wishes to calculate the half-life.

    f) Define the half-life of a radioactive substance.   [1]

    g) Derive a relationship between the decay constant λ and the half-life τ.   [2]

    h) Hence calculate the half-life of this radioactive isotope.   [1]

4. This question is about the quantum concept.

    A biography of Schrödinger contains the following sentence: 'Shortly after de Broglie introduced the concept of *matter waves* in 1924, Schrödinger began to develop a new atomic theory.'

    a) Explain the term '*matter waves*'. State what quantity determines the wavelength of such waves.   [2]

    b) Electron diffraction provides evidence to support the existence of matter waves. What is electron diffraction?   [2]

5. Light is incident on a clean metal surface in a vacuum. The maximum kinetic energy KE_{max} of the electrons ejected from the surface is measured for different values of the frequency f of the incident light.

    The measurements are shown plotted below.

    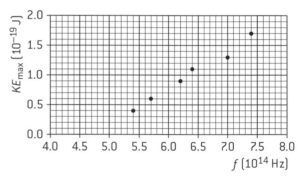

    a) Draw a line of best fit for the plotted data points.   [1]

    b) Use the graph to determine

       (i) the Planck constant   [2]

       (ii) the minimum energy required to eject an electron from the surface of the metal (the *work function*).   [3]

    c) Explain briefly how Einstein's photoelectric theory accounts for the fact that no electrons are emitted from the surface of this metal if the frequency of the incident light is less than a certain value.   [3]

6. Thorium-227 (Th-227) undergoes a-decay with a half-life of 18 days to form radium-223 (Ra-223). A sample of Th-227 has an initial activity of 3.2×10^5 Bq.

    Determine the activity of the remaining thorium-227 after 50 days.   [4]

7. Explain:

    a) The role of angular momentum in the Bohr model for hydrogen   [3]

    b) Pair production and annihilation   [3]

    c) Quantum tunnelling   [3]

Reference frames

OBSERVERS AND FRAMES OF REFERENCE

The proper treatment of large velocities involves an understanding of Einstein's theory of relativity and this means thinking about space and time in a completely different way. The reasons for this change are developed in the following pages, but they are surprisingly simple. They logically follow from two straightforward assumptions. In order to see why this is the case we need to consider what we mean by an object in motion in the first place.

A person sitting in a chair will probably think that they are at rest. Indeed from their point of view this must be true, but this is not the only way of viewing the situation. The Earth is in orbit around the Sun, so from the Sun's point of view the person sitting in the chair must be in motion. This example shows that an object's motion (or lack of it) depends on the observer.

The calculation of relative velocity was considered on page 9. This treatment, like all the mechanics in this book so far, assumes that the velocities are small enough to be able to apply Newton's laws to different frames of reference.

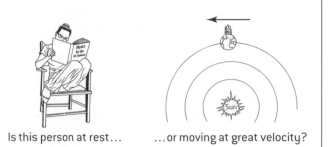

Is this person at rest... ...or moving at great velocity?

GALILEAN TRANSFORMATIONS

It is possible to formalize the relationship between two different frames of reference. The idea is to use the measurement in one frame of reference to work out the measurements that would be recorded in another frame of reference. The equations that do this without taking the theory of relativity into consideration are called **Galilean transformations**.

The simplest situation to consider is two frames of reference (S and S') with one frame (S') moving past the other one (S) as shown below.

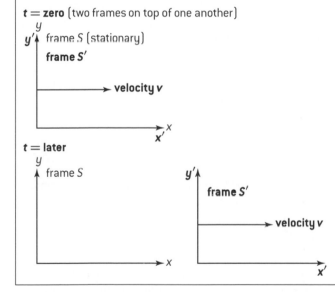

t = zero (two frames on top of one another)

frame S (stationary)
frame S'
velocity v

t = later

frame S
frame S'
velocity v

Each frame of reference can record the position and time of an event. Since the relative motion is along the x-axis, most measurements will be the same:

$$y' = y; z' = z; t' = t$$

If an event is stationary according to one frame, it will be moving according to the other frame – the frames will record different values for the x measurement. The transformation between the two is given by

$$x' = x - vt$$

We can use these equations to formalize the calculation of velocities. The frames will agree on any velocity measured in the y or z direction, but they will disagree on a velocity in the x-direction. Mathematically,

$$u' = u - v$$

For example, if the moving frame is going at 4 m s^{-1}, then an object moving in the same direction at a velocity of 15 m s^{-1} as recorded in the stationary frame will be measured as travelling at 11 m s^{-1} in the moving frame.

Newton's 3 laws of motion describe how an object's motion is effected. An assumption (Newton's Postulates) underlying these laws is that the time interval between two events is the same for all observers. Time is the same for all frames and the separation between events will also be the same in all frames. As a result, the same physical laws will apply in all frames.

PION DECAY EXPERIMENTS

In 1964 an experiment at the European Centre for Nuclear Reseach (CERN) measured the speed of gamma-ray photons that had been produced by particles moving close to the speed of light and found these photons also to be moving at the speed of light. This is consistent with the speed of light being independent of the speed of its source, to a high degree of accuracy.

The experiment analysed the decay of a particle called the neutral pion into two gamma-ray photons. Energy considerations meant that the pions were known to be moving faster than 99.9% of the speed of light and the speed of the photons was measured to be $2.9977 \pm 0.0040 \times 10^8$ m s^{-1}.

FAILURE OF GALILEAN TRANSFORMATION EQUATIONS

If the speed of light has the same value for all observers (see box on left) then the Galilean transformation equations cannot work for light.

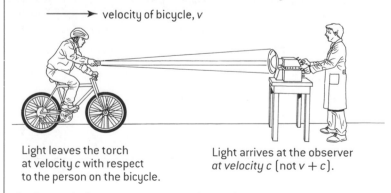

velocity of bicycle, v

Light leaves the torch at velocity c with respect to the person on the bicycle.

Light arrives at the observer *at velocity c (not $v + c$).*

The theory of relativity attempts to work out what has gone wrong.

Maxwell's equations

MAXWELL AND THE CONSTANCY OF THE SPEED OF LIGHT

In 1864 James Clerk Maxwell presented a new theory at the Royal Society in London. His ideas were encapsulated in a mathematical form that elegantly expressed not only what was known at the time about the magnetic field B and the electric field E, but it also proposed a unifying link between the two – electromagnetism. The 'rules' of electromagnetic interactions are summarized in four equations known as Maxwell's equations. These equations predict the nature of electromagnetic waves.

Most people know that light is an electromagnetic wave, but it is quite hard to understand what this actually means. A physical wave involves the oscillation of matter, whereas an electromagnetic wave involves the oscillation of electric and magnetic fields. The diagram below attempts to show this.

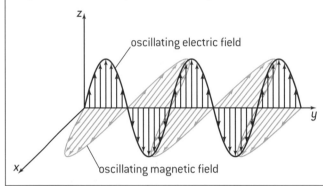

The changing electric and magnetic fields move through space – the technical way of saying this is that the fields **propagate** through space. The physics of how these fields propagate allows the speed of all electromagnetic waves (including light) to be predicted. It turns out that this can be done in terms of the electric and magnetic constants of the medium through which they travel.

$$c = \sqrt{\frac{1}{\varepsilon_0 \mu_0}}$$

This equation does not need to be understood in detail. The only important idea is that the speed of light is **independent** of the velocity of the source of the light. In other words, a prediction from Maxwell's equations is that the speed of light in a vacuum has the same value for all observers.

This prediction of the constancy of the speed of light highlights an inconsistency that cannot be reconciled with Newtonian mechanics (where the resultant speed of light would be equal to the addition of the relative speed of the source and the relative speed of light as measured by the source). Einstein's analysis forced long-held assumptions about the independence of space and time to be rejected.

COMPARING ELECTRIC AND MAGNETIC FIELDS

Electrostatic forces and magnetic forces appear very different to one another. Fundamentally, however, they are just different aspects of one force – the electromagnetic interaction. The nature of which field is observed depends on the observer. For example:

a) A charge moving at right angles to a magnetic field.

   An observer in a frame of reference that is **at rest with respect to the magnetic field** will explain the force acting on the charge (and its acceleration) in terms of a magnetic force ($F_M = Bqv$) that acts on the moving charge.

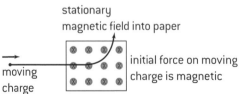

   An observer in a frame of reference that is **at rest with respect to the charge** will explain the initial force acting on the charge (and its initial acceleration) in terms of an induced electric force that results from the cutting of magnetic flux.

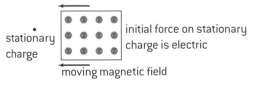

b) Two identically charged particles moving with parallel velocities according to a laboratory frame of reference.

   An observer in a frame of reference that is **moving with the charged particles** will see the particles at rest. Thus this observer sees the force of repulsion between the two charges as solely electrostatic in nature.

   An observer in a frame of reference where **the laboratory is at rest** will see the total force between the two charges as a combination of electrostatic and magnetic. Moving charges are currents and thus each moving charge creates its own magnetic field which is stationary in the laboratory frame. Each charge is moving in the other's stationary magnetic field and will experience a magnetic force.

Special relativity

POSTULATES OF SPECIAL RELATIVITY

The special theory of relativity is based on two fundamental assumptions or **postulates**. If either of these postulates could be shown to be wrong, then the theory of relativity would be wrong. When discussing relativity we need to be even more than usually precise with our use of technical terms.

One important technical phrase is an **inertial frame of reference**. This means a frame of reference in which the laws of inertia (Newton's laws) apply. Newton's laws do not apply in accelerating frames of reference so an inertial frame is a frame that is either stationary or moving with constant velocity.

An important idea to grasp is that there is no fundamental difference between being stationary and moving at constant velocity. Newton's laws link forces and accelerations. If there is no resultant force on an object then its acceleration will be zero.

This could mean that the object is at **rest** or it could mean that the object is **moving at constant velocity**.

The two postulates of special relativity are:

- the speed of light in a vacuum is the same constant for all inertial observers
- the laws of physics are the same for all inertial observers.

The first postulate leads on from Maxwell's equations and can be experimentally verified. The second postulate seems completely reasonable – particularly since Newton's laws do not differentiate between being at rest and moving at constant velocity. If both are accepted as being true then we need to start thinking about space and time in a completely different way. If in doubt, we need to return to these two postulates.

SIMULTANEITY

One example of how the postulates of relativity disrupt our everyday understanding of the world around us is the concept of simultancity. If two events happen together we say that they are simultaneous. We would normally expect that if two events are **simultaneous** to one observer, they should be simultaneous to all observers – but this is not the case! A simple way to demonstrate this is to consider an experimenter in a train.

The experimenter is positioned **exactly** in the middle of a carriage that is moving at constant velocity. She sends out two pulses of light towards the ends of the train. Mounted at the ends are mirrors that reflect the pulses back towards the observer. As far as the experimenter is concerned, the whole carriage is at rest. Since she is in the middle, the experimenter will know that:

- the pulses were sent out simultaneously
- the pulses hit the mirrors simultaneously
- the pulses returned simultaneously.

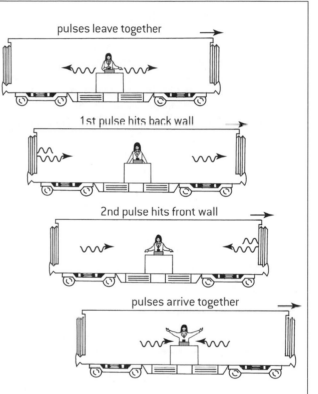

pulses leave together

1st pulse hits back wall

2nd pulse hits front wall

pulses arrive together

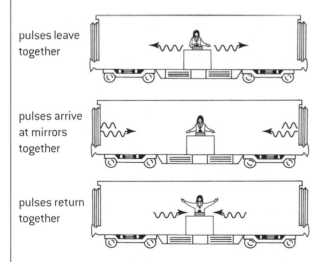

pulses leave together

pulses arrive at mirrors together

pulses return together

The situation will seem very different if watched by a stationary observer (on the platform). This observer knows that light must travel at constant speed – both beams are travelling at the same speed as far as he is concerned, so they must hit the mirrors at different times. The left-hand end of the carriage is moving towards the beam and the right hand end is moving away. This means that the reflection will happen on the left-hand end first.

Interestingly, the observer on the platform does see the beams arriving back at the same time. The observer on the platform will know that:

- the pulses were sent out simultaneously
- the left-hand pulse hit the mirror before the right-hand pulse
- the pulses returned simultaneously.

In general, simultaneous events that take place at the same point in space will be simultaneous to all observers whereas events that take place at different points in space can be simultaneous to one observer but not simultaneous to another!

Do not dismiss these ideas because the experiment seems too fanciful to be tried out. The use of a pulse of light allowed us to rely on the first postulate. This conclusion is valid whatever event is considered.

Lorentz transformations

LORENTZ FACTOR

The formulae for special relativity all involve a factor that depends on the relative velocity between different observers, v.

We define the Lorentz factor, γ as follows:

$$\gamma = \frac{1}{\sqrt{1 - \frac{v^2}{c^2}}}$$

At low velocities, the Lorentz factor is approximately equal to one – relativistic effects are negligible. It approaches infinity near the speed of light.

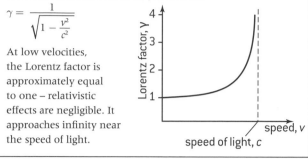

LORENTZ TRANSFORMATIONS

An observer defines a frame of reference and different **events** can be characterized by different coordinates according to the observer's measurements of space and time. In a frame S, an event will be associated with a given position (x, y and z coordinates) and take place at a given time (t). Observers in relative uniform motion disagree on the numerical values for these coordinates.

The Galilean transformations equations (page 131) allowed us to calculate what an observer in a second frame will record if we know the values in one frame but assume that the measurement of time is the same in both frames. Einstein has shown that this is not correct.

clock in frame S and clock in frame S' are synchronized to $t = t' =$ zero when frames coincide.
(two frames on top of one another)

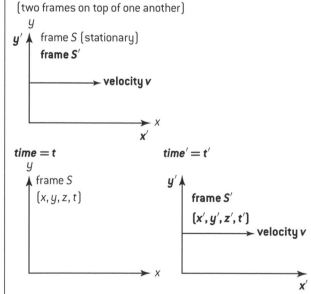

Because the frames were synchronized, the observers agree on the measurements of y and z. To switch between the other measurements made by different observers we need to use the Lorentz transformations. These all involve the Lorentz factor, γ, as defined above. The derivation of these equations is not required.

$x' = \gamma(x - vt); \quad \Delta x' = \gamma(\Delta x - v\Delta t);$
$t' = \gamma\left(t - \frac{vx}{c^2}\right); \quad \Delta t' = \gamma\left(\Delta t - \frac{v\Delta x}{c^2}\right)$

The reverse transformations also apply. These are just a consequence of the relative velocity of frame S (with respect to frame S') being in the opposite direction.

$x = \gamma(x' + vt'); \quad t = \gamma\left(t' + \frac{vx'}{c^2}\right)$

LORENTZ TRANSFORMATION EXAMPLE

We can apply the Lorentz transformation equations to the situation shown on page 133. Suppose the experiment on the train measures the carriage to be 50.0 m long and the observer on the platform measures the speed of the train to be 2.7×10^8 m s^{-1} (0.90 c) to the right. In this situation, we know the times (t) and locations (x) are measured according to the experimenter on the train (frame S) and the experimenter on the platform is frame S'.

1. According to the experimenter on the train (frame S),

    Time taken for each pulse to reach mirror at end of carriage is given by:

    $$\Delta t = \frac{25.0}{3.0 \times 10^8} = 8.33 \times 10^{-8} \text{ s}$$

    Total time taken for each pulse to complete the round journey to the experimenter is:

    $$\Delta t_{\text{total}} = \frac{50.0}{(3.0 \times 10^8)} = 1.67 \times 10^{-7} \text{ s}$$

2. According to the experimenter on the platform (frame S'),

    $$\gamma = \frac{1}{\sqrt{1 - \frac{v^2}{c^2}}} = \frac{1}{\sqrt{1 - \frac{(0.9c)^2}{c^2}}}$$
    $$= \frac{1}{\sqrt{1 - 0.81}} = \frac{1}{\sqrt{0.19}} = 2.29$$

    Time taken for LH pulse to reach mirror at end of carriage is given by:

    $$\Delta t'_{(LH \, pulse)} = \gamma\left(\Delta t - \frac{v\Delta x}{c^2}\right)$$

    where $\Delta t = 8.33 \times 10^{-8}$ s, $v = -2.7 \times 10^8$ m s^{-1} (relative velocity of platform is moving to the left) and $\Delta x = -25.0$ m (pulse moving to left)

    $$\therefore \Delta t'_{(LH \, pulse)} = 2.29\left(8.33 \times 10^{-8} - \frac{(-2.7 \times 10^8) \times (-25.0)}{(3.0 \times 10^8)^2}\right)$$
    $$= 1.91 \times 10^{-7} - 1.72 \times 10^{-7}$$
    $$= 1.9 \times 10^{-8} \text{ s}$$

    Time taken for RH pulse to reach mirror at end of carriage is given by:

    $$\Delta t'_{(RH \, pulse)} = \gamma\left(\Delta t - \frac{v\Delta x}{c^2}\right)$$
    $$= 2.29\left(8.33 \times 10^{-8} - \frac{(-2.7 \times 10^8) \times 25.0}{(3.0 \times 10^8)^2}\right)$$
    $$= 1.91 \times 10^{-7} + 1.72 \times 10^{-7} = 3.63 \times 10^{-7} \text{ s}$$

    Note that the time taken by each pulse is different – they do not arrive simultaneously according to the experimenter on the platform.

    The return time for the LH pulse is the same as the time taken for the RH to initially reach the mirror (in each case, $\Delta x = 25.0$ m and $\Delta t = 8.33 \times 10^{-8}$ s)

    So total time taken for LH pulse to return to centre of carriage is

    total time$'_{(LH \, pulse)} = 1.9 \times 10^{-8} + 3.63 \times 10^{-7} = 3.82 \times 10^{-7}$ s

    This is the same as the total time taken for the RH pulse so both experimenters observe the return of the pulses to be simultaneous.

    Check: The above calculates that for frame S', the total time taken for the round trip is 3.82×10^{-7} s. The Lorentz transformation, can also be applied to the pulse's journey. In this situation, Δx (in frame S) = 0 as the pulse returns to its starting position.

    $$\text{total } \Delta t'_{(either \, pulse)} = \gamma\left(\Delta t - \frac{v\Delta x}{c^2}\right) = \gamma\Delta t$$
    $$= 2.29 \times 1.67 \times 10^{-7} = 3.82 \times 10^{-7} \text{ s}$$

Velocity addition

VELOCITY ADDITION

When two observers measure each other's velocity, they will always agree on the value. The calculation of relative velocity is not, however, normally straightforward. For example, an observer might see two objects approaching one another, as shown below.

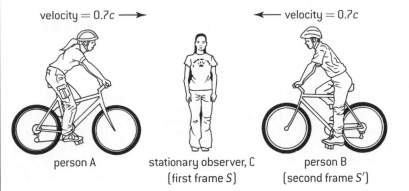

velocity = 0.7c →   ← velocity = 0.7c

person A   stationary observer, C   person B
                  (first frame S)   (second frame S')

If each object has a relative velocity of 0.7 c, the Galilean transformations would predict that the relative velocity between the two objects would be 1.4 c. This cannot be the case as the Lorentz factor can only be worked out for objects travelling at less than the speed of light.

The situation considered is one frame moving relative to another frame at velocity v.

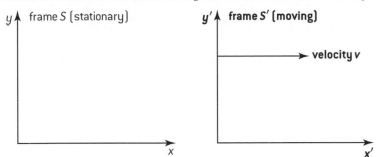

y ↑ frame S (stationary)

y' ↑ frame S' (moving)

→ velocity v

x

x'

Application of the Lorentz transformation gives the equation used to move between frames:

$$u' = \frac{u - v}{1 - \frac{uv}{c^2}}$$

  u' – the velocity under consideration in the x-direction as measured in the second frame, S'

  u – the velocity under consideration in the x-direction as measured in the first frame, S

  v – the velocity of the second frame, S', as measured in the first frame, S

In each of these cases, a positive velocity means motion along the positive x-direction. If something is moving in the negative x-direction then a negative velocity should be substituted into the equation.

Example

In the example above, two objects approached each other with 70% of the speed of light. So u' is person A's velocity as measured in person B's frame of reference.

  u' = relative velocity of approach – to be calculated

  $u = 0.7\,c$

  $v = -0.7\,c$

  $u' = \dfrac{1.4\,c}{(1 + 0.49)}$ *note the sign in the brackets*

  $= \dfrac{1.4\,c}{1.49}$

  $= 0.94\,\text{c}$

COMPARISON WITH GALILEAN EQUATION

The top line of the relativistic addition of velocities equation can be compared with the Galilean equation for the calculation of relative velocities.

  $u' = u - v$

At low values of v these two equations give the same value. The Galilean equation only starts to fail at high velocities.

At high velocities, the Galilean equation can give answers of greater than c, while the relativistic one always gives a relative velocity that is less than the speed of light.

Invariant quantities

SPACETIME INTERVAL

Relativity has shown that our Newtonian ideas of space and time are incorrect. Two inertial observers will generally disagree on their measurements of space and time but they will agree on a measurement of the speed of light. Is there anything else upon which they will agree?

In relativity, a good way of imagining what is going on is to consider everything as different 'events' in something called **spacetime**. From one observer's point of view, three co-ordinates (x, y and z) can define a position in space. One further 'coordinate' is required to define its position in time (t). An event is a given point specified by these four coordinates (x, y, z, t).

As a result of the Lorentz transformation, another observer would be expected to come up with totally different numbers for all of these four measurements – (x', y', z', t'). The amazing thing is that these two observers will agree on something. This is best stated mathematically:

$$(ct)^2 - x^2 - y^2 - z^2 = (ct')^2 - x'^2 - y'^2 - z'^2$$

On normal axes, Pythagoras's theorem shows us that the quantity $\sqrt{(x^2 + y^2 + z^2)}$ is equal to the length of the line from the origin, so $(x^2 + y^2 + z^2)$ is equal to (the length of the line) ^{2}. In other words, it is the separation in space.

$$(\text{Separation in space})^2 = (x^2 + y^2 + z^2)$$

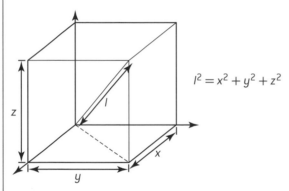

$$l^2 = x^2 + y^2 + z^2$$

The two observers agree about something very similar to this, but it includes a coordinate of time. This can be thought of as the separation in imaginary four-dimensional spacetime.

$$(\text{Separation in spacetime})^2 = (ct)^2 - x^2 - y^2 - z^2$$

or

$(\text{Separation in spacetime})^2$
$$= (\text{time separation})^2 - (\text{space separation})^2$$

In 1 dimension, this is simplified to

$$(ct')^2 - (x')^2 = (ct)^2 - (x)^2$$

OTHER INVARIANT QUANTITIES

In addition to the spacetime interval between two events (see box above), all observers agree on the values of three other quantities associated with the separation between two events or with reference to a given object. These are:

- Proper time interval Δt_0
- Proper length L_0
- Rest mass m_0

These four quantities are said to be **invariant** as they are always constant and do not vary with a change of observer. There are additional quantities, not associated with mechanics, that are also invariant e.g. electric charge.

PROPER TIME, PROPER LENGTH & REST MASS

a) Proper time interval Δt_0

When expressing the time taken between events (for example the length of time that a firework is giving out light), the **proper time** is the time as measured in a frame where the events take place at the same point in space. It turns out to be the shortest possible time that any observer could correctly record for the event.

measuring how long a firework lasts

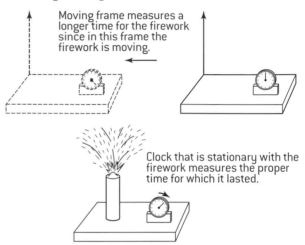

Moving frame measures a longer time for the firework since in this frame the firework is moving.

Clock that is stationary with the firework measures the proper time for which it lasted.

If A is moving past B then B will think that time is running slowly for A. From A's point of view, B is moving past A. This means that A will think that time is running slowly for B. Both views are correct!

b) Proper length L_0

As before, different observers will come up with different measurements for the length of the same object depending on their relative motions. The **proper length** of an object is the length recorded in a frame where the object is at rest.

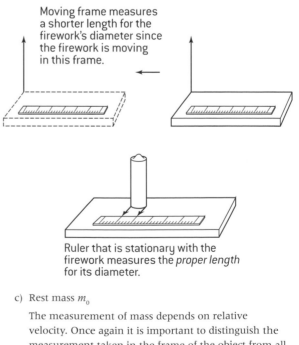

Moving frame measures a shorter length for the firework's diameter since the firework is moving in this frame.

Ruler that is stationary with the firework measures the *proper length* for its diameter.

c) Rest mass m_0

The measurement of mass depends on relative velocity. Once again it is important to distinguish the measurement taken in the frame of the object from all other possible frames. The **rest mass** of an object is its mass as measured in a frame where the object is at rest. A particle's rest mass does not change.

Time dilation

LIGHT CLOCK

A **light clock** is an imaginary device. A beam of light bounces between two mirrors – the time taken by the light between bounces is one 'tick' of the light clock.

As shown in the derivation the path taken by light in a light clock that is moving at constant velocity is longer. We know that the speed of light is fixed so the time between the 'ticks' on a moving clock must also be longer. This effect – that moving clocks run slow – is called **time dilation**.

The time between bounces Δt_0 is the proper time for this clock in the frame where the clock is at rest.

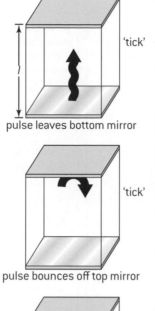

pulse leaves bottom mirror

'tick'

pulse bounces off top mirror

'tick'

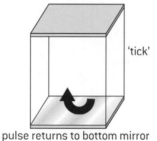

pulse returns to bottom mirror

'tick'

DERIVATION OF EFFECT FROM LORENTZ TRANSFORMATION

If frame S is a frame where two events take place at the same point in space, then the time interval between these two events must be the proper time interval, Δt_0.

Time dilation is then a direct consequence of the Lorentz transformation:

$$\Delta t' = \gamma \left(\Delta t - \frac{v \Delta x}{c^2} \right)$$

Where $\Delta t = \Delta t_0$, (the proper time interval) and $\Delta x = $ zero (same point in space)

$\therefore$ time interval in frame S', $\Delta t' = \gamma \Delta t_0$

DERIVATION OF THE EFFECT FROM FIRST PRINCIPLES

If we imagine a stationary observer with one light clock then t is the time between 'ticks' on their stationary clock. In **this stationary frame**, a moving clock runs **slowly** and t' is the time between 'ticks' on the moving clock: t' is greater than t.

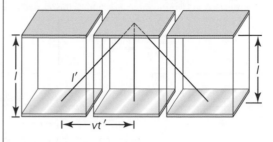

In the time t',

the clock has moved on a distance $= vt'$

Distance travelled by the light, $l' = \sqrt{((vt')^2 + l^2)}$

$$t' = \frac{l'}{c}$$

$$= \frac{\sqrt{(vt')^2 + l^2}}{c}$$

$\therefore \qquad t'^2 = \frac{v^2 t'^2 + l^2}{c^2}$

$\therefore \qquad t'^2 \left(1 - \frac{v^2}{c^2}\right) = \frac{l^2}{c^2}$

but $\qquad \frac{l^2}{c^2} = t^2$

$\therefore \qquad t'^2 \left(1 - \frac{v^2}{c^2}\right) = t^2$

or $\qquad t' = \frac{1}{\sqrt{1 - \frac{v^2}{c^2}}} \times t \quad$ or $\quad t' = \gamma t$

This equation is true for all measurements of time, whether they have been made using a light clock or not.

Length contraction and evidence to support special relativity

EFFECT OF LENGTH CONTRACTION

Time is not the only measurement that is affected by relative motion. There is another relativistic effect called **length contraction**. According to a (stationary) observer, the separation between two points in space contracts if there is relative motion in that direction. The contraction is in the same direction as the relative motion.

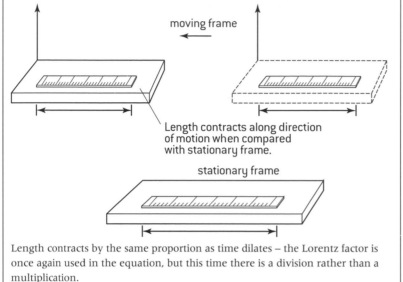

moving frame

Length contracts along direction of motion when compared with stationary frame.

stationary frame

Length contracts by the same proportion as time dilates – the Lorentz factor is once again used in the equation, but this time there is a division rather than a multiplication.

$$L = \frac{L_0}{\gamma}$$

EXAMPLE

An unstable particle has a lifetime of 4.0×10^{-8} s in its own rest frame. If it is moving at 98% of the speed of light calculate:

a) Its lifetime in the laboratory frame.

b) The length travelled in both frames.

a) $\gamma = \sqrt{\dfrac{1}{1 - (0.98)^2}}$

$= 5.025$

$\Delta t = \gamma \Delta t_0$

$= 5.025 \times 4.0 \times 10^{-8}$

$= 2.01 \times 10^{-7}$ s

b) In the laboratory frame, the particle moves

Length = speed × time

$= 0.98 \times 3 \times 10^8 \times 2.01 \times 10^{-7}$

$= 59.1$ m

In the particle's frame, the laboratory moves

$\Delta l = \dfrac{59.1}{\gamma}$

$= 11.8$ m

(alternatively: length = speed × time

$= 0.98 \times 3 \times 10^8 \times 4.0 \times 10^{-8}$

$= 11.8$ m)

DERIVATION OF LENGTH CONTRACTION FROM LORENTZ TRANSFORMATION

When we measure the length of a moving object, then we are recording the position of each end of the object at one given instant of time according to that frame of reference. In other words the time interval measured in frame S between these two events will be zero, $\Delta t = 0$. In this case, the length measured Δx is the length of the moving object L_0.

Length contraction is then a direct consequence of the Lorentz transformation, as, if we move into the frame, S',

where the object is at rest, we will be measuring the proper length L_0:

$$\Delta x' = \gamma(\Delta x - v\Delta t)$$

Where $\Delta x' = L_0$ (the proper length) and

$\Delta t =$ zero (simultaneous measurements of position of end of object)

$\therefore$ Length in frame S', $L_0 = \gamma(L)$

$$L = \frac{L_0}{\gamma}$$

THE MUON EXPERIMENT

Muons are leptons (see page 78) – they can be thought of as a more massive version of an electron. They can be created in the laboratory but they quickly decay. Their average lifetime is 2.2×10^{-6} s as measured in the frame in which the muons are at rest.

Muons are also created high up (10 km above the surface) in the atmosphere. Cosmic rays from the Sun can cause them to be created with huge velocities – perhaps $0.99\,c$. As they travel towards the Earth some of them decay but there is still a detectable number of muons arriving at the surface of the Earth.

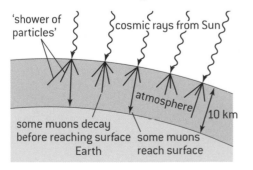

'shower of particles'

cosmic rays from Sun

some muons decay before reaching surface

some muons reach surface

atmosphere 10 km

Earth

Without relativity, no muons would be expected to reach the surface at all. A particle with a lifetime of 2.2×10^{-6} s which is travelling near the speed of light (3×10^8 m s^{-1}) would be expected to travel less than a kilometre before decaying ($2.2 \times 10^{-6} \times 3 \times 10^8 = 660$ m).

The moving muons are effectively moving 'clocks'. Their high speed means that the Lorentz factor is high.

$$\gamma = \sqrt{\frac{1}{1 - 0.99^2}} = 7.1$$

Therefore an average lifetime of 2.2×10^{-6} s in the muons' frame of reference will be time dilated to a longer time as far as a stationary observer on the Earth is concerned. From this frame of reference they will last, on average, 7.1 times longer. Many muons will still decay but some will make it through to the surface – this is exactly what is observed.

In the muons' frame they exist for 2.2×10^{-6} s on average. They make it down to the surface because the atmosphere (and the Earth) is moving with respect to the muons. This means that the atmosphere will be length-contracted. The 10 km distance as measured by an observer on the Earth will only be $\frac{10}{7.1} = 1.4$ km. A significant number of muons will exist long enough for the Earth to travel this distance.

Spacetime diagrams (Minkowski diagrams) 1

SPACETIME DIAGRAMS

Spacetime separation was introduced on page 136. A spacetime diagram is a visual way of representing the geometry. Measurements can be taken from the diagram to calculate actual values.

We cannot represent all four dimensions on the one diagram, so we usually limit the number of dimensions of space that we represent. The simplest representation has only one dimension of space and one of time as shown below.

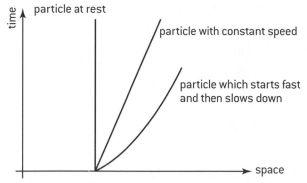

An object (moving or stationary) is always represented as a line in spacetime.

Note that:

- The values on the spacetime diagram are as would be measured by an observer whose worldline is represented by the vertical axis.

- The vertical axis in the above spacetime diagram is time t. An alternative is to plot (speed of light × time), ct. This means that both axes can have the same units (m, light-years or equivalent).

- Whatever axes are being used, by convention, the path of a beam of light is represented by a line at 45° to the axes.

- The advance of proper time for any traveller can be calculated from the overall separation in spacetime. In the traveller's frame of reference, they remained stationary so the separation between two events can be calculated as shown below.

EXAMPLE 1 OF SPACETIME DIAGRAMS

The advance of proper time for the journey between the events A→B→C→D can be calculated from the values on the spacetime diagram.

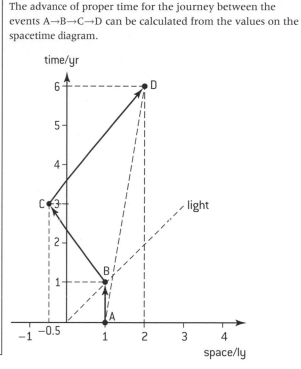

A journey through spacetime

Journey	Space separation (x)/ly	Time separation (t)/yr	(Spacetime separation)^2 $(ct)^2 - (x)^2$/ly^2	Advance of proper time according to traveller / yr $t' = \sqrt{\dfrac{(ct)^2 - (x)^2}{c}}$
A→B	0.0	1.0	$1^2 - 0^2 = 1$	$\sqrt{1.00} = 1.00$
B→C	1.5	2.0	$4 - 2.25 = 1.75$	$\sqrt{1.75} = 1.32$
C→D	2.5	3.0	$9 - 6.25 = 2.75$	$\sqrt{2.75} = 1.66$

The total advance of proper time for the traveller is $1.00 + 1.32 + 1.66 = 3.98$ yr. This compares with the advance of 6.0 years according to an observer whose worldline is a vertical line on this spacetime diagram. This difference is an example of **time dilation** (see page 137).

The alternative journey direct from A → D shows a greater elapsed proper time.

Journey	Space separation (x)/ly	Time separation (t)/yr	(Spacetime separation)^2 $(ct)^2 - (x)^2$/ly^2	Advance of proper time according to traveller / yr $t' = \sqrt{\dfrac{(ct)^2 - (x)^2}{c}}$
A→D	1.0	6.0	$36 - 1 = 35$	$\sqrt{35} = 5.92$

This is always true. A direct worldline always has a greater amount of elapsed proper time than an indirect worldline.

Spacetime diagrams 2

CALCULATION OF TIME DILATION AND LENGTH CONTRACTION

Time dilation and **length contraction** are quantitatively represented on spacetime diagrams. Refer to diagram on page 139.

a) **Time dilation**: In the journey direct from B → C, the relative velocity between the traveller and the stationary observer is $\frac{1.5\,\text{ly}}{2.0\,\text{yrs}} = 0.75$ c. The Lorentz gamma factor is:

$$\gamma = \frac{1}{\sqrt{1 - \frac{v^2}{c^2}}} = \frac{1}{\sqrt{1 - 0.75^2}} = 1.51$$

The journey takes 2 yrs according to the observer at rest. This means the proper time as measured by the traveller will be:

$$\Delta t = \gamma \Delta t_0 \Rightarrow \Delta t_0 = \frac{\Delta t}{\gamma} = \frac{2.0}{1.51} = 1.32 \text{ yr}$$

as shown in the table on page 139.

b) **Length contraction**: The observer at rest measures the journey length from B→C to be 1.5 ly. The journey will be length contracted to be

$$L = \frac{L_0}{\gamma} = \frac{1.5}{1.51} = 0.99 \text{ ly}$$

The relative velocity of travel is 0.75 c, and the time taken to go from from B → C, in the traveller's frame of reference, is 1.32 yr. This makes the distance according to the traveller to be 0.75 c × 1.32 yr = 0.99 ly as shown above.

EXAMPLE 2 – CURVED WORLDLINE

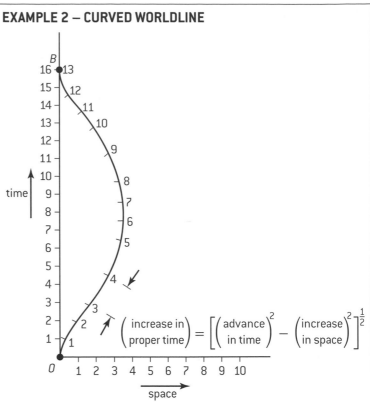

$$\left(\begin{array}{c}\text{increase in}\\\text{proper time}\end{array}\right) = \left[\left(\begin{array}{c}\text{advance}\\\text{in time}\end{array}\right)^2 - \left(\begin{array}{c}\text{increase}\\\text{in space}\end{array}\right)^2\right]^{\frac{1}{2}}$$

Proper time along a curved worldline from event O to event B is smaller than the proper time along the straight line from O to B.

The twin paradox 1

As mentioned on page 136, the theory of relativity gives no preference to different inertial observers – the time dilation effect (moving clocks run slowly) is always the same. This leads to the '**twin paradox**'. In this imaginary situation, two identical twins compare their views of time. One twin remains on Earth while the other twin undergoes a very fast trip out to a distant star and back again.

As far as the twin on the Earth is concerned the other twin is a moving observer. This means that the twin that remains on the Earth will think that time has been running slowly for the other twin. When they meet up again, the returning twin should have aged less.

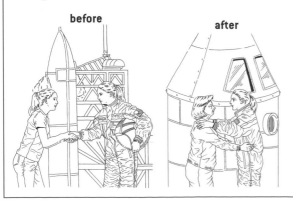

This seems a very strange prediction, but it is correct according to the time dilation formula. Remember that:

- This is a relativistic effect – time is running at different rates because of the **relative velocity** between the two twins and **not** because of the **distance** between them.

- The difference in ageing is relative. Neither twin is getting younger; as far as both of them are concerned, time has been passing at the normal rate. It's just that the moving twin thinks that she has been away for a shorter time than the time as recorded by the twin on the Earth.

The paradox is that, according to the twin who made the journey, the twin on the Earth was moving all the time and so the twin left on the Earth should have aged less. Whose version of time is correct?

The solution to the paradox comes from the realization that the equations of special relativity are only symmetrical when the two observers are in constant relative motion. For the twins to meet back up again, one of them would have to turn around. This would involve external forces and acceleration. If this is the case then the situation is no longer symmetrical for the twins. The twin on the Earth has not accelerated so her view of the situation must be correct.

The resolution of the twin paradox using a spacetime diagram is on page 141.

Twin paradox 2

RESOLVING THE TWIN PARADOX USING SPACETIME DIAGRAMS

The diagram below is a spacetime diagram for a journey to a distant planet followed by an immediate return.

According to the twin remaining on Earth:

- the distance to the planet = 3.0 ly
- relative velocity of traveller is 0.6 c
- each leg of the journey takes $\frac{3.0}{0.6} = 5.0$ yr
- Total journey time = 10.0 yr

The gamma factor is

$$\gamma = \frac{1}{\sqrt{1 - \frac{v^2}{c^2}}} = \frac{1}{\sqrt{1 - 0.6^2}} = 1.25$$

So according to the twin undertaking the journey:

- each leg of the journey takes $\frac{5.0}{1.25} = 4.0$ yr
- Total journey time = 8.0 yr
- the distance to the planet = $\frac{3.0}{1.25} = 2.4$ ly
- relative velocity of Earth = $\frac{4.8}{8.0} = 0.6$ c

In order to check whose version of time 'is correct', **they agree to send light signals every year**. The spacetime diagram for this situation in the Earth's frame of reference is shown below (left).

Note that there is no paradox; they agree on the number of signals sent and received; the travelling twin has aged less than the twin that stayed on Earth.

A more complicated spacetime diagram can be drawn for the reference frame of the outbound traveller (below right). Note that:

- The first four years has the travelling twin's worldline vertical i.e. stationary.
- When the travelling twin turns round, she leaves her original frame of reference and changes to a frame where the Earth is moving towards her at $\frac{3}{5} c$ ($= 0.6\ c$).
- Her relative velocity towards the Earth with respect to her original frame of reference can be calculated from the velocity transformation equations as $\frac{15}{17} c$ ($= 0.88\ c$) back.
- In this frame of reference, the total time for the round trip would be measured as 12.5 yr

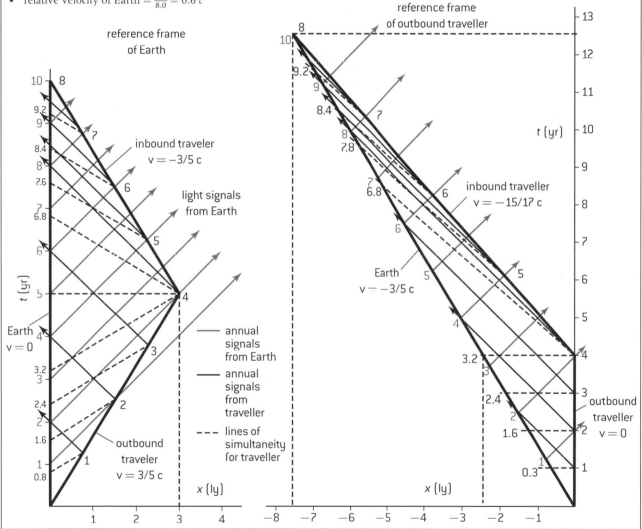

Spacetime diagrams 3

REPRESENTING MORE THAN ONE INERTIAL FRAME ON THE SAME SPACETIME DIAGRAM

The Lorentz transformations describe how measurements of space and time in one frame can be converted into the measurements observed in another frame of reference. The situation in each frame of reference can be visualized by using separate spacetime diagrams for each frame of reference (see page 141 for examples).

It is also possible to represent two inertial frames on the same spacetime diagram. A frame S' (coordinates x' and ct') is moving at relative constant velocity $+v$ according to a frame S (coordinates x and ct). The principles are as follows:

- The same worldline applies to both sets of coordinate axes (that is, to x and ct, as well as to x' and ct').

- The Lorentz transformation is made by *changing the coordinate system for frame S'* rather than the position of the worldline.

- The spacetime axes for frame S has x and ct at right angles to one another as normal.

- The spacetime axes for frame S' has its x' and ct' axes both angled in towards the $x = ct$ line (which represents a path of a beam of light.

- The coordinates of a spacetime event in S are read from the x and ct axes directly.

- The coordinates of a spacetime event in S' are measured by drawing lines parallel to the ct' and x' axes until they hit the x' and ct' axes.

Mathematically for the above process to agree with the Lorentz transformation calculations, the following must apply:

- The angle between the ct' axis (the worldline for the origin of S') and the ct axis is the same as the angle between the x' axis and the x axis. It is:

$$\theta = \tan^{-1}\left(\frac{v}{c}\right)$$

- The scales used by the axes in S' are different to the scales used by the axes in S.

- A given value is represented by a greater length on the ct' axis when compared with the ct axis.

- A given value is represented by a greater length on the x' axis when compared with the x axis.

- The ratio of the measurements on the axes depends on the relative velocity between the frames. The equation (which does not need to be recalled) is:

$$\text{ratio of units } \frac{ct'}{ct} = \sqrt{\frac{1 + \frac{v^2}{c^2}}{1 - \frac{v^2}{c^2}}}$$

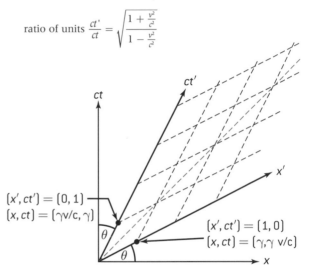

$[x', ct'] = [0, 1]$
$[x, ct] = [\gamma v/c, \gamma]$

$[x', ct'] = [1, 0]$
$[x, ct] = [\gamma, \gamma \, v/c]$

Summary

- At greater speed:
  ◊ the S' axes swing towards the $x = ct$ line as the angle θ increases.
  ◊ the ct' and x' axes are more stretched when compared with the ct and x axes.

- Events that are simultaneous in S are on the same horizontal line.

- Events that are simultaneous in S' are on a line parallel to the x' axis.

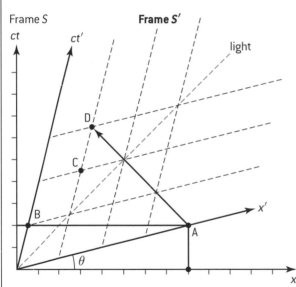

Frame S Frame S'

1. Events A & B are simultaneous in frame S but are *not* simultaneous in frame S' (A occurs before B)

$$\tan \theta = \frac{2}{8} = 0.25$$

∴ relative velocity of frames S' and $S = 0.25\,c$

2. Events C & D occur at same location in frame S'.

   Events C & D occur at different locations in frame S.

3. A pulse of light emitted by event A arrives at event D according to both frames of reference. It cannot arrive at events B or C.

ⓗⓛ Mass and energy

$E = MC^2$

The most famous equation in all of physics is surely Einstein's mass–energy relationship $E = mc^2$, but where does it come from? By now it should not be a surprise that if time and length need to be viewed in a different way, then so does energy.

According to Newton's laws, a constant force produces a constant acceleration. If this was always true then any velocity at all should be achievable – even faster than light. All we have to do is apply a constant force and wait.

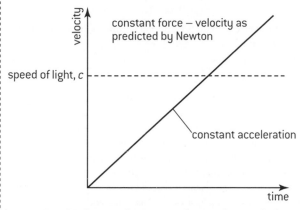

In practice, this does not happen. As soon as the speed of an object starts to approach the speed of light, the acceleration gets less and less even if the force is constant.

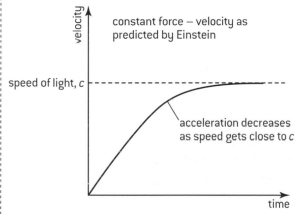

The force is still doing work (= force × distance), therefore the object must still be gaining kinetic energy and a new relativistic equation is needed for energy:

$E = \gamma m_0 c^2$

Note that some textbooks compare this equation with the definition of rest energy ($E_0 = m_0 c^2$) in order to define a concept of relativistic mass that varies with speed ($m = \gamma m_0$). The current IB syllabus does not encourage this approach.

The preferred approach is to see rest mass as invariant and to adopt a new relativistic formula for kinetic energy:

Total energy = rest energy + kinetic energy = $\gamma m_0 c^2$

rest energy = $m_0 c^2$

so, kinetic energy $E_K = (\gamma - 1)m_0 c^2$

MASS AND ENERGY

Mass and energy are equivalent. This means that energy can be converted into mass and vice versa. Einstein's mass–energy equation can always be used, but one needs to be careful about how the numbers are substituted. Newtonian equations (such as $KE = \frac{1}{2}mv^2$ or momentum $= mv$) will take different forms when relativity theory is applied.

The energy needed to create a particle at rest is called the rest energy E_0 and can be calculated from the rest mass:

$E_0 = m_0 c^2$

If this particle is given a velocity, it will have a greater total energy.

$E = \gamma m_0 c^2$

HL Relativistic momentum and energy

EQUATIONS

The laws of conservation of momentum and conservation of energy still apply in relativistic situations. However the concepts often have to be refined to take into account the new ways of viewing space and time.

For example, in Newtonian mechanics, momentum p is defined as the product of mass and velocity.

$$p = mv$$

In relativity it has a similar form, but the Lorentz factor needs to be taken into consideration.

$$p = \gamma m_0 v$$

The momentum of an object is related to its total energy. In relativistic mechanics, the relationship can be stated as

$$E^2 = p^2 c^2 + m_0^2 c^4$$

In Newtonian mechanics, the relationship between energy and momentum is

$$E = \frac{p^2}{2m}$$

Do not be tempted to use the standard Newtonian equations – if the situation is relativistic, then you need to use the relativistic equations.

UNITS

SI units can be applied in these equations. Sometimes, however, it is useful to use other units instead.

At the atomic scale, the joule is a huge unit. Often the electronvolt (eV) is used. One electronvolt is the energy gained by one electron if it moves through a potential difference of 1 volt. Since

$$\text{Potential difference} = \frac{\text{energy difference}}{\text{charge}}$$

$$1 \text{ eV} = 1 \text{ V} \times 1.6 \times 10^{-19} \text{ C}$$

$$= 1.6 \times 10^{-19} \text{ J}$$

In fact the electronvolt is too small a unit, so the standard SI multiples are used

$$1 \text{ keV} = 1000 \text{ eV}$$

$$1 \text{ MeV} = 10^6 \text{ eV} \quad \text{etc.}$$

Since mass and energy are equivalent, it makes sense to have comparable units for mass. The equation that links the two ($E = mc^2$) defines a new unit for mass – the MeV c^{-2}. The speed of light is included in the unit so that no change of number is needed when switching between mass and energy – If a particle of mass of 5 MeV c^{-2} is converted completely into energy, the energy released would be 5 MeV. It would also be possible to use keV c^{-2} or GeV c^{-2} as a unit for mass.

In a similar way, the easiest unit for momentum is the MeV c^{-1}. This is the best unit to use if using the equation which links relativistic energy and momentum.

EXAMPLE

The Large Electron / Positron (LEP) collider at the European Centre for Nuclear Research (CERN) accelerates electrons to total energies of about 90 GeV. These electrons then collide with *positrons* moving in the opposite direction as shown below. Positrons are identical in rest mass to electrons but carry a positive charge. The positrons have the same energy as the electrons.

Electron	Electron
$\bullet \longrightarrow$	$\longleftarrow \bullet$
Total energy = 90 GeV	Total energy = 90 GeV

a) Use the equations of special relativity to calculate,

(i) the velocity of an electron (with respect to the laboratory);

Total energy = 90 GeV = 90000 MeV

Rest mass = 0.5 MeVc^{-2} $\therefore \gamma = 18000$ (huge)

$\therefore v \simeq c$

(ii) the momentum of an electron (with respect to the laboratory).

$p^2 c^2 = E^2 - m_0^2 c^4$

$\simeq E^2$

$p \simeq 90 \text{ GeV}c^{-1}$

b) For these two particles, estimate their relative velocity of approach.

since γ so large

relative velocity $\simeq c$

c) What is the total momentum of the system (the two particles) before the collision?

zero

d) The collision causes new particles to be created.

(i) Estimate the maximum total rest mass possible for the new particles.

Total energy available = 180 GeV

$\therefore$ max total rest mass possible = 180 GeVc^{-2}

(ii) Give one reason why your answer is a *maximum*.

Above assumes that particles were created at rest

Ⓗ Relativistic mechanics examples

PARTICLE ACCELERATION AND ELECTRIC CHARGE

In a particle accelerator (e.g. a linear accelerator or cyclotron), charged particles are accelerated up to very high energies. The basic principle is to pass the charged particles through a series of potential differences and each time, the particle's total energy increases as a result. The increase in kinetic energy (ΔE_K) as a result of a charge q passing through a potential difference V is given by:

$$qV = \Delta E_K$$

EXAMPLE

An electron is accelerated through a pd of 1.0×10^6 V. Calculate its velocity.

$$\text{Energy gained} = 1.0 \times 10^6 \times 1.6 \times 10^{-19} \text{ J}$$
$$= 1.6 \times 10^{-13} \text{ J}$$
$$E_0 = m_0c^2 = 9.11 \times 10^{-31} \times (3 \times 10^8)^2$$
$$= 8.2 \times 10^{-14} \text{ J}$$
$$\therefore \quad \text{Total energy} = 1.6 \times 10^{-13} + 8.2 \times 10^{-14}$$
$$= 2.42 \times 10^{-13} \text{ J}$$
$$\therefore \gamma = \frac{2.42 \times 10^{-13}}{8.2 \times 10^{-14}} = 2.95$$
$$\text{velocity} = \sqrt{1 - \frac{1}{\gamma^2}} \cdot c$$
$$= 0.94\, c$$

PHOTONS

Photons are particles that have a zero rest mass and travel at the speed of light, c. Their total energy and their frequency f is linked by Planck's constant h:

$$E = hf$$

The relativistic equation that links total energy, E and momentum, p, must also apply to photons:

$$E^2 - p^2c^2 + m_0^2c^4$$

The rest mass of a photon is zero so the momentum of a photon is:

$$p = \frac{E}{c} = \frac{hf}{c} = \frac{h}{\lambda}$$

EXAMPLE: DECAY OF A PION

A neutral pion (π^0) is a meson of rest mass $m_0 = 135.0$ MeV c^{-2}. A typical mode of decay is to convert into two photons:

$$\pi^0 \to 2\gamma$$

The wavelength of these photons can be calculated:

a) Decay at rest

If the pion was at rest when it decayed, each photon would have half the total energy of the pion:

$$E = 67.5 \text{ MeV} = 67.5 \times 10^6 \times 1.6 \times 10^{-19} \text{ J}$$
$$- 1.08 \times 10^{-11} \text{ J}$$

Planck's constant can be used to calculate the wavelength of one of the photons:

$$E = h\frac{c}{\lambda}$$
$$\lambda = h\frac{c}{E} = 6.63 \times 10^{-34} \times \frac{3.0 \times 10^8}{1.08 \times 10^{-11}}$$
$$= 1.84 \times 10^{-14} \text{ m}$$

The momentum of the pion was initially zero as it was at rest. Conservation of momentum means that the photons will be emitted in opposite directions. The total momentum of each photon add together to give a total, once again, of zero.

b) Decay while moving

Suppose the pion was moving forward when it decayed with a total energy 270.0 MeV c^{-2}; the photons will be emitted as shown below:

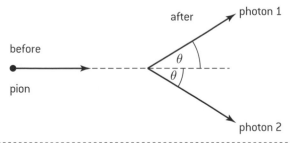

Note that in this example, total energy $2m_0c^2$, so $\gamma = 2$ so $v = 0.866\, c$

Each photon will have a total energy of 135 MeV $= 2.16 \times 10^{-11}$ J

and a momentum of 135 MeV c^{-1}. The wavelengths of the photons will be:

$$\lambda = h\frac{c}{E} = 6.63 \times 10^{-34} \times \frac{3.0 \times 10^8}{2.16 \times 10^{-11}}$$
$$= 9.21 \times 10^{-15} \text{ m}$$

Initial total momentum for the pion in the forward direction can be calculated from

$$E^2 = p^2c^2 + m_0^2c^4$$
$$p^2c^2 - E^2 - m_0^2c^4 = (4 - 1)m_0^2c^4$$
$$p = \sqrt{3}\, m_0 c = 1.73 \times 135.0 = 233.8 \text{ MeV } c^{-1}$$

So conservation of momentum in forward direction is:

$$233.8 = 2 \times 135 \times \cos\theta$$
$$\therefore \cos\theta = \frac{233.8}{270} = 0.866$$
$$\therefore \quad \theta = 30°$$

HL General relativity – the equivalence principle

PRINCIPLE OF EQUIVALENCE

One of Einstein's 'thought experiments' considers how an observer's view of the world would change if they were accelerating. The example below considers an observer inside a closed spaceship.

There are two possible situations to compare.

- The rocket could be far away from any planet but accelerating forwards.
- The rocket could be at rest on the surface of a planet.

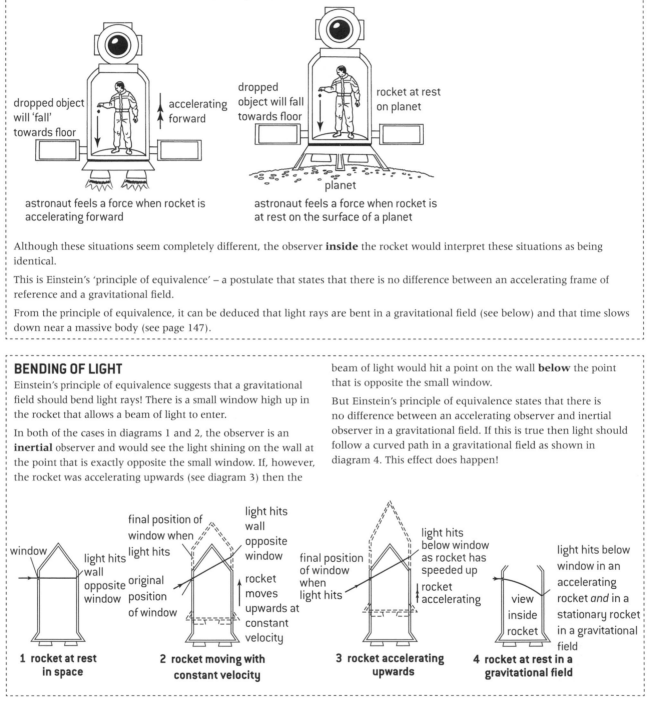

astronaut feels a force when rocket is accelerating forward

astronaut feels a force when rocket is at rest on the surface of a planet

Although these situations seem completely different, the observer **inside** the rocket would interpret these situations as being identical.

This is Einstein's 'principle of equivalence' – a postulate that states that there is no difference between an accelerating frame of reference and a gravitational field.

From the principle of equivalence, it can be deduced that light rays are bent in a gravitational field (see below) and that time slows down near a massive body (see page 147).

BENDING OF LIGHT

Einstein's principle of equivalence suggests that a gravitational field should bend light rays! There is a small window high up in the rocket that allows a beam of light to enter.

In both of the cases in diagrams 1 and 2, the observer is an **inertial** observer and would see the light shining on the wall at the point that is exactly opposite the small window. If, however, the rocket was accelerating upwards (see diagram 3) then the

beam of light would hit a point on the wall **below** the point that is opposite the small window.

But Einstein's principle of equivalence states that there is no difference between an accelerating observer and inertial observer in a gravitational field. If this is true then light should follow a curved path in a gravitational field as shown in diagram 4. This effect does happen!

1 rocket at rest in space

2 rocket moving with constant velocity

3 rocket accelerating upwards

4 rocket at rest in a gravitational field

(HL) Gravitational red shift

CONCEPT

The general theory of relativity makes other predictions that can be experimentally tested. One such effect is **gravitational red shift** – clocks slow down in a gravitational field. In other words a clock on the ground floor of a building will run slowly when compared with a clock in the attic – the attic is further away from the centre of the Earth.

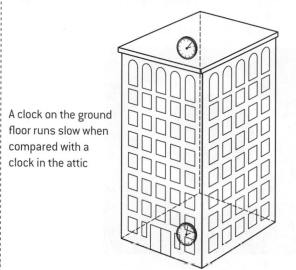

A clock on the ground floor runs slow when compared with a clock in the attic

The same effect can be imagined in a different way. We have seen that a gravitational field affects light. If light is shone away from a mass (for example the Sun), the photons of light must be increasing their gravitational potential energy as they move away. This means that they must be decreasing their total energy. Since frequency is a measure of the energy of a photon, the observed frequency away from the source must be less than the emitted frequency.

At the top of the building, the photon has less energy, and so a lower frequency, than when it was at the bottom.

The oscillations of the light can be imagined as the pulses of a clock. An observer at the top of the building would perceive the clock on the ground floor to be running slowly.

MATHEMATICS

This gravitational time dilation effect can be mathematically worked out for a uniform gravitational field g. The change in frequency Δf is given by

$$\frac{\Delta f}{f} = \frac{g\Delta h}{c^2}$$

where

f is the frequency emitted at the source

g is the gravitational field strength (assumed to be constant)

Δh is the height difference and

c is the speed of light.

EXAMPLE

A UFO travels at such a speed to remain above one point on the Earth at a height of 200 km above the Earth's surface. A radio signal of frequency of 110 MHz is sent to the UFO.

(i) What is the frequency received by the UFO?

(ii) If the signal was reflected back to Earth, what would be the observer frequency of the return signal? Explain your answer.

(i) $f = 1.1 \times 10^8$ Hz

          $g = 10$ m s^{-2}

          $\Delta h = 2.0 \times 10^5$ m

  $\therefore \Delta f = \dfrac{10 \times 2.0 \times 10^5}{(3 \times 10^8)^2} \times 1.1 \times 10^8$ Hz

          $= 2.4 \times 10^{-3}$ Hz

  $\therefore f$ received $= 1.1 \times 10^8 - 2.4 \times 10^{-3}$

          $= 109999999.998$ Hz

          $\approx 1.1 \times 10^8$ Hz

(ii) The return signal will be gravitationally blue shifted. Therefore it will arrive back at **exactly** the same frequency as emitted.

EVIDENCE TO SUPPORT GENERAL RELATIVITY

Bending of star light

The predictions of general relativity, just like those of special relativity, seem so strange that we need strong experimental evidence. One main prediction was the bending of light by a gravitational field. One of the first experiments to check this effect was done by a physicist called Arthur Eddington in 1919.

The idea behind the experiment was to measure the deflection of light (from a star) as a result of the Sun's mass. During the day, the stars are not visible because the Sun is so bright. During a solar eclipse, however, stars are visible during the few minutes when the Moon blocks all of the light from the Sun. If the positions of the stars during the total eclipse were compared with the positions of the same stars recorded at a different time, the stars that appeared near the edge of the Sun would appear to have moved.

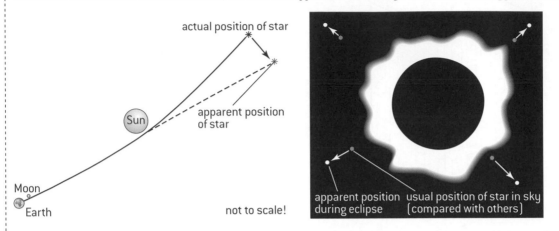

The angle of the shift of these stars turned out to be exactly the angle as predicted by Einstein's general theory of relativity.

Gravitational lensing

The bending of the path of light or the warping of spacetime (depending on which description you prefer) can also produce some very extreme effects. Massive galaxies can deflect the light from quasars (or other very distance sources of light) so that the rays bend around the galaxy as shown below.

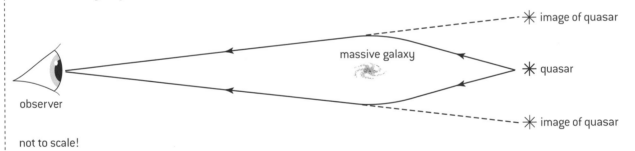

In this strange situation, the galaxy is acting like a lens and we can observe multiple images of the distant quasar.

EVIDENCE TO SUPPORT GRAVITATIONAL RED SHIFT

Pound–Rebka–Snider experiment

The decrease in the frequency of a photon as it climbs out of a gravitational field can be measured in the laboratory. The measurements need to be very sensitive, but they have been successfully achieved on many occasions. One of the experiments to do this was done in 1960 and is called the **Pound–Rebka** experiment. The frequencies of gamma-ray photons were measured after they ascended or descended Jefferson Physical Laboratory Tower at Harvard University.

The original **Pound–Rebka** experiment was repeated with greater accuracy by Pound and Snider.

Atomic clock frequency shift

Because they are so sensitive, comparing the difference in time recorded by two identical atomic clocks can provide a direct measurement of gravitational red shift. One of the clocks is taken to high altitude by a rocket, whereas a second one remains on the ground. The clock that is at the higher altitude will run faster.

Global positioning system

For the global positioning system to be so accurate, general relativity must be taken into account in calculating the details of the satellite's orbit.

HL Curvature of spacetime

EFFECT OF GRAVITY ON SPACETIME

The Newtonian way of describing gravity is in terms of the forces between two masses. In general relativity the way of thinking about gravity is not to think of it as a force, but as changes in the shape (warping) of spacetime. The warping of spacetime is caused by mass. Think about two travellers who both set off from different points on the Earth's equator and travel north.

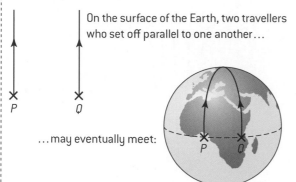

On the surface of the Earth, two travellers who set off parallel to one another...

...may eventually meet:

As they travel north they will get closer and closer together. They could explain this coming together in terms of a force of attraction between them or they could explain it as a consequence of the surface of the Earth being curved. The travellers have to move in straight lines across the surface of the Earth so their paths come together.

Einstein showed how spacetime could be thought of as being curved by mass. The more matter you have, the more curved spacetime becomes. Moving objects follow the curvature of spacetime or in other words, they take the shortest path in spacetime. As has been explained, it is very hard to imagine the four dimensions of spacetime. It is easier to picture what is going on by representing spacetime as a flat two-dimensional sheet.

spacetime represented by flat sheet

Any mass present warps (or bends) spacetime. The more mass you have the greater the warping that takes place. This warping of spacetime can be used to describe the orbit of the Earth around the Sun. The diagram below represents how Einstein would explain the situation. The Sun warps spacetime around itself. The Earth orbits the Sun because it is travelling along the shortest possible path in spacetime. This turns out to be a curved path.

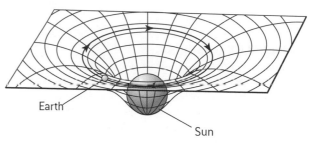

Earth

Sun

- Mass 'tells' spacetime how to curve.
- Spacetime 'tells' matter how to move.

APPLICATIONS OF GENERAL RELATIVITY TO THE UNIVERSE AS A WHOLE

General relativity is now fundamental to understanding how the objects in the Universe interact with spacetime and thus how they affect each other. This allows far-reaching predictions to be created about the future development and fate of the Universe – see cosmology sections of the astrophysics option (option D).

The development of the Universe can be modelled in detail. Many current aspects (e.g. its large-scale structure, the creation of the elements and the presence of cosmic background radiation) are predicted.

Very large mass black holes may exist at the centres of many galaxies. General relativity predicts how these may interact with matter and astronomers are searching for appropriate evidence.

General relativity predicts the existence of gravitational waves associated with high energy events such as the collision of two black holes. Experimental evidence for the existence of these waves is being sought.

Black holes

DESCRIPTION

When a star has used up all of its nuclear fuel, the force of gravity makes it collapse down on itself (see the astrophysics option for more details). The more it contracts the greater the density of matter and thus the greater the gravitational field near the collapsing star. In terms of general relativity, this would be described in terms of the spacetime near a collapsing star becoming more and more curved. The curvature of spacetime becomes more and more severe depending on the mass of the collapsing star.

If the collapsing star is less than about 1.4 times the mass of the Sun, then the electrons play an important part in eventually stopping this contraction. The star that is left is called a white dwarf. If the collapsing star is greater than this, the electrons cannot halt the contraction. A contracting mass of up to three times the mass of the Sun can also be stopped – this time the neutrons play an important role and the star that is left is called a neutron star. The curvature of spacetime near a neutron star is more extreme than the curvature near a white dwarf.

At masses greater than this we do not know of any process that can stop the contraction. Spacetime around the mass becomes more and more warped until eventually it becomes so great that it folds in over itself. What is left is called a black hole. All the mass is concentrated into a point – the **singularity**.

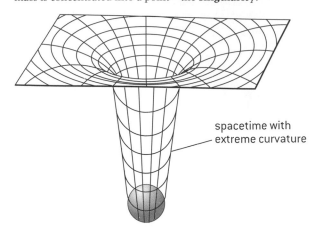

spacetime with extreme curvature

SCHWARZCHILD RADIUS

The curvature of spacetime near a black hole is so extreme that nothing, not even light, can escape. Matter can be attracted into the hole, but nothing can get out since nothing can travel faster than light. The gravitational forces are so extreme that light would be severely deflected near a black hole.

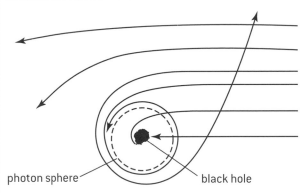

photon sphere   black hole

If you were to approach a black hole, the gravitational forces on you would increase. The first thing of interest would be the **photon sphere**. This consists of a very thin shell of light photons captured in orbit around the black hole. As we fall further in, the gravitational forces increase and so the escape velocity at that distance also increases.

EXAMPLE

Calculate the size of a black hole that has the same mass as our Sun (1.99×10^{30} kg).

$$R_{\text{Sch}} = \frac{2 \times 6.67 \times 10^{-11} \times 1.99 \times 10^{30}}{(3 \times 10^8)^2}$$

$$= 2949.6 \text{ m}$$

$$= 2.9 \text{ km}$$

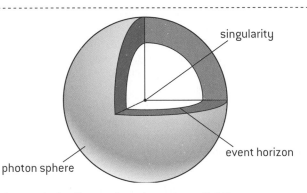

singularity

event horizon

photon sphere

At a particular distance from the centre, called the **Schwarzchild radius**, we get to a point where the escape velocity is equal to the speed of light. Newtonian mechanics predicts that the escape velocity v from a mass M of radius r is given by the formula

$$v = \sqrt{\frac{2GM}{r}}$$

If the escape velocity is the speed of light, c, then the Schwarzchild radius would be given by

$$R_{\text{s}} = \frac{2GM}{c^2}$$

It turns out that this equation is also correct if we use the proper equations of general relativity. If we cross the Schwarzchild radius and get closer to the singularity, we would no longer be able to communicate with the Universe outside. For this reason crossing the Schwarzchild radius is sometimes called crossing the **event horizon**. An observer watching an object approaching a black hole would see time slowing down for the object.

The observed time dilation is worked out from

$$\Delta t = \frac{\Delta t_0}{\sqrt{1 - \frac{R_{\text{s}}}{r}}}$$

where r is the distance from the black hole.

IB Questions – option A – relativity

1. In the laboratory frame of reference, a slow moving alpha particle travels parallel to stationary metal wire that carries an electric current. In this frame of reference, the velocity of the alpha particle and the drift velocity of the electrons in the wire are identical. Explain the origin of the force on the alpha particle in the frame of reference of

   a) The alpha particle [2]

   b) The laboratory [2]

2. Two identical rockets are moving along the same straight line as viewed from Earth. Rocket 1 is moving away from the Earth at speed 0.80 c **relative to the Earth** and rocket 2 is moving away from rocket 1 at speed 0.60 c **relative to rocket 1**.

   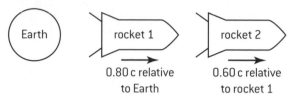

   a) Calculate the velocity of rocket 2 relative to the Earth, using the

      (i) Galilean transformation equation. [1]

      (ii) relativistic transformation equation. [2]

   b) Comment on your answers in (a). [2]

   c) The rest mass of rocket 1 is 1.0×10^3 kg. Determine the **relativistic** kinetic energy of rocket 1, as measured by an observer on Earth. [3]

3. The spacetime diagram below shows two events, A and B, as observed in a reference frame S. Each event emits a light signal.

   Use the diagram to calculate, according to frame S,

   a) The time between event A and event B [2]

   b) The time taken for the light signal leaving event A to arrive at the position of event B. [2]

   c) The location of a stationary observer who receives the light signal from events A simultaneously with receiving the light signal from event B. [2]

   d) The velocity of a moving frame of reference in which event A and event B occurred simultaneously. [4]

   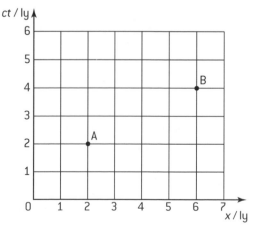

4. *Relativity and simultaneity*

   a) State two postulates of the special theory of relativity. [2]

   Einstein proposed a 'thought experiment' along the following lines. Imagine a train of proper length 100 m passing through a station at half the speed of light. There are two lightning strikes, one at the front and one at the rear of the train, leaving scorch marks on both the train

and the station platform. Observer S is standing on the station platform midway between the two strikes, while observer T is sitting in the middle of the train. Light from each strike travels to both observers.

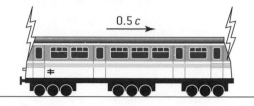

   b) If observer S on the station concludes from his observations that the two lightning strikes occurred simultaneously, explain why observer T on the train will conclude that they did **not** occur simultaneously. [4]

   c) Which strike will T conclude occurred first? [1]

   d) What will be the distance between the scorch marks on the *train*, according to T and according to S? [3]

   e) What will be the distance between the scorch marks on the *platform*, according to T and according to S? [2]

HL

5. In a laboratory experiment two identical particles (P and Q), each of rest mass m_0, collide. In the **laboratory frame of reference**, they are both moving at a velocity of 2/3 c. The situation before the collision is shown in the diagram below.

   Before:

   $$2/3\,c \qquad\qquad\qquad 2/3\,c$$
   $$\bullet\!\!\longrightarrow \qquad\qquad\qquad \longleftarrow\!\!\bullet$$
   $$P \qquad\qquad\qquad\qquad\qquad\qquad Q$$

   a) In the laboratory frame of reference,

      (i) what is the total momentum of P and Q? [1]

      (ii) what is the total energy of P and Q? [3]

   The same collision can be viewed according to **P's frame of reference** as shown in the diagrams below.

   $$\qquad\qquad\qquad\qquad \text{velocity} = v$$
   $$\bullet \qquad\qquad\qquad\qquad \longleftarrow$$
   $$P\,(\text{rest}) \qquad\qquad\qquad\qquad Q$$

   b) In P's frame of reference,

      (i) what is Q's velocity, v? [3]

      (ii) what is the total momentum of P and Q? [3]

      (iii) what is the **total energy** of P and Q? [3]

   c) As a result of the collision, many particles and photons are formed, but the total energy of the particles depends on the frame of reference. Do the observers in each frame of reference agree or disagree on the number of particles and photons formed in the collision? Explain your answer. [2]

6. The concept of gravitational red-shift indicates that clocks run slower as they approach a black hole.

   a) Describe what is meant by

      (i) gravitational red-shift. [2]

      (ii) spacetime. [1]

      (iii) a black hole with reference to the concept of spacetime. [2]

   b) A particular black hole has a Schwarzschild radius R. A person at a distance of 2R from the event horizon of the black hole measures the time between two events to be 10 s. Deduce that for a person a very long way from the black hole the time between the events will be measured as 12 s. [1]

Translational and rotational motion

CONCEPTS

The complex motion of a rigid body can be analysed as a **combination** of two types of motion: *translation* and *rotation*. Both these types of motion are studied separately in this study guide (pages 9 and 65).

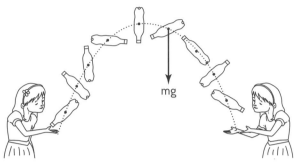

A bottle thrown through the air – the centre of mass of the bottle follows a path as predicted by projectile motion. In addition the bottle rotates about one (or more) axes.

Translational motion is described using displacements, velocities and linear accelerations; all these quantities apply to the **centre of mass** of the object. Rotational motion is described using angles (angular displacement), angular velocities and angular accelerations; all these quantities apply to circular motion about a given axis of rotation.

The concept of angular velocity, ω, has already been introduced with the mechanics of circular motion (see page 66) and is linked to the frequency of rotation by the following formula:

$$\omega = 2\pi f$$

angular velocity   frequency

EQUATIONS OF UNIFORM ANGULAR ACCELERATION

The definitions of average linear velocity and average linear acceleration can be rearranged to derive the constant acceleration equations (page 11). An equivalent rearrangement derives the equations of constant angular acceleration.

Translational motion		Rotational motion	
Displacement	s	Angular displacement	θ
Initial velocity	u	Initial angular velocity	ω_i
Final velocity	v	Final angular velocity	ω_f
Time taken	t	Time taken	t
Acceleration	a	Angular acceleration	α
	[constant]		[constant]
$v = u + at$		$\omega_f = \omega_i + \alpha t$	
$s = ut + \frac{1}{2}at^2$		$\theta = \omega_i t + \frac{1}{2}\alpha t^2$	
$v^2 = u^2 + 2as$		$\omega_f^2 = \omega_i^2 + 2\alpha\theta$	
$s = \dfrac{(v+u)t}{2}$		$\theta = \dfrac{(\omega_f + \omega_i)t}{2}$	

Translational motion	Rotational motion
Every particle in the object has the same instantaneous velocity	Every particle in the object moves in a circle around the same axis of rotation
Displacement, s, measured in m	Angular displacement, θ, measured in radians [rad]
Velocity, v, is the rate of change of displacement measured in m s^{-1} $v = \dfrac{ds}{dt}$	Angular velocity, ω, is the rate of change of angle measured in rad s^{-1} $\omega = \dfrac{d\theta}{dt}$
Acceleration, a, is the rate of change of velocity measured in m s^{-2} $a = \dfrac{dv}{dt}$	Angular acceleration, α, is the rate of change of angular velocity measured in rad s^{-2} $\alpha = \dfrac{d\omega}{dt}$

Comparison of linear and rotational motion

EXAMPLE: BICYCLE WHEEL

When a bicycle is moving forward at constant velocity v, the different points on the wheel each have different velocities. The motion of the wheel can be analysed as the addition of the translational and the rotational motion.

a) Translational motion

The bicycle is moving forward at velocity v so the wheel's centre of mass has forward translational motion of velocity v. All points on the wheel's rim have a translational component forward at velocity v.

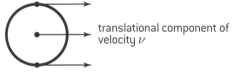

translational component of velocity v

b) Rotational motion

The wheel is rotating around the central axis of rotation at a constant angular velocity ω. All points on the wheel's rim have a tangential component of velocity $v \, (= r\omega)$

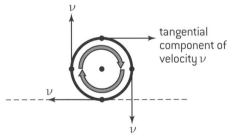

tangential component of velocity v

c) Combined motion

The motion of the different points on the wheel's rim is the vector addition of the above two components:

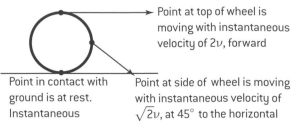

Point at top of wheel is moving with instantaneous velocity of $2v$, forward

Point in contact with ground is at rest. Instantaneous velocity is zero

Point at side of wheel is moving with instantaneous velocity of $\sqrt{2}v$, at 45° to the horizontal

Translational and rotational relationships

RELATIONSHIP BETWEEN LINEAR AND ROTATIONAL QUANTITIES

When an object is just rotating about a fixed axis, and there is no additional translational motion of the object, all the individual particles that make up that object have different instantaneous values of linear displacement, linear velocity and linear acceleration. They do, however, all share the same instantaneous values of angular displacement, angular velocity and angular acceleration. The link between these values involves the distance from the axis of rotation to the particle.

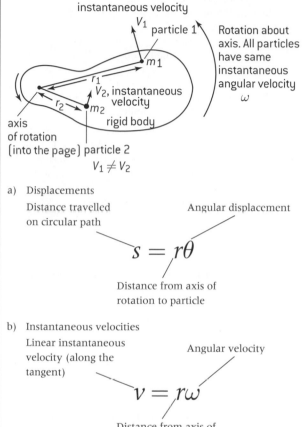

a) Displacements

Distance travelled on circular path   Angular displacement

$$s = r\theta$$

Distance from axis of rotation to particle

b) Instantaneous velocities

Linear instantaneous velocity (along the tangent)   Angular velocity

$$v = r\omega$$

Distance from axis of rotation to particle

c) Accelerations

The total linear acceleration of any particle is made up of two components:

a) The **centripetal acceleration**, a_r (towards the axis of rotation – see page 65), also known as the **radial acceleration**.

Tangential velocity   Angular velocity

$$a_r = \frac{v^2}{r} = r\omega^2$$

Centripetal acceleration (along the radius)   Distance from axis of rotation to particle

b) An additional **tangential acceleration**, a_t, which results from an angular acceleration taking place. If $\alpha = 0$, then $a_t = 0$.

Instantaneous acceleration (along the tangent)   Angular acceleration

$$a_t = r\alpha$$

Distance from axis of rotation to particle

The total acceleration of the particle can be found by vector addition of these two components: $a = r\sqrt{\omega^4 + \alpha^2}$

Translational and rotational equilibrium

THE MOMENT OF A FORCE: THE TORQUE Γ

A particle is in equilibrium if its acceleration is zero. This occurs when the vector sum of all the external forces acting on the particle is zero (see page 16). In this situation, all the forces pass through a single point and sum to zero. The forces on real objects do not always pass through the same point and can create a turning effect about a given axis. The turning effect is called the **moment of the force** or the **torque**. The symbol for torque is the Greek uppercase letter gamma, Γ.

The moment or torque Γ of a force, F about an axis is defined as the product of the force and the perpendicular distance from the axis of rotation to the line of action of the force.

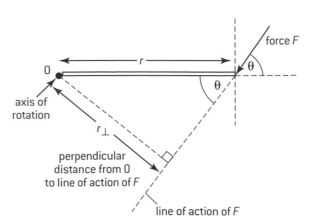

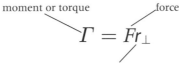

$$\Gamma = Fr_\perp$$

perpendicular distance

$$\Gamma = Fr \sin \theta$$

Note:

- The torque and energy are both measured in N m, but only energy can also be expressed as joules.

- The direction of any torque is clockwise or anticlockwise about the axis of rotation that is being considered. For the purposes of calculations, this can be treated as a vector quantity with the direction of the torque vector considered to be along the axis of rotation. In the example above, the torque vector is directed into the paper. If the force F was applied in the opposite direction, the torque vector would be directed out of the paper.

COUPLES

A **couple** is a system of forces that has no resultant force but which does produce a turning effect. A common example is a pair of equal but anti-parallel forces acting with different points of application. In this situation, the resultant torque is the same about all axes drawn perpendicular to the plane defined by the forces.

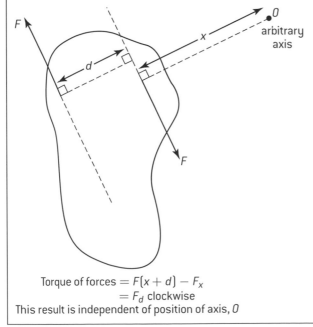

Torque of forces $= F(x + d) - F_x$
$= F_d$ clockwise
This result is independent of position of axis, O

ROTATIONAL AND TRANSLATIONAL EQUILIBRIUM

If a resultant force acts on an object then it must accelerate (page 17). When there is no resultant force acting on an object then we know it to be in translational equilibrium (page 16) as this means its acceleration must be zero.

Similarly, if there is a resultant torque acting on an object then it must have an angular acceleration, α. Thus an object will be in **rotational equilibrium** only if the vector sum of all the external torques acting on the object is zero.

If an object is not moving and not rotating then it is said to be in **static equilibrium**. This must mean that the object is in both rotational and translational equilibrium.

For rotational equilibrium:

$$\alpha = 0 \therefore \sum \Gamma = 0$$

In 2D problems (in the x-y plane), it is sufficient to show that there is no torque about any **one** axis perpendicular to the plane being considered (parallel to the z-axis). In 3D problems, three axis directions (x, y and z) would need to be considered.

For translational equilibrium:

$$a = 0 \therefore \sum F = 0$$

In 2D problems, it is sufficient to show that there is no resultant force in **two** different directions. In 3D problems three axis directions (x, y and z) would need to be considered.

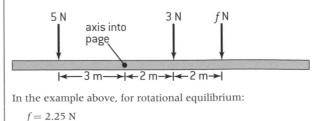

In the example above, for rotational equilibrium:

$$f = 2.25 \text{ N}$$

Equilibrium examples

CENTRE OF GRAVITY

The effect of gravity on all the different parts of the object can be treated as a single force acting at the object's **centre of gravity**.

If an object is of uniform shape and density, the centre of gravity will be in the middle of the object. If the object is not uniform, then finding its position is not trivial – it is possible for an object's centre of gravity to be outside the object. Experimentally, if you suspend an object from a point and it is free to move, then the centre of gravity will always end up below the point of suspension.

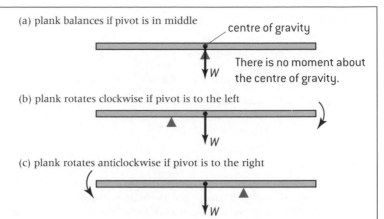

(a) plank balances if pivot is in middle

centre of gravity

There is no moment about the centre of gravity.

(b) plank rotates clockwise if pivot is to the left

(c) plank rotates anticlockwise if pivot is to the right

EXAMPLE 1

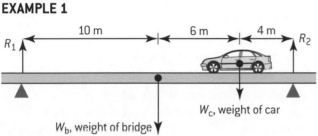

W_c, weight of car

W_b, weight of bridge

When a car goes across a bridge, the forces (on the bridge) are as shown.

Taking moments about right-hand support:
clockwise moment = anticlockwise moment

$$(R_1 \times 20\,m) = (W_b \times 10\,m) + (W_c \times 4\,m)$$

Taking moments about left-hand support:

$$(R_2 \times 20\,m) = (W_b \times 10\,m) + (W_c \times 16\,m)$$

Also, since bridge is not accelerating:

$$R_1 + R_2 = W_b + W_c$$

When solving problems to do with rotational equilibrium remember:

- All forces at an axis have zero moment about that axis.
- You do not have to choose the pivot as the axis about which you calculate torques, but it is often the simplest thing to do (for the reason above).
- You need to remember the sense (clockwise or anticlockwise).
- When solving two-dimensional problems it is sufficient to show that an object is in rotational equilibrium about any ONE axis.
- Newton's laws still apply. Often an object is in rotational AND in translational equilibrium. This can provide a simple way of finding an unknown force.
- The weight of an object can be considered to be concentrated at its centre of gravity.
- If the problem only involves three non-parallel forces, the lines of action of all the forces must meet at a single point in order to be in rotational equilibrium.

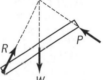

3 forces must meet at a point if in equilibrium

EXAMPLE 2

A ladder of length 5.0 m leans against a smooth wall (no friction) at an angle of 30° to the vertical.

a) Explain why the ladder can only stay in place if there is friction between the ground and the ladder.

b) What is the minimum coefficient of static fraction between the ladder and the ground for the ladder to stay in place?

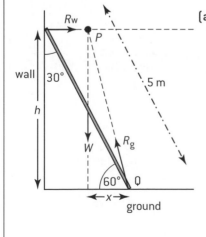

(a) The reaction from the wall, R_W and the ladder's weight meet at point P. For equilibrium the force from the ground, R_g must also pass through this point (for zero torque about P). ∴ R_g is as shown and has a horizontal component (i.e. friction must be acting)

(b) Equilibrium conditions:-

$(\uparrow)$ $W = R_v$ ①

$(\rightarrow)$ $R_H = R_w$ ②

moments about Q $R_w h = Wx$ ③

$$F_f \leq \mu_s R$$

$$\therefore R_H \leq \mu_s R_v$$

using ① & ② $\Rightarrow \mu_s \geq \dfrac{R_w}{W}$

③ $\Rightarrow \mu_s \geq \dfrac{x}{h} = \dfrac{2.5 \cos 60}{5.0 \sin 60}$

$$\therefore \mu_s \geq 0.29$$

Newton's second law – moment of inertia

NEWTON'S SECOND LAW – DEFINITION OF MOMENT OF INERTIA

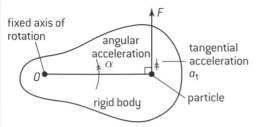

fixed axis of rotation

angular acceleration α

tangential acceleration a_t

particle

rigid body

Newton's second law as applied to one particle in a rigid body

Newton's second law applies to every particle that makes up a large object and must also apply if the object is undergoing rotational motion. In the diagram above, the object is made up of lots of small particles each with a mass m. F is the *tangential component* of the resultant force that acts *on one particle*. The other component, the radial component, cannot produce angular acceleration so it is not included. For this particle we can apply Newton's second law:

$F = m\,a_t = mr\alpha$

so torque $\Gamma = (mr\alpha)r = mr^2\,\alpha$

Similar equations can be created for all the particles that make up the object and summed together:

$\sum \Gamma = \sum mr^2\,\alpha$

or $\sum \Gamma_{ext} = \alpha \sum mr^2$ (1)

Note that:

- Newton's third law applies and, when summing up all the torques, the internal torques (which result from the internal forces between particles) must sum to zero. Only the external torques are left.

- Every particle in the object has the same angular acceleration, α.

The moment of inertial, I, of an object about a particular axis is defined by the summation below:

the distance of the particle from the axis or rotation

moment of inertia

$$I = \sum mr^2$$

mass of an individual particle in the object

Note that moment of inertia, I, is

- A scalar quantity

- Measured in kg m^2 (not kg m^{-2})

- Dependent on:

  ◊ The mass of the object

  ◊ The way this mass is distributed

  ◊ The axis of rotation being considered.

Using this definition, equation 1 becomes:

resultant external torque in N m

angular acceleration in rad s^{-2}

$$\Gamma = I\alpha$$

moment of inertia in kg m^2

This is Newton's second law for rotational motion and can be compared to $F = ma$

MOMENTS OF INERTIA FOR DIFFERENT OBJECTS

Equations for moments of inertia in different situations do not need to be memorized.

Object	Axis of rotation	moment of inertia	Object	Axis of rotation	moment of inertia
thin ring (simple wheel)	through centre, perpendicular to plane	mr^2	Sphere	through centre	$\frac{2}{5}\,mr^2$
thin ring	through a diameter	$\frac{1}{2}\,mr^2$			
disc and cylinder (solid flywheel)	through centre, perpendicular to plane	$\frac{1}{2}\,mr^2$	Rectangular lamina	Through the centre of mass, perpendicular to the plane of the lamina	$m\left(\frac{l^2 + h^2}{12}\right)$
thin rod, length d	through centre, perpendicular to rod	$\frac{1}{12}\,md^2$			

EXAMPLE

A torque of 30 N m acts on a wheel with moment of inertia 600 kg m^2. The wheel starts off at rest.

a) What angular acceleration is produced?

b) The wheel has a radius of 40 cm. After 1.5 minutes:

i. what is the angular velocity of the wheel?

ii. how fast is a point on the rim moving?

a) $\Gamma = I\alpha \Rightarrow \alpha = \frac{\Gamma}{I} = \frac{30}{600} = 5.0 \times 10^{-2}$ rad s^{-2}

b) i. $\omega = \alpha t = 5.0 \times 10^{-2} \times 90 = 4.5$ rad s^{-1}

ii. $v = r\omega = 0.4 \times 4.5 = 1.8$ m s^{-1}

Rotational dynamics

ENERGY OF ROTATIONAL MOTION

Energy considerations often provide simple solutions to complicated problems. When a torque acts on an object, work is done. In the absence of any resistive torque, the work done on the object will be stored as rotational kinetic energy.

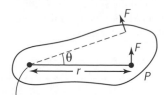

axis of rotation

Calculation of work done by a torque

In the situation above, a force F is applied and the object rotates. As a result, an angular displacement of θ occurs. The work done, W, is calculated as shown below:

$W = F \times$ (distance along arc) $= F \times r\theta = \Gamma\theta$

Using $\Gamma = I\alpha$ we know that $W = I\alpha\theta$

We can apply the constant angular acceleration equation to substitute for $\alpha\theta$:

$\omega_f^2 = \omega_i^2 + 2\alpha\theta$

$\therefore W = I\left(\dfrac{\omega_f^2}{2} - \dfrac{\omega_i^2}{2}\right) = \frac{1}{2}I\omega_f^2 - \frac{1}{2}I\omega_i^2$

This means that we have an equation for rotational KE:

$E_{K_{rot}} = \frac{1}{2}I\omega^2$

Work done by the torque acting on object = change in rotational KE of object

The total KE is equal to the sum of translational KE and the rotational KE:

Total KE = translational KE + rotational KE

Total KE $= \frac{1}{2}Mv^2 + \frac{1}{2}I\omega^2$

ANGULAR MOMENTUM

For a single particle

The linear momentum, p, of a particle of mass m which has a tangential speed v is $m\,v$.

The angular momentum, L, is defined as the *moment* of the linear momentum about the axis of rotation

Angular momentum, $L = (mv)r = (mr\omega)r = (mr^2)\omega$

For a larger object

The angular momentum L of an object about an axis of rotation is defined as

Angular momentum, $L = \sum(mr^2)\omega$

$L = I\omega$

Note that total angular momentum, L, is:

- a vector (in the same way that a torque is considered to be a vector for calculations)
- measured in kg m² s⁻¹ or N m s
- dependent on all rotations taking place. For example, the total angular momentum of a planet orbiting a star would involve:
  ◊ the spinning of the planet about an axis through the planet's centre of mass and
  ◊ the orbital angular momentum about an axis through the star.

CONSERVATION OF ANGULAR MOMENTUM

In exactly the same way that Newton's laws can be applied to linear motion to derive:

- the concept of the impulse of a force
- the relationship between impulse and change in momentum
- the law of conservation of linear momentum,

then Newton's laws can be applied to angular situations to derive:

- The concept of the **angular impulse**:

  Angular impulse is the product of torque and the time for which the torque acts:

  angular impulse $= \Gamma\Delta t$

  If the torque varies with time then the total angular impulse given to an object can be estimated from the area under the graph showing the variation of torque with time. This is analogous to estimating the total impulse given to an object as a result of a varying force (see page 23).

- The relationship between angular impulse and change in angular momentum:

  angular impulse applied to an object = change of angular momentum experienced by the object

- The law of conservation of angular momentum.

The total angular momentum of a system remains constant provided no resultant external torque acts.

Examples:

a) A skater who is spinning on a vertical axis down their body can reduce their moment of inertia by drawing in their arms. This allows their mass to be redistributed so that the mass of the arms is no longer at a significant distance from the axis of rotation thus reducing Σmr^2.

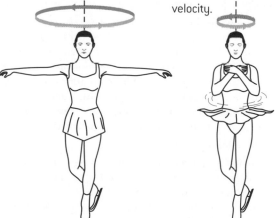

Extended arms mean larger radius and smaller velocity of rotation.

Bringing in her arms decreases her moment of inertia and therefore increases her rotational velocity.

b) The Earth–Moon system produces tides in the oceans. As a result of the relative movement of water, friction exists between the oceans and Earth. This provides a torque that acts to reduce the Earth's spin on its own axis and thus reduces the Earth's angular momentum. The conservation of angular momentum means that there must be a corresponding increase in the orbital angular momentum of the Earth–Moon system. As a result, the Earth–Moon separation is slowly increasing.

Solving rotational problems

SUMMARY COMPARISON OF EQUATIONS OF LINEAR AND ROTATIONAL MOTION

Every equation for linear motion has a corresponding angular equivalent:

	Linear motion	Rotational motion
Physics principles	A resultant external force on a point object causes acceleration. The value of the acceleration is determined by the mass and the resultant force.	A resultant external torque on an extended object causes rotational acceleration. The value of the angular acceleration is determined by the moment of inertia and the resultant torque.
Newton's second law	$F = m\,a$	$\Gamma = I\,\alpha$
Work done	$W = F\,s$	$W = \Gamma\,\theta$
Kinetic energy	$E_K = \frac{1}{2}\,m\,v^2$	$E_{K_{rot}} = \frac{1}{2}\,I\,\omega^2$
Power	$P = F\,v$	$P = \Gamma\,\omega$
Momentum	$p = m\,v$	$L = I\,\omega$
Conservation of momentum	The total linear momentum of a system remains constant provided no resultant external force acts.	The total angular momentum of a system remains constant provided no resultant external torque acts.
Symbols used	Resultant force $\qquad F$ Mass $\qquad m$ Acceleration $\qquad a$ Displacement $\qquad s$ Velocity $\qquad v$ Linear momentum $\qquad p$	Resultant torque $\qquad \Gamma$ Moment of inertia $\qquad I$ Angular acceleration $\qquad \alpha$ Angular displacement $\qquad \theta$ Angular velocity $\qquad \omega$ Angular momentum $\qquad L$

PROBLEM SOLVING AND GRAPHICAL WORK

When analysing any rotational situation, the simplest approach is to imagine the equivalent linear situation and use the appropriate equivalent relationships.

a) Graph of angular displacement vs time

This graph is equivalent to a graph of linear displacement vs time. In the linear situation, the area under the graph does not represent any useful quantity and the gradient of the line at any instant is equal to the instantaneous velocity (see page 10). **Thus the gradient of an angular displacement vs time graph gives the instantaneous angular velocity.**

b) Graph of angular velocity vs time

This graph is equivalent to a graph of linear velocity vs time. In the linear situation, the area under the graph represents the distance gone and the gradient of the line at any instant is equal to the instantaneous acceleration (see page 10). **Thus the area under an angular velocity vs time graph gives the total angular displacement and the gradient of an angular velocity vs time graph gives the instantaneous angular acceleration.**

c) Graph of torque vs time

This graph is equivalent to a graph of force vs time. In the linear situation, the area under the graph represents the total impulse given to the object which is equal to the change of momentum of the object (see page 23). **Thus the area under the torque vs time graph represents the total angular impulse given to the object which is equal to the change of angular momentum.**

EXAMPLE

A solid cylinder, initially at rest, rolls down a 2.0 m long slope of angle 30° as shown in the diagram below:

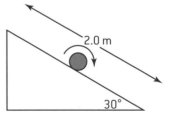

The mass of the cylinder is m and the radius of the cylinder is R. Calculate the velocity of the cylinder at the bottom of the slope.

Answer:

Vertical height fallen by cylinder $= 2.0 \sin 30 = 1.0$ m

$$PE \text{ lost} = mgh$$

$$KE \text{ gained} = \frac{1}{2}\,mv^2 + \frac{1}{2}\,I\omega^2$$

$$\text{but} \quad I = \frac{1}{2}\,mR^2 \quad \text{(cylinder) see page 156}$$

$$\text{and } \omega = \frac{v}{R}$$

$$\Rightarrow \quad KE \text{ gained} = \frac{1}{2}\,mv^2 + \frac{1}{2}\,\frac{mR^2}{2}\cdot\frac{v^2}{R^2}$$

$$= \frac{1}{2}\,mv^2 + \frac{1}{4}\,mv^2$$

$$= \frac{3}{4}\,mv^2$$

Conservation of energy

$$\Rightarrow \quad mgh = \frac{3}{4}\,mv^2$$

$$\therefore \quad v = \sqrt{4\frac{gh}{3}}$$

$$= \sqrt{\frac{4 \times 9.8 \times 1.0}{3}}$$

$$= 3.61 \text{ m s}^{-1}$$

Thermodynamic systems and concepts

DEFINITIONS

Historically, the study of the behaviour of ideal gases led to some very fundamental concepts that are applicable to many other situations. These laws, otherwise known as the laws of **thermodynamics,** provide the modern physicist with a set of very powerful intellectual tools.

The terms used need to be explained.

Thermodynamic system	Most of the time when studying the behaviour of an ideal gas in particular situations, we focus on the macroscopic behaviour of the gas as a whole. In terms of work and energy, the gas can gain or lose thermal energy and it can do work or work can be done on it. In this context, the gas can be seen as a **thermodynamic system**.
The surroundings	If we are focusing our study on the behaviour of an ideal gas, then everything else can be called its **surroundings**. For example the expansion of a gas means that work is done by the gas on the surroundings (see below).

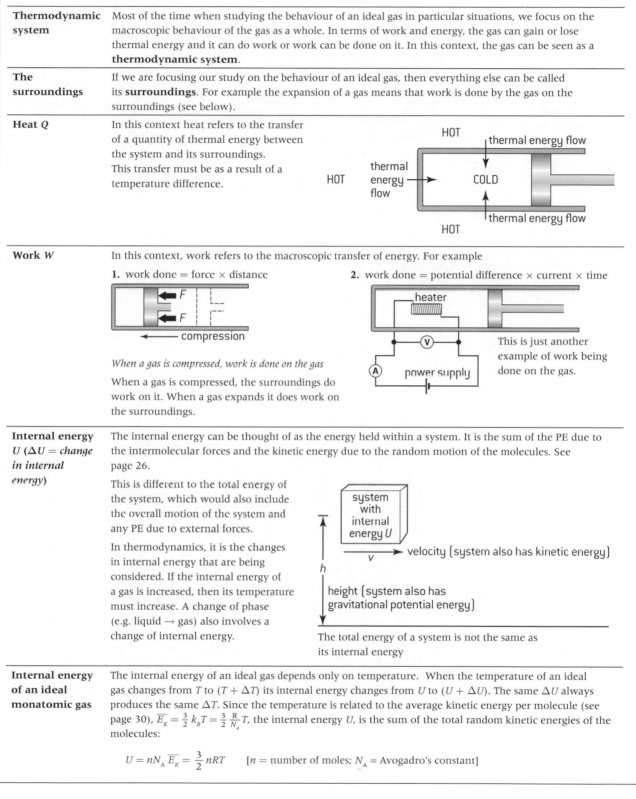

Heat Q	In this context heat refers to the transfer of a quantity of thermal energy between the system and its surroundings. This transfer must be as a result of a temperature difference.
Work W	In this context, work refers to the macroscopic transfer of energy. For example

1. work done = force × distance

When a gas is compressed, work is done on the gas

When a gas is compressed, the surroundings do work on it. When a gas expands it does work on the surroundings.

2. work done = potential difference × current × time

This is just another example of work being done on the gas.

Internal energy U (ΔU = *change in internal energy*)	The internal energy can be thought of as the energy held within a system. It is the sum of the PE due to the intermolecular forces and the kinetic energy due to the random motion of the molecules. See page 26. This is different to the total energy of the system, which would also include the overall motion of the system and any PE due to external forces. In thermodynamics, it is the changes in internal energy that are being considered. If the internal energy of a gas is increased, then its temperature must increase. A change of phase (e.g. liquid → gas) also involves a change of internal energy.

The total energy of a system is not the same as its internal energy

Internal energy of an ideal monatomic gas	The internal energy of an ideal gas depends only on temperature. When the temperature of an ideal gas changes from T to $(T + \Delta T)$ its internal energy changes from U to $(U + \Delta U)$. The same ΔU always produces the same ΔT. Since the temperature is related to the average kinetic energy per molecule (see page 30), $\overline{E_K} = \frac{3}{2} k_B T = \frac{3}{2} \frac{R}{N_A} T$, the internal energy U, is the sum of the total random kinetic energies of the molecules:

$$U = nN_A \overline{E_K} = \frac{3}{2} nRT \qquad [n = \text{number of moles}; \ N_A = \text{Avogadro's constant}]$$

Work done by an ideal gas

WORK DONE DURING EXPANSION AT CONSTANT PRESSURE

Whenever a gas expands, it is doing work on its surroundings. If the pressure of the gas is changing all the time, then calculating the amount of work done is complex. This is because we cannot assume a constant force in the equation of work done (work done = force × distance). If the pressure changes then the force must also change. If the pressure is constant then the force is constant and we can calculate the work done.

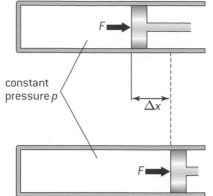

constant pressure p

Work done W = force × distance

$$= F\Delta x$$

Since pressure $= \dfrac{\text{force}}{\text{area}}$

$$F = pA$$

therefore

$$W = pA\Delta x$$

$$but \ A\Delta x = \Delta V$$

$$so \ work \ done = p\Delta V$$

So if a gas increases its volume (ΔV is positive) then the gas does work (W is positive)

pV DIAGRAMS AND WORK DONE

It is often useful to represent the changes that happen to a gas during a thermodynamic process on a pV diagram. An important reason for choosing to do this is that the area under the graph represents the work done. The reasons for this are shown below.

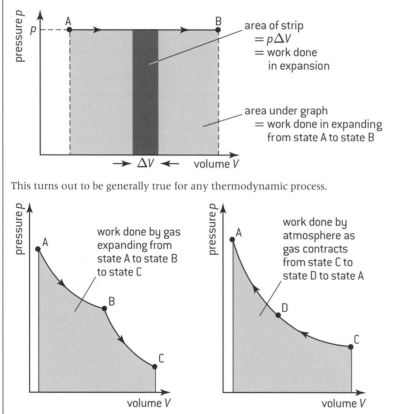

area of strip
$= p\Delta V$
$=$ work done in expansion

area under graph
$=$ work done in expanding from state A to state B

This turns out to be generally true for any thermodynamic process.

work done by gas expanding from state A to state B to state C

work done by atmosphere as gas contracts from state C to state D to state A

The first law of thermodynamics

FIRST LAW OF THERMODYNAMICS

There are three fundamental laws of thermodynamics. The first law is simply a statement of the principle of energy conservation as applied to the system. If an amount of thermal energy Q is given to a system, then one of two things must happen (or a combination of both). The system can increase its internal energy ΔU or it can do work W.

As energy is conserved

$$Q = \Delta U + W$$

It is important to remember what the signs of these symbols mean. They are all taken from the system's 'point of view'.

Q If this is **positive**, then thermal energy is going into the system.
If it is **negative**, then thermal energy is going out of the system.

ΔU If this is **positive**, then the internal energy of the system is **increasing**. (The temperature of the gas is increasing.)
If it is **negative**, the internal energy of the system is **decreasing**.(The temperature of the gas is decreasing.)

W If this is **positive**, then the **system is doing work** on the surroundings.(The gas is expanding.)
If it is **negative**, the **surroundings are doing work** on the system. (The gas is contracting.)

IDEAL GAS PROCESSES

A gas can undergo any number of different types of change or process. Four important processes are considered below. In each case the changes can be represented on a pressure–volume diagram and the first law of thermodynamics must apply. To be precise, these diagrams represent a type of process called a reversible process.

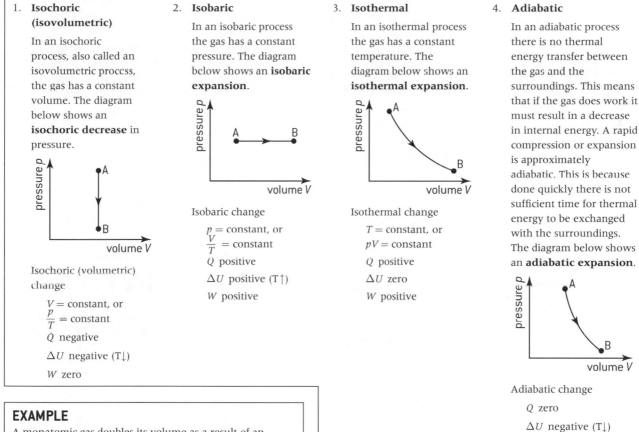

1. **Isochoric (isovolumetric)**

   In an isochoric process, also called an isovolumetric proccss, the gas has a constant volume. The diagram below shows an **isochoric decrease** in pressure.

   Isochoric (volumetric) change

   $V = $ constant, or
   $\frac{p}{T} = $ constant
   Q negative
   ΔU negative (T↓)
   W zero

2. **Isobaric**

   In an isobaric process the gas has a constant pressure. The diagram bclow shows an **isobaric expansion**.

   Isobaric change

   $p = $ constant, or
   $\frac{V}{T} = $ constant
   Q positive
   ΔU positive (T↑)
   W positive

3. **Isothermal**

   In an isothermal process the gas has a constant temperature. The diagram below shows an **isothermal expansion**.

   Isothermal change

   $T = $ constant, or
   $pV = $ constant
   Q positive
   ΔU zero
   W positive

4. **Adiabatic**

   In an adiabatic process there is no thermal energy transfer between the gas and the surroundings. This means that if the gas does work it must result in a decrease in internal energy. A rapid compression or expansion is approximately adiabatic. This is because done quickly there is not sufficient time for thermal energy to be exchanged with the surroundings. The diagram below shows an **adiabatic expansion**.

   Adiabatic change

   Q zero
   ΔU negative (T↓)
   W positive

   For a monatomic gas, the equation for an adiabatic process is

   $pV^{\frac{5}{3}} = $ constant

EXAMPLE

A monatomic gas doubles its volume as a result of an adiabatic expansion. What is the change in pressure?

$$p_1 V_1^{\frac{5}{3}} = p_2 V_2^{\frac{5}{3}}$$

$$\frac{p_2}{p_1} = \left(\frac{V_1}{V_2}\right)^{\frac{5}{3}}$$

$$= 0.5^{\frac{5}{3}}$$

$$= 0.31$$

∴ final pressure = 31% of initial pressure

Second law of thermodynamics and entropy

SECOND LAW OF THERMODYNAMICS

Historically the **second law of thermodynamics** has been stated in many different ways. All of these versions can be shown to be equivalent to one another.

In principle there is nothing to stop the complete conversion of thermal energy into useful work. In practice, a gas can not continue to expand forever – the apparatus sets a physical limit. Thus **the continuous conversion of thermal energy into work requires a cyclical process – a heat engine.**

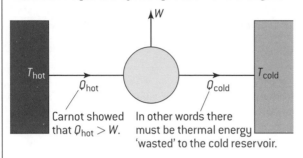

Carnot showed that $Q_{hot} > W$.

In other words there must be thermal energy 'wasted' to the cold reservoir.

This realization leads to possibly the simplest formulation of the second law of thermodynamics (the **Kelvin–Planck** formulation).

> **No heat engine, operating in a cycle, can take in heat from its surroundings and totally convert it into work.**

Other possible formulations include the following:

> **No heat pump can transfer thermal energy from a low-temperature reservoir to a high-temperature reservoir without work being done on it** (Clausius).

> **Heat flows from hot objects to cold objects.**

The concept of **entropy** leads to one final version of the second law.

> **The entropy of the Universe can never decrease.**

EXAMPLES

The first and second laws of thermodynamics both must apply to all situations. Local decreases of entropy are possible so long as elsewhere there is a corresponding increase.

1. A refrigerator is an example of a heat pump.

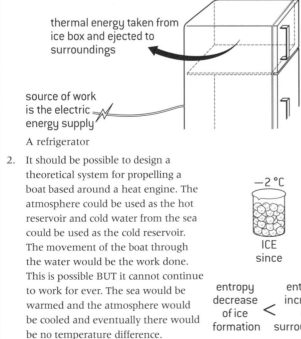

thermal energy taken from ice box and ejected to surroundings

source of work is the electric energy supply

A refrigerator

2. It should be possible to design a theoretical system for propelling a boat based around a heat engine. The atmosphere could be used as the hot reservoir and cold water from the sea could be used as the cold reservoir. The movement of the boat through the water would be the work done. This is possible BUT it cannot continue to work for ever. The sea would be warmed and the atmosphere would be cooled and eventually there would be no temperature difference.

ENTROPY AND ENERGY DEGRADATION

Entropy is a property that expresses the disorder in the system.

The details are not important but the entropy S of a system is linked to the number of possible arrangements W of the system. [$S = k_B \ln(W)$]

Because molecules are in random motion, one would expect roughly equal numbers of gas molecules in each side of a container.

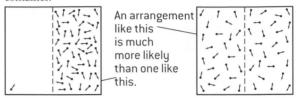

An arrangement like this is much more likely than one like this.

The number of ways of arranging the molecules to get the set-up on the right is greater than the number of ways of arranging the molecules to get the set-up on the left. This means that the entropy of the system on the right is greater than the entropy of the system on the left.

In any random process the amount of disorder will tend to increase. In other words, the total entropy will always increase. The entropy change ΔS is linked to the thermal energy change ΔQ and the temperature T. ($\Delta S = \frac{\Delta Q}{T}$)

thermal energy flow

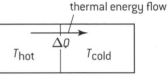

$$\text{decrease of entropy} = \frac{\Delta Q}{T_{hot}} \qquad \text{increase of entropy} = \frac{\Delta Q}{T_{cold}}$$

When thermal energy flows from a hot object to a colder object, overall the total entropy has increased.

In many situations the idea of energy **degradation** is a useful concept. The more energy is shared out, the more degraded it becomes – it is harder to put it to use. For example, the internal energy that is 'locked' up in oil can be released when the oil is burned. In the end, all the energy released will be in the form of thermal energy – shared among many molecules. It is not feasible to get it back.

3. Water freezes at 0 °C because this is the temperature at which the entropy increase of the surroundings (when receiving the latent heat) equals the entropy decrease of the water molecules becoming more ordered. It would not freeze at a higher temperature because this would mean that the overall entropy of the system would decrease.

increasing temperature of surroundings

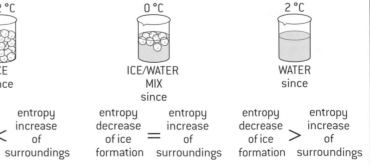

−2 °C	0 °C	2 °C
ICE since	ICE/WATER MIX since	WATER since
entropy decrease of ice formation < entropy increase of surroundings	entropy decrease of ice formation = entropy increase of surroundings	entropy decrease of ice formation > entropy increase of surroundings

Heat engines and heat pumps

HEAT ENGINES

A central concept in the study of thermodynamics is the **heat engine**. A heat engine is any device that uses a source of thermal energy in order to do work. It converts heat into work. The internal combustion engine in a car and the turbines that are used to generate electrical energy in a power station are both examples of heat engines. A block diagram representing a generalized heat engine is shown below.

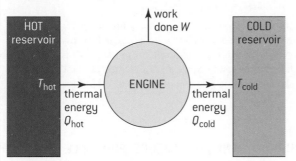

Heat engine

In this context, the word **reservoir** is used to imply a constant temperature source (or sink) of thermal energy. Thermal energy can be taken from the hot reservoir without causing the temperature of the hot reservoir to change. Similarly thermal energy can be given to the cold reservoir without increasing its temperature.

An ideal gas can be used as a heat engine. The pV diagram right represents a simple example. The four-stage cycle returns the gas to its starting conditions, but the gas has done work. The area enclosed by the cycle represents the amount of work done.

In order to do this, some thermal energy must have been taken from a hot reservoir (during the isovolumetric increase in pressure and the isobaric expansion). A different amount of thermal energy must have been ejected to a cold reservoir (during the isovolumetric decrease in pressure and the isobaric compression).

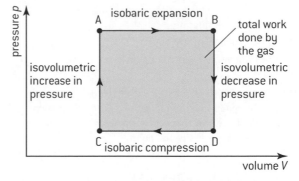

The thermal efficiency of a heat engine is defined as

$$\eta = \frac{\text{work done}}{\text{(thermal energy taken from hot reservoir)}}$$

This is equivalent to

$$\eta = \frac{\text{rate of doing work}}{\text{(thermal power taken from hot reservoir)}}$$

$$\eta = \frac{\text{useful work done}}{\text{energy input}}$$

The cycle of changes that results in a heat engine with the maximum possible efficiency is called the **Carnot cycle**.

HEAT PUMPS

A **heat pump** is a heat engine being run in reverse. A heat pump causes thermal energy to be moved from a cold reservoir to a hot reservoir. In order for this to be achieved, mechanical work must be done.

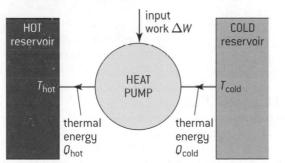

Heat pump

Once again an ideal gas can be used as a heat pump. The thermodynamic processes can be exactly the same ones as were used in the heat engine, but the processes are all opposite. This time an anticlockwise circuit will represent the cycle of processes.

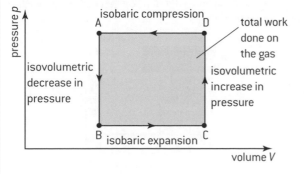

CARNOT CYCLES AND CARNOT THEOREM

The Carnot cycle represents the cycle of processes for a theoretical heat engine with the maximum possible efficiency. Such an idealized engine is called a **Carnot engine**.

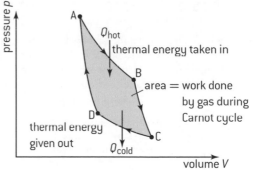

Carnot cycle

It consists of an ideal gas undergoing the following processes.

- Isothermal expansion (A → B)
- Adiabatic expansion (B → C)
- Isothermal compression (C → D)
- Adiabatic compression (D → A)

The temperatures of the hot and cold reservoirs fix the maximum possible efficiency that can be achieved.

The efficiency of a Carnot engine can be shown to be

$$\eta_{\text{Carnot}} = 1 - \frac{T_{\text{cold}}}{T_{\text{hot}}} \quad \text{(where T is in kelvin)}$$

An engine operates at 300 °C and ejects heat to the surroundings at 20 °C. The maximum possible theoretical efficiency is

$$\eta_{\text{Carnot}} = 1 - \frac{293}{573} = 0.49 = 49\%$$

ⓗ Fluids at rest

DEFINITIONS OF DENSITY AND PRESSURE

The symbol representing density is the Greek letter rho, ρ. The average density of a substance is defined by the following equation:

$$\underset{\text{average density}}{\rho} = \frac{\overset{\text{mass}}{m}}{\underset{\text{volume}}{V}}$$

- Density is a scalar quantity.
- The SI units of density are kg m^{-3}.
- Densities can also be quoted in g cm^{-3} (see conversion factor below)
- The density of water is $1\,\text{g cm}^{-3} = 1{,}000\,\text{kg m}^{-3}$

Pressure at any point in a fluid (a gas or a liquid) is defined in terms of the force, ΔF, that acts normally (at 90°) to a small area, ΔA, that contains the point.

$$\overset{\text{pressure}}{p} = \frac{\overset{\text{normal force}}{\Delta F}}{\underset{\text{area}}{\Delta A}}$$

- Pressure is a scalar quantity – the force has a direction but the pressure does not. Pressure acts equally in all directions.
- The SI unit of pressure is N m^{-2} or pascals (Pa). $1\,\text{Pa} = 1\,\text{N m}^{-2}$
- Atmospheric pressure $\approx 10^5\,\text{Pa}$
- Absolute pressure is the actual pressure at a point in a fluid. Pressure gauges often record the **difference** between absolute pressure and atmospheric pressure. Thus if a difference pressure gauge gives a reading of $2 \times 10^5\,\text{Pa}$ for a gas, the absolute pressure of the gas is $3 \times 10^5\,\text{Pa}$.

VARIATION OF FLUID PRESSURE

The pressure in a fluid increases with depth. If two points are separated by a vertical distance, d, in a fluid of constant density, ρ_f, then the pressure difference, Δp, between these two points is:

$$\underset{\text{pressure difference due to depth}}{\Delta p} = \rho_f \overset{\text{density of fluid}\quad\text{gravitational field strength}}{g} \underset{\text{depth}}{d}$$

The total pressure at a given depth in a liquid is the addition of the pressure acting at the surface (atmospheric pressure) and the additional pressure due to the depth:

$$\underset{\text{Total pressure}}{P} = \overset{\text{Atmospheric pressure}}{P_0} + \rho_f \underset{\text{gravitational field strength}}{g} \overset{\text{density of fluid}}{d}^{\text{depth}}$$

Note that:

- Pressure can be expressed in terms of the equivalent depth (or head) in a known liquid. Atmospheric pressure is approximately the same as exerted by a 760 mm high column of mercury (Hg) or a 10 m column of water.
- As pressure is dependent on depth, the pressures at two points that are at the same horizontal level in the same liquid must be the same provided they are connected by that liquid and the liquid is static.

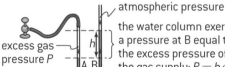

atmospheric pressure

the water column exerts a pressure at B equal to the excess pressure of the gas supply: $P = h\rho g$

excess gas pressure P

- The pressure is independent of the cross-sectional area – this means that liquids will always find their own level.

HYDROSTATIC EQUILIBRIUM

A fluid is in **hydrostatic equilibrium** when it is at rest. This happens when all the forces on a given volume of fluid are balanced. Typically external forces (e.g. gravity) are balanced by a pressure gradient across the volume of fluid (pressure increases with depth – see above).

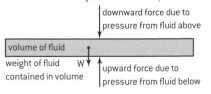

downward force due to pressure from fluid above

volume of fluid

weight of fluid W contained in volume

upward force due to pressure from fluid below

BUOYANCY AND ARCHIMEDES' PRINCIPLE

Archimedes' principle states that when a body is immersed in a fluid, it experiences a buoyancy upthrust equal in magnitude to the weight of the fluid displaced. $B = \rho_f V_f g$

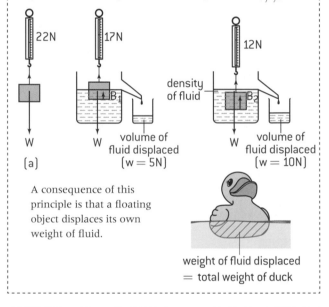

A consequence of this principle is that a floating object displaces its own weight of fluid.

weight of fluid displaced = total weight of duck

PASCAL'S PRINCIPLE

Pascal's principle states that the pressure applied to an enclosed liquid is transmitted to every part of the liquid, whatever the shape it takes. This principle is central to the design of many **hydraulic** systems and is different to how solids respond to forces.

When a solid object (e.g. an incompressible stick) is pushed at one end and its other end is held in place, then the same force will be exerted on the restraining object.

Incompressible solids transmit forces whereas incompressible liquids transmit pressures.

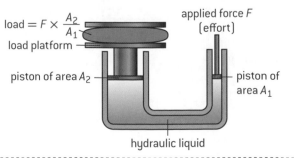

$\text{load} = F \times \dfrac{A_2}{A_1}$

applied force F (effort)

load platform

piston of area A_2

piston of area A_1

hydraulic liquid

ⓗ Fluids in motion – Bernoulli effect

THE IDEAL FLUID

In most real situations, fluid flow is extremely complicated. The following properties define an ideal fluid that can be used to create a simple model. This simple model can be later refined to be more realistic.

An ideal fluid:

- Is **incompressible** – thus its density will be constant.

- Is **non-viscous** – as a result of fluid flow, no energy gets converted into thermal energy. See page 167 for the definition of the viscosity of a real fluid.

- Involves a **steady flow** (as opposed to a **turbulent**, or chaotic, flow) of fluid. Under these conditions the flow is laminar (see box below). See page 167 for an analysis of turbulent flow.

- Does **not have angular momentum** – it does not rotate.

LAMINAR FLOW, STREAMLINES AND THE CONTINUITY EQUATION

When the flow of a liquid is steady or **laminar**, different parts of the fluid can have different instantaneous velocities. The flow is said to be laminar if every particle that passes through a given point has the same velocity whenever the observation is made. The opposite of **laminar flow**, turbulent flow, takes place when the particles that pass through a given point have a wide variation of velocities depending on the instant when the observation is made (see page 167).

A **streamline** is the path taken by a particle in the fluid and laminar flow means that all particles that pass through a given point in the fluid must follow the same streamline. The direction of the tangent to a streamline gives the direction of the instantaneous velocity that the particles of the fluid have at that point. No fluid ever crosses a streamline. Thus a collection of streamlines can together define a **tube of flow**. This is tubular region of fluid where fluid only enters and leaves the tube through its ends and never through its sides.

In a time Δt, the mass, m_1, entering the cross-section A_1 is

$$m_1 = \rho_1 A_1 v_1 \Delta t$$

Similarly the mass, m_2, leaving the cross-section A_2 is

$$m_2 = \rho_2 A_2 v_2 \Delta t$$

Conservation of mass applies to this tube of flow, so

$$\rho_1 A_1 v_1 = \rho_2 A_2 v_2$$

This is an ideal fluid and thus incompressible meaning $\rho_1 = \rho_2$, so

$$A_1 v_1 = A_2 v_2 \text{ or } Av = \text{constant}$$

This is the **continuity equation**.

THE BERNOULLI EFFECT

When a fluid flows into a narrow section of a pipe:

- The fluid must end up moving at a higher speed (continuity equation).

- This means the fluid must have been accelerated forwards.

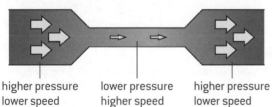

| higher pressure lower speed | lower pressure higher speed | higher pressure lower speed |

- This means there must be a pressure difference forwards with a lower pressure in the narrow section and a higher pressure in the wider section.

Thus an increase in fluid speed must be associated with a decrease in fluid pressure. This is the Bernoulli effect – the greater the speed, the lower the pressure and vice versa.

THE BERNOULLI EQUATION

The Bernoulli equation results from a consideration of the work done and the conservation of energy when an ideal fluid changes:

- its speed (as a result of a change in cross-sectional area)

- its vertical height as a result of work done by the fluid pressure.

The equation identifies a quantity that is always constant along any given streamline:

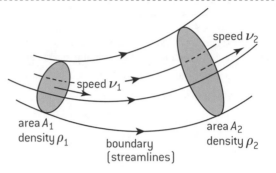

$$\frac{1}{2}\rho v^2 + \rho g z + p = \text{constant}$$

Note that:

- The first term ($\frac{1}{2}\rho v^2$), can be thought of as the dynamic pressure.

- The last two terms ($\rho g z + p$), can be thought of as the static pressure.

- Each term in the equation has several possible units: N m^{-2}, Pa, J m^{-3}.

- The last of the above units leads to a new interpretation for the Bernoulli equation:

$$\text{KE per unit volume} + \text{gravitational PE per unit volume} + \text{pressure} = \text{constant}$$

ⓗ Bernoulli – examples

APPLICATIONS OF THE BERNOULLI EQUATION

a) Flow out of a container

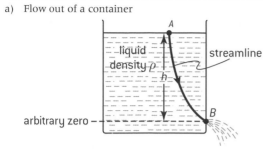

To calculate the speed of fluid flowing out of a container, we can apply Bernoulli's equation to the streamline shown above.

At A, p = atmospheric and v = zero

At B, p = atmospheric and v = ?

$$\frac{1}{2}\rho v^2 + \rho gz + p = \text{constant}$$

$$\therefore 0 + h\rho g + p = \frac{1}{2}\rho v^2 + 0 + p$$

$$v = \sqrt{2gh}$$

b) Venturi tubes

A Venturi meter allows the rate of flow of a fluid to be calculated from a measurement of pressure difference between two different cross-sectional areas of a pipe.

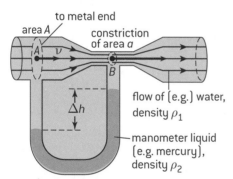

* The pressure difference between A and B can be calculated by taking readings of Δh and ρ_2 from the attached manometer:

$$P_A - P_B = \Delta h \rho_2 g$$

* This value and measurements of A, a and ρ_1 allows the fluid speed at A to be calculated by using Bernoulli's equation and the equation of continuity

$$v = \sqrt{\frac{2\Delta h \rho_2 g}{\left[\rho_1\left(\frac{A}{a}\right)^2 - 1\right]}}$$

* The rate of flow of fluid through the pipe is equal to $A \times v$

c) Fragrance spray

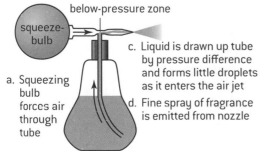

b. Constriction in tube causes low pressure region as air travels faster in this section

below-pressure zone

squeeze-bulb

c. Liquid is drawn up tube by pressure difference and forms little droplets as it enters the air jet

a. Squeezing bulb forces air through tube

d. Fine spray of fragrance is emitted from nozzle

d) Pitot tube to determine the speed of a plane

A pitot tube is attached facing forward on a plane. It has two separate tubes:

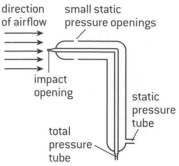

* The front hole (impact opening) is placed in the airstream and measures the total pressure (sometimes called the stagnation pressure), P_T.
* The side hole(s) measures the static pressure, P_s.
* The difference between P_T and P_s, is the dynamic pressure. The Bernoulli equation can be used to calculate airspeed:

$$P_T - P_s = \frac{1}{2}\rho v^2$$

$$v = \sqrt{\frac{2(P_T - P_s)}{\rho}}$$

e) Aerofoil (aka airfoil)

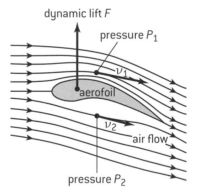

Note that:

* Streamlines closer together above the aerofoil imply a decrease in cross-sectional area of equivalent tubes of flow above the aerofoil.
* Decrease in cross-sectional area of tube of flow implies increased velocity of flow above the aerofoil (equation of continuity). $v_1 > v_2$
* Since $v_1 > v_2$, $P_1 < P_2$
* Bernoulli equation can be used to calculate the pressure different (height difference not relevant) which can support the weight of the aeroplane.
* When angle of attack is too great, the flow over the upper surface can become turbulent. This reduces the pressure difference and leads to the plane 'stalling'.

 **Viscosity**

DEFINITION OF VISCOSITY

An ideal fluid does not resist the relative motion between different layers of fluid. As a result there is no conversion of work into thermal energy during laminar flow and no external forces are needed to maintain a steady rate of flow. Ideal fluids are non-viscous whereas real fluids are viscous. In a viscous fluid, a steady external force is needed to maintain a steady rate of flow (no acceleration). Viscosity is an internal friction between different layers of a fluid which are moving with different velocities.

The definition of the viscosity of a fluid, η, (Greek letter Nu) is in terms of two new quantities, the **tangential stress**, τ, and the **velocity gradient**, $\frac{\Delta v}{\Delta y}$ (see RH side).

The coefficient of viscosity η is defined as:

$$\eta = \frac{\text{tangential stress}}{\text{velocity gradient}} = \frac{F/A}{\Delta v/\Delta y}$$

- The units of η are N s m^{-2} or kg m^{-1} s^{-1} or Pa s
- Typical values at room temperature:
  ◊ Water: 1.0×10^{-3} Pa s
  ◊ Thick syrup: 1.0×10^{2} Pa s
- Viscosity is very sensitive to changes of temperature.

For a class of fluid, called **Newtonian fluids**, experimental measurements show that tangential stress is proportional to velocity gradient (e.g. many pure liquids). For these fluids the coefficient of viscosity is constant provided external conditions remain constant.

A) Tangential stress

The tangential stress is defined as:

$$\tau = \frac{F}{A}$$

- Units of tangential stress are N m^{-2} or Pa

B) Velocity gradient

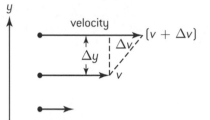

The velocity gradient is defined as:

$$\text{velocity gradient} = \frac{\Delta v}{\Delta y}$$

- Units of velocity gradient are s^{-1}

STOKES' LAW

Stokes' law predicts the viscous drag force F_D that acts on a perfect sphere when it moves through a fluid:

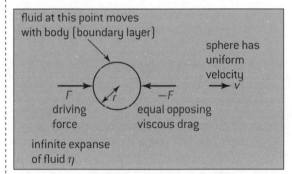

Drag force acting on sphere in N   viscosity of fluid in Pa s

$$F_D = 6\pi\eta r v$$

radius of sphere in m   velocity of sphere in m s^{-1}

Note Stokes' law assumes that:

- The speed of the sphere is small so that:
  ◊ the flow of fluid past the sphere is streamlined
  ◊ there is no slipping between the fluid and the sphere

- The fluid is infinite in volume. Real spheres falling through columns of fluid can be affected by the proximity of the walls of the container.
- The size of the particles of the fluid is very much smaller than the size of the sphere.

The forces on a sphere falling through a fluid at terminal velocity are as shown below:

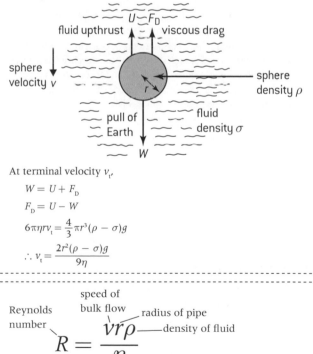

At terminal velocity v_t,

$$W = U + F_D$$
$$F_D = U - W$$
$$6\pi\eta r v_t = \frac{4}{3}\pi r^3(\rho - \sigma)g$$
$$\therefore v_t = \frac{2r^2(\rho - \sigma)g}{9\eta}$$

TURBULENT FLOW – THE REYNOLDS NUMBER

Streamline flow only occurs at low fluid flow rates. At high flow rates the flow becomes turbulent:

laminar   turbulent

It is extremely difficult to predict the exact conditions when fluid flow becomes turbulent. When considering fluid flow down a pipe, a useful number to consider is the Reynolds number, R, which is defined as:

Reynolds number   speed of bulk flow   radius of pipe / density of fluid

$$R = \frac{vr\rho}{\eta}$$

viscosity of fluid

Note that:

- The Reynolds number does not have any units – it is just a ratio.
- Experimentally, fluid flow is often laminar when R < 1000 and turbulent when R > 2000 but precise predictions are difficult.

Ⓗ Forced oscillations and resonance (1)

DAMPING

Damping involves a frictional force that is always in the opposite direction to the direction of motion of an oscillating particle. As the particle oscillates, it does work against this resistive (or dissipative) force and so the particle loses energy. As the total energy of the particle is proportional to the (amplitude)^2 of the SHM, the amplitude decreases exponentially with time.

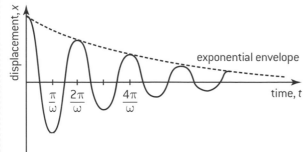

The above example shows the effect of **light damping** (the system is said to be **underdamped**) where the resistive force is small so a small fraction of the total energy is removed each cycle. The time period of the oscillations is not affected and the oscillations continue for a significant number of cycles. The time taken for the oscillations to 'die out' can be long.

Heavy damping or **overdamping** involves large resistive forces (e.g. the SHM taking place in a viscous liquid) and can completely prevent the oscillations from taking place. The time taken for the particle to return to zero displacement can again be long.

Critical damping involves an intermediate value for resistive force such that the time taken for the particle to return to zero displacement is a minimum. Effectively there is no 'overshoot'. Examples of critically damped systems include electric meters with moving pointers and door closing mechanisms.

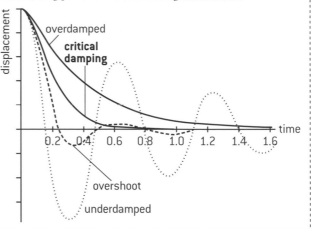

NATURAL FREQUENCY AND RESONANCE

If a system is temporarily displaced from its equilibrium position, the system will oscillate as a result. This oscillation will be at the **natural frequency of vibration** of the system. For example, if you tap the rim of a wine glass with a knife, it will oscillate and you can hear a note for a short while. Complex systems tend to have many possible modes of vibration each with its own natural frequency.

It is also possible to force a system to oscillate at any frequency that we choose by subjecting it to a changing force that varies with the chosen frequency. This periodic driving force must be provided from outside the system. When this **driving frequency** is first applied, a combination of natural and forced oscillations take place which produces complex **transient** oscillations. Once the amplitude of the transient oscillations 'die down', a steady condition is achieved in which:

- The system oscillates at the driving frequency.
- The amplitude of the forced oscillations is fixed. Each cycle energy is dissipated as a result of damping and the driving force does work on the system. The overall result is that the energy of the system remains constant.

- The amplitude of the forced oscillations depends on:
  ◊ the comparative values of the natural frequency and the driving frequency
  ◊ the amount of damping present in the system.

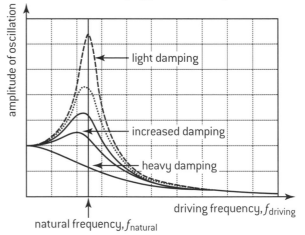

Resonance occurs when a system is subject to an oscillating force at exactly the same frequency as the natural frequency of oscillation of the system.

Q FACTOR AND DAMPING

The degree of damping is measured by a quantity called the quality factor or Q factor. It is a ratio (no units) and the definition is:

$$Q = 2\pi \frac{\text{energy stored}}{\text{energy lost per cycle}}$$

Since the energy stored is proportional to the square of amplitude of the oscillation, measurements of decreasing amplitude with time can be used to calculate the Q factor. The Q factor is approximately equal to the number of oscillations that are completed before damping stops the oscillation.

Typical orders of magnitude for different Q-factors:

Car suspension:	1
Simple pendulum:	10^3
Guitar string:	10^3
Excited atom:	10^7

When a system is in resonance and its amplitude is constant, the energy provided by the driving frequency during one cycle is all used to overcome the resistive forces that cause damping. In this situation, the Q factor can be calculated as:

$$Q = 2\pi \times \text{resonant frequency} \times \frac{\text{energy stored}}{\text{power loss}}$$

⒣ Resonance (2)

PHASE OF FORCED OSCILLATIONS

After transient oscillations have died down, the frequency of the forced oscillations equals the driving frequency. The phase relationship between these two oscillations is complex and depends on how close the driven system is to resonance:

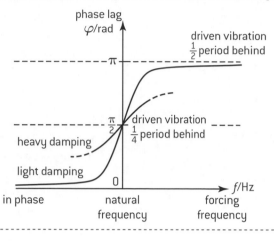

EXAMPLES OF RESONANCE

	Comment
Vibrations in machinery	When in operation, the moving parts of machinery provide regular driving forces on the other sections of the machinery. If the driving frequency is equal to the natural frequency, the amplitude of a particular vibration may get dangerously high. e.g. at a particular engine speed a truck's rear view mirror can be seen to vibrate.
Quartz oscillators	A quartz crystal feels a force if placed in an electric field. When the field is removed, the crystal will oscillate. Appropriate electronics are added to generate an oscillating voltage from the mechanical movements of the crystal and this is used to drive the crystal at its own natural frequency. These devices provide accurate clocks for microprocessor systems.
Microwave generator	Microwave ovens produce electromagnetic waves at a known frequency. The changing electric field is a driving force that causes all charges to oscillate. The driving frequency of the microwaves provides energy, which means that water molecules in particular are provided with kinetic energy – i.e. the temperature is increased.
Radio receivers	Electrical circuits can be designed (using capacitors, resistors and inductors) that have their own natural frequency of electrical oscillations. The free charges (electrons) in an aerial will feel a driving force as a result of the frequency of the radio waves that it receives. Adjusting the components of the connected circuit allows its natural frequency to be adjusted to equal the driving frequency provided by a particular radio station. When the driving frequency equals the circuit's natural frequency, the electrical oscillations will increase in amplitude and the chosen radio station's signal will dominate the other stations.
Musical instruments	Many musical instruments produce their sounds by arranging for a column of air or a string to be driven at its natural frequency which causes the amplitude of the oscillations to increase.
Greenhouse effect	The natural frequency of oscillation of the molecules of greenhouse gases is in the infra-red region. Radiation emitted from the Earth can be readily absorbed by the greenhouse gases in the atmosphere. See page 92 for more details.

IB Questions – option B – engineering physics

1. A sphere of mass m and radius r rolls, without slipping, from rest down an inclined plane. When it reaches the base of the plane, it has fallen a vertical distance h. Show that the speed of the sphere, v, when it arrives at the base of the incline is given by:

   $$v = \sqrt{\frac{10gh}{7}}$$ [4]

2. A flywheel of moment of inertia 0.75 kg m² is accelerated uniformly from rest to an angular speed of 8.2 rad s⁻¹ in 6.5 s.

   a) Calculate the resultant torque acting on the flywheel during this time. [2]

   b) Calculate the rotational kinetic energy of the flywheel when it rotates at 8.2 rad s⁻¹ [2]

   c) The radius of the flywheel is 15 cm. A breaking force applied on the circumference and brings it to rest from an angular speed of 8.2 rad s⁻¹ in exactly 2 revolutions. Calculate the value of the breaking force. [2]

3. A fixed mass of a gas undergoes various changes of temperature, pressure and volume such that it is taken round the p–V cycle shown in the diagram below.

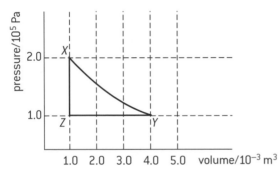

   The following sequence of processes takes place during the cycle.

   X → Y the gas expands at constant temperature and the gas absorbs energy from a reservoir and does 450 J of work.

   Y → Z the gas is compressed and 800 J of thermal energy is transferred from the gas to a reservoir.

   Z → X the gas returns to its initial stage by absorbing energy from a reservoir.

   a) Is there a change in internal energy of the gas during the processes X → Y? Explain. [2]

   b) Is the energy absorbed by the gas during the process X → Y less than, equal to or more than 450 J? Explain. [2]

   c) Use the graph to determine the work done on the gas during the process Y → Z. [3]

   d) What is the change in internal energy of the gas during the process Y → Z? [2]

   e) How much thermal energy is absorbed by the gas during the process Z → X? Explain your answer. [2]

   f) What quantity is represented by the area enclosed by the graph? Estimate its value. [2]

   g) The overall efficiency of a heat engine is defined as

   $$\text{Efficiency} = \frac{\text{net work done by the gas during a cycle}}{\text{total energy absorbed during a cycle}}$$

   If this p–V cycle represents the cycle for a particular heat engine determine the efficiency of the heat engine. [2]

4. In a **diesel** engine, air is initially at a pressure of 1×10^5 Pa and a temperature of 27 °C. The air undergoes the cycle of changes listed below. At the end of the cycle, the air is back at its starting conditions.

   1 An **adiabatic compression** to 1/20th of its original volume.

   2 A brief **isobaric expansion** to 1/10th of its original volume.

   3 An **adiabatic expansion** back to its original volume.

   4 A cooling down at constant volume.

   a) Sketch, with labels, the cycle of changes that the gas undergoes. Accurate values are not required. [3]

   b) If the pressure after the **adiabatic compression** has risen to 6.6×10^6 Pa, calculate the temperature of the gas. [2]

   c) In which of the four processes:

   (i) is work done **on** the gas? [1]

   (ii) is work done **by** the gas? [1]

   (iii) does ignition of the air-fuel mixture take place? [1]

   d) Explain how the 2nd law of thermodynamics applies to this cycle of changes. [2]

HL

5. With the aid of diagrams, explain

   a) What is meant by laminar flow

   b) The Bernoulli effect

   c) Pascal's principle

   d) An ideal fluid [8]

6. Oil, of viscosity 0.35 Pa s and density 0.95 g cm⁻³, flows through a pipe of radius 20 cm at a velocity of 2.2 m s⁻¹. Deduce whether the flow is laminar or turbulent. [4]

7. A pendulum clock maintains a constant amplitude by means of an electric power supply. The following information is available for the pendulum:

   Maximum kinetic energy: 5×10^{-2} J

   Frequency of oscillation: 2 Hz

   Q factor: 30

   Calculate:

   a) The driving frequency of the power supply [3]

   b) The power needed to drive the clock. [3]

Image formation

RAY DIAGRAMS

If an object is placed in front of a plane mirror, an image will be formed.

The process is as follows:

- Light sets off in all directions from every part of the object. (This is a result of diffuse reflections from a source of light.)
- Each ray of light that arrives at the mirror is reflected according to the law of reflection.
- These rays can be received by an observer.
- The location of the image seen by the observer arises because the rays are assumed to have travelled in straight lines.

In order to find the location and nature of this image a ray diagram is needed.

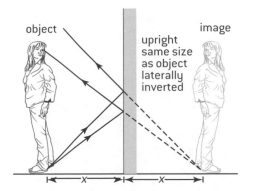

The image formed by reflection in a plane mirror is always:

- **the same distance behind the mirror as the object is in front**
- **upright** (as opposed to being inverted)
- **the same size as the object** (as opposed to being magnified or diminished)
- **laterally inverted** (i.e. left and right are interchanged)
- **virtual** (see below).

REAL AND VIRTUAL IMAGES

The image formed by reflection in a plane mirror is described as a **virtual image**. This term is used to describe images created when rays of light **seem** to come from a single point but in fact they do not pass through that point. In the example above, the rays of light seem to be coming from behind the mirror. They do not, of course, actually pass behind the mirror at all.

The opposite of a virtual image is a **real image**. In this case, the rays of light do actually pass through a single point. Real images cannot be formed by plane mirrors, but they can be formed by concave mirrors or by lenses. For example, if you look into the concave surface of a spoon, you will see an image of yourself. This particular image is

- Upside down
- Diminished
- Real.

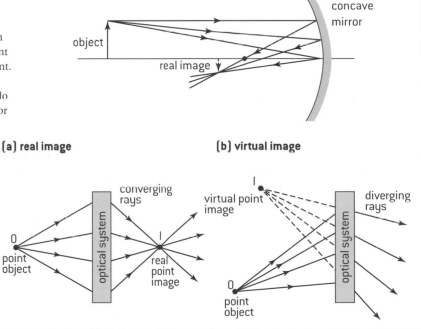

(a) real image

(b) virtual image

STICK IN WATER

The image formed as a result of the refraction of light leaving water is so commonly seen that most people forget that the objects are made to seem strange. A straight stick will appear bent if it is placed in water. The brain assumes that the rays that arrive at one's eyes must have been travelling in a straight line.

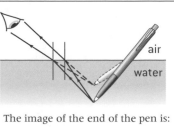

A straight stick appears bent when placed in water

The image of the end of the pen is:

- Nearer to the surface than the pen actually is.
- Virtual.

Converging lenses

CONVERGING LENSES

A converging lens brings parallel rays into one focus point.

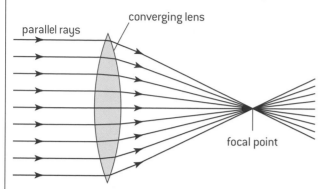

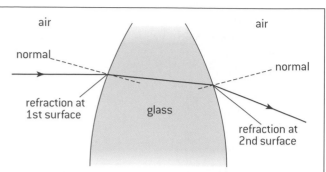

The reason that this happens is the refraction that takes place at both surfaces of the lens.

The rays of light are all brought together in one point because of the particular shape of the lens. Any one lens can be thought of as a collection of different-shaped glass blocks. It can be shown that any thin lens that has surfaces formed from sections of spheres will converge light into one focus point.

A converging lens will always be thicker at the centre when compared with the edges.

POWER OF A LENS

The power of a lens measures the extent to which light is bent by the lens. A higher power lens bends the light more and thus has a smaller focal length. The definition of the power of a lens, P, is the reciprocal of the focal length, f:

$$P = \frac{1}{f}$$

f is the focal length measured in m

P is the power of the lens measured in m^{-1} or **dioptres** (dpt)

A lens of power = +5 dioptre is converging and has a focal length of 20 cm. When two thin lenses are placed close together their powers approximately add.

WAVE MODEL OF IMAGE FORMATION

Formation of a real image by refraction (ignoring diffraction)

DEFINITIONS

When analysing lenses and the images that they form, some technical terms need to be defined.

- The curvature of each surface of a lens makes it part of a sphere. The **centre of curvature** for the lens surface is the centre of this sphere.

- The **principal axis** is the line going directly through the middle of the lens. Technically it joins the centres of curvature of the two surfaces.

- The **focal point** (principal focus) of a lens is the point on the principal axis to which rays that were parallel to the principal axis are brought to focus after passing through the lens. A lens will thus have a focal point on each side.

- The **focal length** is the distance between the centre of the lens and the focal point.

- The **linear magnification**, m, is the ratio between the size (height) of the image and the size (height) of the object. It has no units.

$$\text{linear magnification, } m = \frac{\text{image size}}{\text{object size}} = \frac{h_i}{h_o}$$

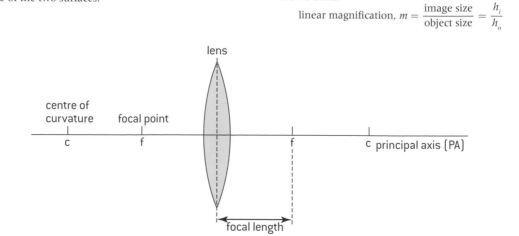

Image formation in convex lenses

IMPORTANT RAYS

In order to determine the nature and position of the image created of a given object, we need to construct a scaled **ray diagram** of the set-up. In order to do this, we concentrate on the paths taken by three particular rays. As soon as the paths taken by two of these rays have been constructed, the paths of all the other rays can be inferred. These important rays are described below.

Converging lens

1. Any ray that was travelling parallel to the principal axis will be refracted towards the focal point on the other side of the lens.

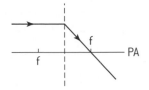

2. Any ray that travelled through the focal point will be refracted parallel to the principal axis.

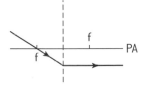

3. Any ray that goes through the centre of the lens will be undeviated.

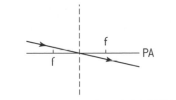

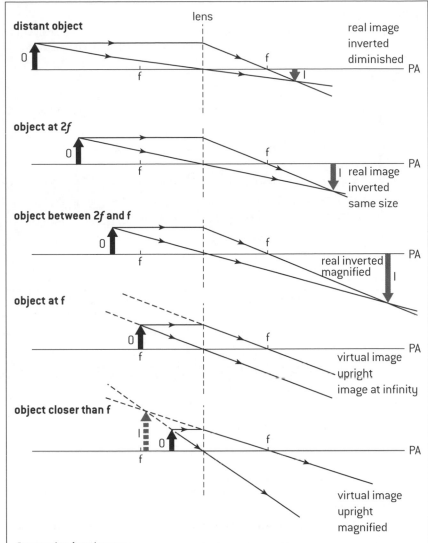

Converging lens images

POSSIBLE SITUATIONS

A ray diagram can be constructed as follows:

*   An upright arrow on the principal axis represents the object.
*   The paths of two important rays from the top of the object are constructed.
*   This locates the position of the top of the image.
*   The bottom of the image must be on the principal axis directly above (or below) the top of the image.

A full description of the image created would include the following information:

*   if it is real or virtual
*   if it is upright or inverted
*   if it is magnified or diminished
*   its exact position.

It should be noted that the important rays are just used to locate the image. The real image also consists of all the other rays from the object. In particular, the image will still be formed even if some of the rays are blocked off.

An observer receiving parallel rays sees an image located in the far distance (at infinity).

Thin lens equation

LENS EQUATION

There is a mathematical method of locating the image formed by a lens. An analysis of the angles involved shows that the following equation can be applied to thin spherical lenses:

$$\frac{1}{f} = \frac{1}{v} + \frac{1}{u}$$

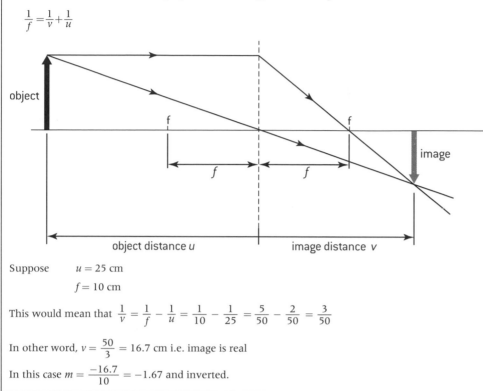

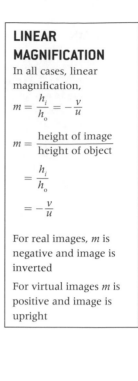

Suppose $u = 25$ cm

$f = 10$ cm

This would mean that $\frac{1}{v} = \frac{1}{f} - \frac{1}{u} = \frac{1}{10} - \frac{1}{25} = \frac{5}{50} - \frac{2}{50} = \frac{3}{50}$

In other word, $v = \frac{50}{3} = 16.7$ cm i.e. image is real

In this case $m = \frac{-16.7}{10} = -1.67$ and inverted.

REAL IS POSITIVE

Care needs to be taken with virtual images. The equation does work but for this to be the case, the following convention has to be followed:

- Distances are taken to be **positive** if actually traversed by the light ray (i.e. distances to real object and image).
- Distances are taken to be **negative** if apparently traversed by the light ray (distances to virtual objects and images).
- Thus a virtual image is represented by a negative value for v – in other words, it will be on the same side of the lens as the object.

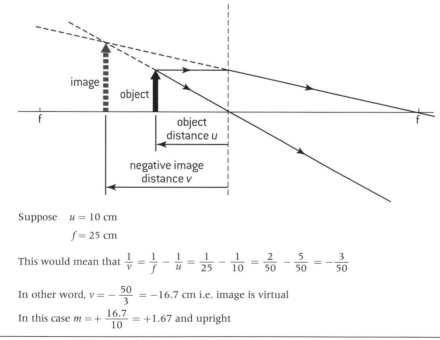

Suppose $u = 10$ cm

$f = 25$ cm

This would mean that $\frac{1}{v} = \frac{1}{f} - \frac{1}{u} = \frac{1}{25} - \frac{1}{10} = \frac{2}{50} - \frac{5}{50} = -\frac{3}{50}$

In other word, $v = -\frac{50}{3} = -16.7$ cm i.e. image is virtual

In this case $m = +\frac{16.7}{10} = +1.67$ and upright

Diverging lenses

DIVERGING LENSES

A diverging lens spreads parallel rays apart. These rays appear to all come from one focus point on the other side of the lens.

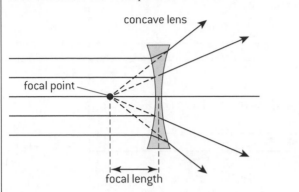

The reason that this happens is the refraction that takes place at both surfaces. A diverging lens will always be thinner at the centre when compared with the edges.

DEFINITIONS AND IMPORTANT RAYS

Diverging lenses have the same analogous definitions as converging lenses for all of the following terms.

Centre of curvature, principal axis, focal point, focal length, linear magnification.

Note that:

- The focal point is the point on the principal axis **from which** rays that were parallel to the principal axis **appear to come** after passing through the lens.

- As the focal point is behind the diverging lens, **the focal length of a diverging lens is negative**.

When constructing ray diagrams for diverging lenses, the important rays whose paths are known (and from which all other ray paths can be inferred) are:

1. Any ray that was travelling parallel to the principal axis will be refracted away from a focal point on the incident side of the lens.

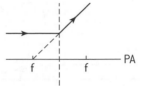

2. Any ray that is heading towards the focal point on the other side of the lens, will be refracted so as to be parallel to the principal axis.

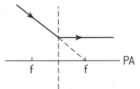

3. Any ray that goes through the centre of the lens will be undeviated.

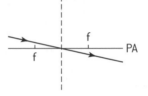

IMAGES CREATED BY A DIVERGING LENS

Whatever the position of the object, a diverging lens will always create an upright, diminished and virtual image located between the focal point and the lens on the same side of the lens as the object.

If you look at an object through a concave lens, it will look smaller and closer.

If you move the object further out, the image will not move as much.

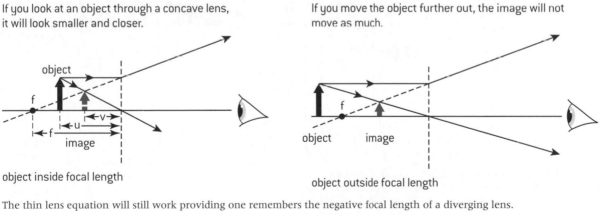

object inside focal length

object outside focal length

The thin lens equation will still work providing one remembers the negative focal length of a diverging lens.

For example, if an object is placed at a distance $2l$ away from a diverging lens of focal length l, the image can be calculated as follows:

Given: $u = 2l, f = -l, v = ?$

$$\frac{1}{u} + \frac{1}{v} = \frac{1}{f}$$

$$\frac{1}{v} = \frac{1}{f} - \frac{1}{u} = \frac{1}{-l} - \frac{1}{2l} = \frac{-3}{2l}$$

$$\therefore v = -\frac{2l}{3}$$

This is a virtual diminished and upright image with $m = +\frac{1}{3}$

Converging and diverging mirrors

GEOMETRY OF MIRRORS AND LENSES

The geometry of the paths of rays after reflection by a spherical concave or convex mirror is exactly analogous to the paths of rays through converging or diverging lenses. The only difference is that mirrors reflect all rays backwards whereas rays pass through lenses and continue forwards.

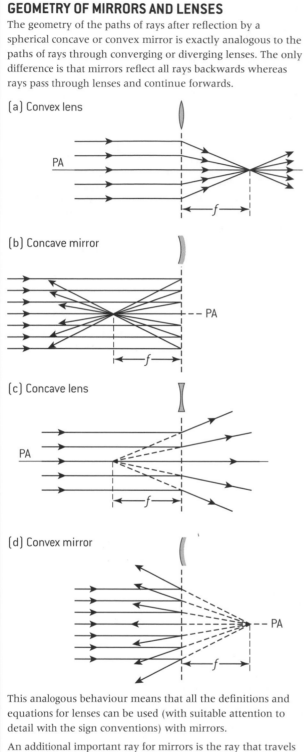

(a) Convex lens

(b) Concave mirror

(c) Concave lens

(d) Convex mirror

This analogous behaviour means that all the definitions and equations for lenses can be used (with suitable attention to detail with the sign conventions) with mirrors.

An additional important ray for mirrors is the ray that travels through (or towards) the centre of curvature of the mirror (located at twice the focal length). This ray will be reflected back along the same path.

IMAGE FORMATION IN MIRRORS

(1) Concave
object at infinity

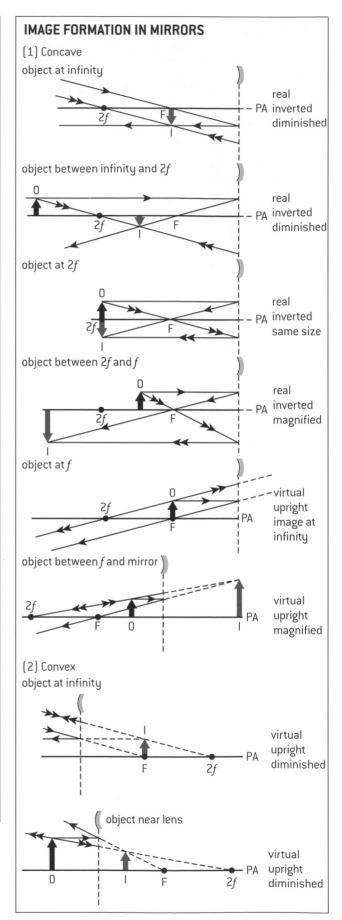

real inverted diminished

object between infinity and 2f

real inverted diminished

object at 2f

real inverted same size

object between 2f and f

real inverted magnified

object at f

virtual upright image at infinity

object between f and mirror

virtual upright magnified

(2) Convex
object at infinity

virtual upright diminished

object near lens

virtual upright diminished

The simple magnifying glass

NEAR AND FAR POINT

The human eye can focus objects at different distances from the eye. Two terms are useful to describe the possible range of distances – the **near point** and the **far point** distance.

- The distance to the **near point** is the distance between the eye and the nearest object that can be brought into clear focus (without strain or help from, for example, lenses). It is also known as the 'least distance of distinct vision'. By convention it is taken to be 25 cm for normal vision.

- The distance to the **far point** is the distance between the eye and the furthest object that can be brought into focus. This is taken to be infinity for normal vision.

ANGULAR SIZE

If we bring an object closer to us (and our eyes are still able to focus on it) then we see it in more detail. This is because, as the object approaches, it occupies a bigger visual angle. The technical term for this is that the object **subtends** a larger angle.

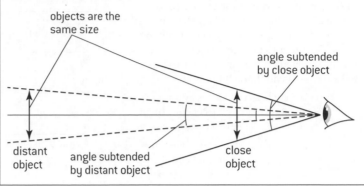

ANGULAR MAGNIFICATION

The angular magnification, M, of an optical instrument is defined as the ratio between the angle that an object subtends normally and the angle that its image subtends as a result of the optical instrument. The 'normal' situation depends on the context. It should be noted that the angular magnification is not the same as the linear magnification.

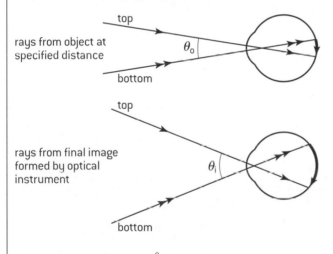

Angular magnification, $M = \frac{\theta_i}{\theta_o}$

The largest visual angle that an object can occupy is when it is placed at the near point. This is often taken as the 'normal' situation.

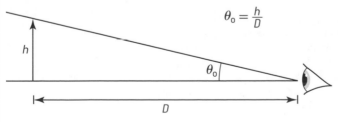

$\theta_o = \frac{h}{D}$

A simple lens can increase the angle subtended. It is usual to consider two possible situations.

1. Image formed at infinity

In this arrangement, the object is placed at the focal point. The resulting image will be formed at infinity and can be seen by the relaxed eye.

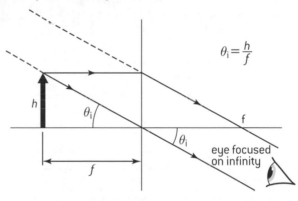

$\theta_i = \frac{h}{f}$

eye focused on infinity

In this case the angular magnification would be

$$M_{\text{infinity}} = \frac{\theta_i}{\theta_o} = \frac{\frac{h}{f}}{\frac{h}{D}} = \frac{D}{f}$$

This is the smallest value that the angular magnification can be.

2. Image formed at near point

In this arrangement, the object is placed nearer to the lens. The resulting virtual image is located at the near point. This arrangement has the largest possible angular magnification.

$$M = \frac{\theta_i}{\theta_o} = \frac{h_i/D}{h/D} = \frac{h_i}{h} = \frac{D}{a}$$

$$\frac{1}{u} + \frac{1}{v} = \frac{1}{f}$$

$$\Rightarrow \frac{1}{a} - \frac{1}{D} = \frac{1}{f}$$

$$\therefore \frac{D}{a} = \frac{D}{f} + 1$$

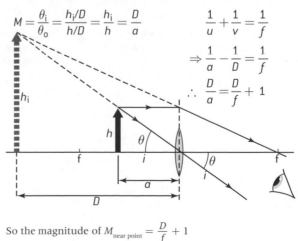

So the magnitude of $M_{\text{near point}} = \frac{D}{f} + 1$

Aberrations

SPHERICAL

A lens is said to have an **aberration** if, for some reason, a point object does not produce a perfect point image. In reality, lenses that are spherical do not produce perfect images. **Spherical aberration** is the term used to describe the fact that rays striking the outer regions of a spherical lens will be brought to a slightly different focus point from those striking the inner regions of the same lens. This is not to be confused with **barrel distortion**.

In general, a point object will focus into a small circle of light, rather than a perfect point. There are several possible ways of reducing this effect:

- the shape of the lens could be altered in such a way as to correct for the effect. The lens would, of course, no longer be spherical. A particular shape only works for objects at a particular distance away.

- the effect can be reduced for a given lens by decreasing the aperture. The technical term for this is **stopping down** the aperture. The disadvantage is that the total amount of light is reduced and the effects of diffraction (see page 46) would be made worse.

The effect for mirrors can be eliminated for all point objects on the axis by using a parabolic (as opposed to a spherical) mirror. For mirrors, the effect can again be reduced by using a smaller aperture.

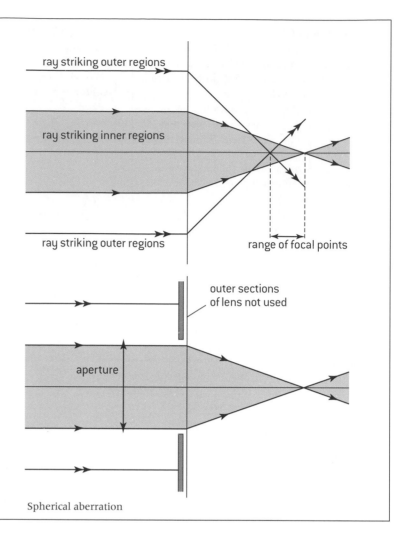

Spherical aberration

CHROMATIC

Chromatic aberration is the term used to describe the fact that rays of different colours will be brought to a slightly different focus point by the same lens. The refractive index of the material used to make the lens is different for different frequencies of light.

A point object will produce a blurred image of different colours.

The effect can be eliminated for two given colours (and reduced for all) by using two different materials to make up a compound lens. This compound lens is called an **achromatic doublet**. The two types of glass produce equal but opposite dispersion.

Mirrors do not suffer from chromatic aberration.

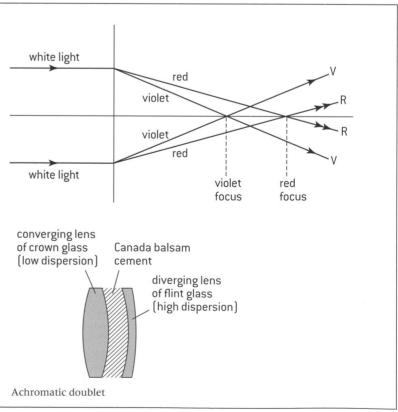

Achromatic doublet

The compound microscope and astronomical telescope

COMPOUND MICROSCOPE

A compound microscope consists of two lenses – the **objective lens** and the **eyepiece lens**. The first lens (the objective lens) forms a **real magnified** image of the object being viewed. This real image can then be considered as the object for the second lens (the eyepiece lens) which acts as a magnifying lens. The rays from this real image travel into the eyepiece lens and they form a **virtual magnified** image. In normal adjustment, this virtual image is arranged to be located at the near point so that maximum angular magnification is obtained.

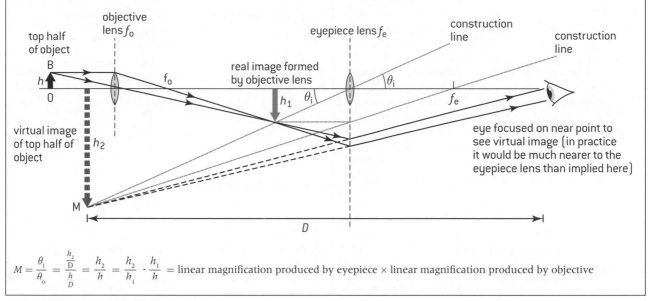

$$M = \frac{\theta_i}{\theta_o} = \frac{\frac{h_2}{D}}{\frac{h}{D}} = \frac{h_2}{h} = \frac{h_2}{h_i} \cdot \frac{h_1}{h} = \text{linear magnification produced by eyepiece} \times \text{linear magnification produced by objective}$$

ASTRONOMICAL TELESCOPE

An astronomical telescope also consists of two lenses. In this case, the objective lens forms a **real** but **diminished** image of the distant object being viewed. Once again, this real image can then be considered as the object for the eyepiece lens acting as a magnifying lens. The rays from this real image travel into the eyepiece lens and they form a **virtual magnified** image. In normal adjustment, this virtual image is arranged to be located at infinity.

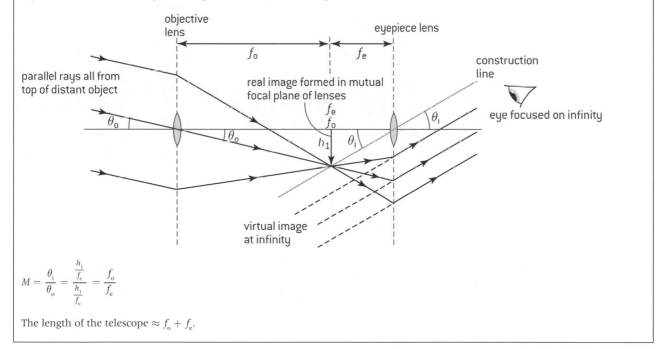

$$M = \frac{\theta_i}{\theta_o} = \frac{\frac{h_1}{f_e}}{\frac{h_1}{f_o}} = \frac{f_o}{f_e}$$

The length of the telescope $\approx f_o + f_e$.

Astronomical reflecting telescopes

COMPARISON OF REFLECTING AND REFRACTING TELESCOPES

A refracting telescope uses an objective (converging) lens to form a real diminished image of a distant object. This image is then viewed by the eyepiece lens (converging) which, acting as a simple magnifying glass, produces a virtual but magnified final image.

In an analogous way, a reflecting telescope uses a concave mirror set up so as to form a real, diminished image of a distant object. This image, however, would be difficult to view as it would be produced in front of the concave mirror. Thus mirrors are used to produce a viewable image that can, like the refracting telescope, be viewed by the eyepiece lens (converging). Once again the eyepiece acts as a simple magnifying glass and produces the virtual, but magnified, final image. Two common mountings for reflecting telescopes are the **Newtonian mounting** and the **Cassegrain mounting.**

All telescopes are made to have large apertures in order to:

a)   reduce diffraction effects, and

b)   collect enough light to make bright images of low power sources.

Large telescopes are reflecting because:

* Mirrors do not suffer from chromatic aberration

* It is difficult to get a uniform refractive index throughout a large volume of glass

* Mounting a large lens is harder to achieve than mounting a large mirror.

* Only one surface needs to be the right shape.

Reflecting telescopes can easily suffer damage to the mirror surface.

NEWTONIAN MOUNTING

A small flat mirror is placed on the principal axis of the mirror to reflect the image formed to the side:

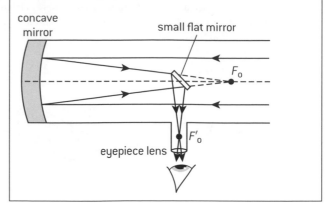

CASSEGRAIN MOUNTING

A small convex mirror is mounted on the principal axis of the mirror. The mirror has a central hole to allow the image to be viewed.

The convex mirror will add to the angular magnification achieved.

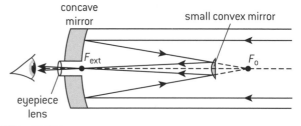

Radio telescopes

SINGLE DISH RADIO TELESCOPES

A single dish radio telescope operates in a very similar way to a reflecting telescope. Rather than reflecting visible light to form an image, the much longer wavelengths of radio waves are reflected by the curved receiving dish. The antenna that is the receiver of the radio waves can be tuned to pick up specific wavelengths under observation and are used to study naturally occurring radio emission from stars, galaxies, quasars and other astronomical objects between wavelengths of about 10 m and 1mm.

Radio telescope

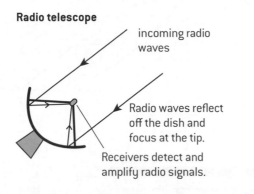

incoming radio waves

Radio waves reflect off the dish and focus at the tip.

Receivers detect and amplify radio signals.

Diffraction effects can significantly limit the accuracy with which a radio telescope can locate individual sources of radio signals. Increasing the diameter of a radio telescope improves the telescope's ability to resolve different sources and ensure that more power can be received (see resolution on page 101).

RADIO INTERFEROMETRY TELESCOPES

The angular resolution of a radio telescope can be improved using a principle called interferometry. This process analyses signals received at two (or more) radio telescopes that are some distance apart but pointing in the same direction. This effectively creates a virtual radio telescope that is much larger than any of the individual telescopes.

The technique is complex as it involves collecting signals from two or more radio telescopes (an **array telescope**) in one central location. The arrival of each signal at an individual antenna needs to be carefully calibrated against a single shared reference signal so that different signals can be combined as though they arrived at one single antenna. When the signals from the different telescopes are added together, they interfere. The result is to create a combined telescope that is equivalent in resolution (though not in sensitivity) to a single radio telescope whose diameter is approximately equal to the maximum separations of the antennae.

The principle can be extended, in a process called **Very Long Baseline Interferometry**, to allow recordings of radio signals (originally made hundreds of km apart) to be synchronized to within a millionth of a second thus allowing scientists from different countries to collaborate to create a virtual radio telescope of huge size and high resolving power.

COMPARATIVE PERFORMANCE OF EARTH-BOUND AND SATELLITE-BORNE TELESCOPES

The following points about Earth-based (EB) and satellite-borne (SB) telescopes can be made:

- SB observations are free from interference and/or absorptions due to the Earth's atmosphere that hinder EB observations, giving better resolution for SB telescopes.

- Modern computer techniques can effectively correct for many atmospheric effects making new ground-based telescopes similar in resolution to some SB telescopes.

- Many significant wavelengths of EM radiation (UV, IR and long wavelength radio) are absorbed by the Earth's atmosphere so SB telescopes are the only possibility in their wavelengths.

- SB observations do not suffer from light pollution / radio interference as a result of nearby human activity.

- SB facilities are not subject to continual wear and tear as a result of the Earth's atmosphere (storms etc.).

- The possibility of damage from space debris exists for SB telescopes.

- There is a great deal of added cost in getting the telescope into orbit and controlling it remotely, meaning that SB telescopes are significantly more expensive to build and this places a limit on their size and weight.

- There is an added difficulty of effecting repairs / alterations to a SB telescope once operational.

- SB telescopes need to withstand wider temperature variations than EB telescopes.

- EB optical telescopes can only operate at night whereas SB telescopes can operate at all times.

Fibre optics

OPTIC FIBRE

Optic fibres use the principle of total internal reflection (see page 45) to guide light along a certain path. The idea is to make a ray of light travel along a transparent fibre by bouncing between the walls of the fibre. So long as the incident angle of the ray on the wall of the fibre is always greater than the critical angle, the ray will always remain within the fibre even if the fibre is bent (see right).

As shown on page 45, the relation between critical angle, c, and refractive index n is given by

$$n = \frac{1}{\sin c}$$

Two important uses of optic fibres are:

* In the communication industry. Digital data can be encoded into pulses of light that can then travel along the fibres. This is used for telephone communication, cable TV etc.

* In the medical world. Bundles of optic fibres can be used to carry images back from inside the body. This instrument is called an endoscope.

* This type of optic fibre is known as a step-index optic fibre. Cladding of a material with a lower refractive index surrounds the fibre. This cladding protects and strengthens the fibre.

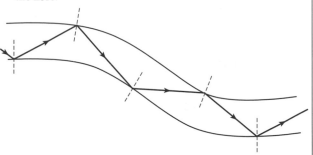

TYPES OF OPTIC FIBRES

The simplest fibre optic is a step-index fibre. Technically this is known as a **multimode step-index fibre**. Multimode refers to the fact that light can take different paths down the fibre which can result in some distortion of signals (see waveguide dispersion, page 183). The (multimode) **graded-index** fibre is an improvement. This uses a graded refractive index profile in the fibre meaning that rays travel at different speeds depending on their distance from the centre. This has the effect of reducing the spreading out of the pulse. Most fibres used in data communications have a graded index. The optimum solution is to have a very narrow core – a **singlemode step-index fibre**.

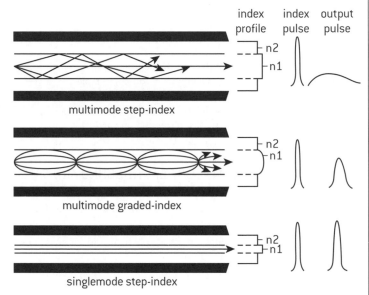

index profile   index pulse   output pulse

n2
n1

multimode step-index

n2
n1

multimode graded-index

n2
n1

singlemode step-index

Dispersion, attenuation and noise in optical fibres

MATERIAL DISPERSION

The refractive index of any substance depends on the frequency of electromagnetic radiation considered. This is the reason that white light is dispersed into different colours when it passes through a triangular prism.

As light travels along an optical fibre, different frequencies will travel at slightly different speeds. This means that if the source of light involves a range of frequencies, then a pulse that starts out as a square wave will tend to spread out as it travels along the fibre. This process is known as **material dispersion**.

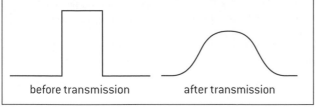

before transmission   after transmission

WAVEGUIDE DISPERSION

If the optical fibre has a significant diameter, another process called **waveguide dispersion** that can cause the stretching of a pulse is **multipath** or **modal dispersion**. The path length along the centre of a fibre is shorter than a path that involves multiple reflections. This means that rays from a particular pulse will not all arrive at the same time because of the different distances they have travelled.

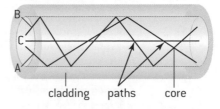

cladding   paths   core

The problems caused by modal dispersion have led to the development of **monomode** (or **singlemode**) **step-index fibres**. These optical fibres have very narrow cores (of the same order of magnitude as the wavelength of the light being used (approximately 5 μm) so that there is only one effective transmission path – directly along the fibre.

ATTENUATION

As light travels along an optic fibre, some of the energy can be scattered or absorbed by the glass. The intensity of the light energy that arrives at the end of the fibre is less than the intensity that entered the fibre. The signal is said to be **attenuated**.

The amount of attenuation is measured on a logarithmic scale in decibels (dB). The attenuation is given by

$$\text{attenuation (dB)} = 10\log\frac{I}{I_o}$$

I is the intensity of the output power measured in W

I_o is the intensity of the original input power measured in W

A negative attenuation means that the signal has been reduced in power. A positive attenuation would imply that the signal has been amplified.

See page 188 for another example of the use of the decibel scale.

It is common to quote the attenuation per unit length as measured in dB km^{-1}. For example, 5 km of fibre optic cable causes an input power of 100 mW to decrease to 1 mW. The attenuation per unit length is calculated as follows:

$$\text{attenuation} = 10 \log (10^{-3}/10^{-1}) = 10 \log (10^{-2})$$
$$= -20 \text{ dB}$$

$$\text{attenuation per unit length} = -20 \text{ dB}/5 \text{ km}$$
$$= -4 \text{ dB km}^{-1}$$

The attenuation of a 10 km length of this fibre optic cable would therefore be −40 dD. The overall attenuation resulting from a series of factors is the algebraic sum of the individual attenuations.

The attenuation in an optic fibre is a result of several processes: those caused by impurities in the glass, the general scattering that takes place in the glass and the extent to which the glass absorbs the light. These last two factors are affected by the wavelength of light used. A typical the overall attenuation is shown below:

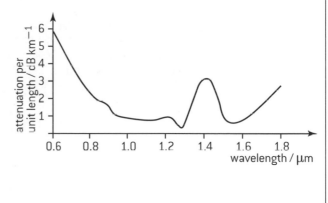

CAPACITY

Attenuation causes an upper limit to the amount of digital information that can be sent along a particular type of optical fibre. This is often stated in terms of its capacity.

$$\text{capacity of an optical fibre} = \text{bit rate} \times \text{distance}$$

A fibre with a capacity of 80 Mbit s^{-1} km can transmit 80 Mbit s^{-1} along a 1 km length of fibre but only 20 Mbit s^{-1} along a 4 km length.

NOISE, AMPLIFIERS AND RESHAPERS

Noise is inevitable in any electronic circuit. Any dispersions or scatterings that take place within an optical fibre will also add to the noise.

An amplifier increases the signal strength and thus will tend to correct the effect of attenuation – these are also sometimes called regenerators. An amplifier will also increase any noise that has been added to the electrical signal.

A reshaper can reduce the effects of noise on a digital signal by returning the signal to a series of 1s and 0s with sharp transitions between the allowed levels.

Channels of communication

The table below shows some common communication links.

	Options for communication	Uses	Advantages and disadvantages
Wire pairs (twisted pair) copper wire insulation	Two wires can connect the sender and receiver of information. For example a simple link between a microphone, an amplifier and a loudspeaker.	Very simple communication systems e.g. intercom	Very simple and cheap. Susceptible to noise and interference. Unable to transfer information at the highest rates.
Coaxial cables copper wire insulation copper mesh outside insulation	This arrangement of two wires reduces electrical interference. A central wire is surrounded by the second wire in the form of an outer cylindrical copper tube or mesh. An insulator separates the two wires. Wire links can carry frequencies up to about 1 GHz but the higher frequencies will be attenuated more for a given length of wire. A typical 100 MHz signal sent down low-loss cable would need repeaters at intervals of approximately 0.5 km. The upper limit for a single coaxial cable is approximately 140 Mbit s^{-1}.	Coaxial cables are used to transfer signals from TV aerials to TV receivers. Historically they are standard for underground telephone links.	Simple and straightforward. Less susceptible to noise compared to simple wire pair but noise still a problem.
Optical fibres	Laser light can be used to send signals down optical fibres with approximately the same frequency limit as cables – 1 GHz. The attenuation in an optical fibre is less than in a coaxial cable. The distance between repeaters can easily be tens (or even hundreds) of kilometres.	Long-distance telecommunication and high volume transfer of digital data including video data.	Compared to coaxial cables with equivalent capacity, optical fibres: • have a higher transmission capacity • are much smaller in size and weight • cost less • allow for a wider possible spacing of regenerators • offer immunity to electromagnetic interference • suffer from negligible cross talk (signals in one channel affecting another channel) • are very suitable for digital data transmission • provide good security • are quiet – they do not hum even when carrying large volumes of data. There are some disadvantages: • the repair of fibres is not a simple task • regenerators are more complex and thus potentially less reliable.

INTENSITY, QUALITY AND ATTENUATION

The effects of X-rays on matter depend on two things, the **intensity** and the **quality** of the X-rays.

- The intensity, I, is the amount of energy per unit area that is carried by the X-rays.

- The quality of the X-ray beam is the name given to the spread of wavelengths that are present in the beam. Low-energy photons will be absorbed by all tissues and potentially cause harm without contributing to forming the image. It is desirable to remove these from the beam.

If the energy of the beam is absorbed, then it is said to be **attenuated**. If there is nothing in the way of an X-ray beam, it will still be attenuated as the beam spreads out. Two processes of attenuation by matter, **simple scattering** and the **photoelectric effect** are the dominant ones for low-energy X-rays.

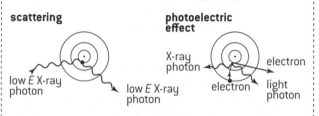

Simple scattering affects X-ray photons that have energies between zero and 30 keV.

- In the photoelectric effect, the incoming X-ray has enough energy to cause one of the inner electrons to be ejected from the atom. It will result in one of the outer electrons 'falling down' into this energy level. As it does so, it releases some light energy. This process affects X-ray photons that have energies between zero and 100 keV.

Both attenuation processes result in a near exponential transmission of radiation as shown in the diagram below. For a given energy of X-rays and given material there will be a certain thickness that reduces the intensity of the X-ray by 50%. This is known as the **half-value thickness**.

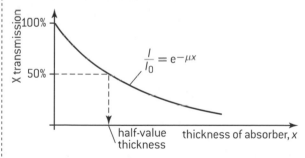

The **attenuation coefficient** μ is a constant that mathematically allows us to calculate the intensity of the X-rays given any thickness of material. The equation is as follows:

$$I = I_0 e^{-\mu x}$$

The relationship between the attenuation coefficient and the half-value thickness is

$$\mu x_{\frac{1}{2}} = \ln 2$$

$x_{\frac{1}{2}}$ The half-value thickness of the material (in m)

$\ln 2$ The natural log of 2. This is the number 0.6931

μ The attenuation coefficient (in m^{-1})

μ depends on the wavelength of the X-rays – short wavelengths are highly penetrating and these X-rays are **hard**. **Soft** x-rays are easily attenuated and have long wavelengths.

BASIC X-RAY DISPLAY TECHNIQUES

The basic principle of X-ray imaging is that some body parts (for example bones) will attenuate the X-ray beam much more than other body parts (for example skin and tissue). Photographic film darkens when a beam of X-rays is shone on them so bones show up as white areas on an X-ray picture.

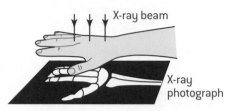

The sharpness of an X-ray image is a measure of how easy it is to see edges of different organs or different types of tissue.

X-ray beams will be scattered in the patient being scanned and the result will be to blur the final image and to reduce the contrast and sharpness. To help reduce this effect, a metal filter grid is added below the patient:

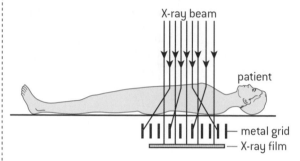

Alternatively computer software can be used to detect and enhance edges.

Since X-rays cause ionizations, they are dangerous. This means that the intensity used needs to be kept to an absolute minimum. This can be done by introducing something to **intensify** (to enhance) the image. There are two simple techniques of **enhancement**:

- When X-rays strike an intensifying screen the energy is re-radiated as visible light. The photographic film can absorb this extra light. The overall effect is to darken the image in the areas where X-rays are still present (see page 187).

- In an image-intensifier tube, the X-rays strike a fluorescent screen and produce light. This light causes electrons to be emitted from a photocathode. These electrons are then accelerated towards an anode where they strike another fluorescent screen and give off light to produce an image.

MASS ATTENUATION COEFFICIENT

An alternative way of writing the equation for the attenuation coefficient is shown below:

$$I = I_0 e^{-\left(\frac{\mu}{\rho}\right)\rho x}$$

Where ρ is the density of the substance. In this format, $\frac{\mu}{\rho}$ is known as the **mass attenuation coefficient** $\frac{\mu}{\rho}$, and ρx is known as the **area density** or **mass thickness**.

Units of mass attenuation coefficient, $\frac{\mu}{\rho} = m^2\ kg^{-1}$

Units of area density, $\rho x = kg\ m^{-2}$

$$I = I_0 e^{-\left(\frac{\mu}{\rho}\right)\rho x}$$

HL X-ray imaging techniques

1) Intensifying screens

The arrangement of the intensifying screens described on page 185 are shown below.

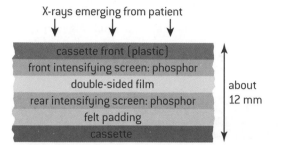

X-rays emerging from patient

| cassette front (plastic) |
| front intensifying screen: phosphor |
| double-sided film |
| rear intensifying screen: phosphor |
| felt padding |
| cassette |

about 12 mm

With a simple X-ray photograph it is hard to identify problems within soft tissue, for example in the gut. There are two general techniques aimed at improving this situation.

2) Barium meals

In a **barium meal**, a dense substance is swallowed and its progress along the gut can be monitored. The contrast between the gut and surrounding tissue is increased. Typically the patient is asked to swallow a harmless solution of barium sulfate. The result is an increase in the sharpness of the image.

3) Tomography

Tomography is a technique that makes the X-ray photograph focus on a certain region or 'slice' through the patient. All other regions are blurred out of focus. This is achieved by moving the source of X-rays and the film together.

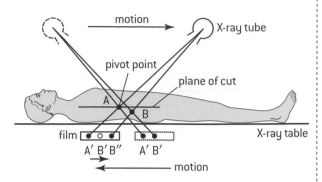

An extension of basic tomography is the **computed tomography scan** or **CT scan**. In this set-up a tube sends out a pulse of X-rays and a set of sensitive detectors collects information about the levels of X-radiation reaching each detector. The X-ray source and the detectors are then rotated around a patient and the process is repeated. A computer can analyse the information recorded and is able to reconstruct a 3-dimensional 'map' of the inside of the body in terms of X-ray attenuation.

(HL) Ultrasonic imaging

ULTRASOUND

The limit of human hearing is about 20 kHz. Any sound that is of higher frequency than this is known as **ultrasound**. Typically ultrasound used in medical imaging is just within the MHz range. The velocity of sound through soft tissue is approximately 1500 m s^{-1} meaning that typical wavelengths used are of the order of a few millimetres.

Unlike X-rays, ultrasound is not ionizing so it can be used very safely for imaging inside the body – with pregnant women for example. The basic principle is to use a probe that is capable of emitting and receiving pulses of ultrasound. The ultrasound is reflected at any boundary between different types of tissue. The time taken for these reflections allows us to work out where the boundaries must be located.

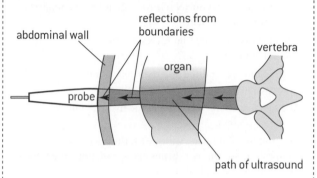

ACOUSTIC IMPEDANCE

The acoustic impedance of a substance is the product of the density, ρ, and the speed of sound, c.

$$Z = \rho c$$

unit of Z = kg m^{-2} s^{-1}

Very strong reflections take place when the boundary is between two substances that have very different acoustic impedances. This can cause some difficulties.

- In order for the ultrasound to enter the body in the first place, there needs to be no air gap between the probe and the patient's skin. An air gap would cause almost all of the ultrasound to be reflected straight back. The transmission of ultrasound is achieved by putting a gel or oil (of similar density to the density of tissue) between the probe and the skin.

- Very dense objects (such as bones) can cause a strong reflection and multiple images can be created. These need to be recognized and eliminated.

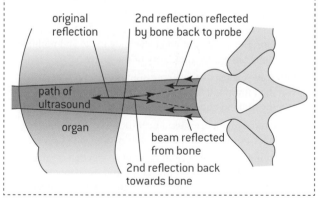

PIEZOELECTRIC CRYSTALS

These quartz crystals change shape when an electric current flows and can be used with an alternating pd to generate ultrasound. They also generate pds when receiving sound pressure waves so one crystal is used for generation and detection.

A- AND B-SCANS

There are two ways of presenting the information gathered from an ultrasound probe, the **A-scan** or the **B-scan**. The A-scan (amplitude-modulated scan) presents the information as a graph of signal strength versus time. The B-scan (brightness-modulated scan) uses the signal strength to affect the brightness of a dot of light on a screen.

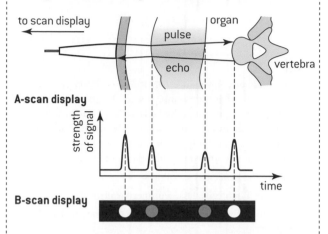

A-scans are useful where the arrangement of the internal organs is well known and a precise measurement of distance is required. If several B-scans are taken of the same section of the body at one time, all the lines can be assembled into an image which represent a section through the body. This process can be achieved using a large number of transducers.

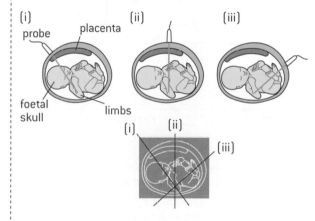

Building a picture from a series of B-scan lines

CHOICE OF FREQUENCY

The choice of frequency of ultrasound to use can be seen as the choice between resolution and attenuation.

- Here, the resolution means the size of the smallest object that can be imaged. Since ultrasound is a wave motion, diffraction effects will be taking place. In order to image a small object, we must use a small wavelength. If this was the only factor to be considered, the frequency chosen would be as large as possible.

- Unfortunately attenuation increases as the frequency of ultrasound increases. If very high frequency ultrasound is used, it will all be absorbed and none will be reflected back. If this was the only factor to be considered, the frequency chosen would be as small as possible.

On balance the frequency chosen has to be somewhere between the two extremes. It turns out that the best choice of frequency is often such that the part of the body being imaged is about 200 wavelengths of ultrasound away from the probe.

RELATIVE INTENSITY LEVELS OF ULTRASOUND

The relative intensity levels of ultrasound between two points are compared using the decibel scale (dB). As its name suggests, the decibel unit is simply one tenth of a base unit that is called the bel (B). The decibel scale is logarithmic.

Mathematically,

Relative intensity level in bels,

$$L_I = \log \frac{\text{intensity level of ultrasound at measurement point}}{\text{intensity level of ultrasound at reference point}}$$

or Relative intensity level in bels $= \log \frac{I_1}{I_0}$

Since 1 bel = 10 dB,

Relative intensity level in decibels, $L_I = 10 \log \frac{I_1}{I_0}$

NMR

Nuclear Magnetic Resonance (NMR) is a very complicated process but one that is extremely useful. It can provide detailed images of sections through the body without any invasive or dangerous techniques. It is of particular use in detecting tumours in the brain. It involves the use of a non-uniform magnetic field in conjuction with a large uniform field.

In outline, the process is as follows:

- The nuclei of atoms have a property called spin.

- The spin of these nuclei means that they can act like tiny magnets.

- These nuclei will tend to line up in a strong magnetic field.

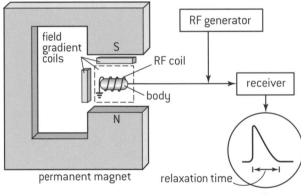

- They do not, however, perfectly line up – they oscillate in a particular way that is called **precession**. This happens at a very high frequency – the same as the frequency of radio waves.

- The particular frequency of precession depends on the magnetic field and the particular nucleus involved. It is called the **Larmor frequency**.

- If a pulse of radio waves is applied at the Larmor frequency, the nuclei can absorb this energy in a process called **resonance**. The protons make a spin transition.

- After the pulse, the nuclei return to their lower energy state by emitting radio waves.

- The time over which radio waves are emitted is called the **relaxation time**.

- The radio waves emitted and their relaxation times can be processed to produce the NMR scan image.

- The signal analysis is targeted at the hydrogen nuclei (protons) present.

- The number of H nuclei varies with the chemical composition so different tissues extract different amounts of energy from the applied signal.

- Thus RF signal forces protons to make a spin transition and
  ◊ The gradient field allows determination of the point from which the photons are emitted.
  ◊ The proton spin relaxation time depends on the type of tissue at the point where the radiation is emitted.

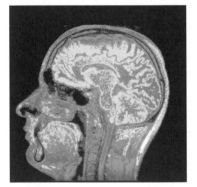

COMPARISON BETWEEN ULTRASOUND AND NMR

The following points can be noted:

- NMR imaging is very expensive when compared with ultrasound equipment and is very bulky – patient needs to be brought to the NMR machine and process is time consuming.

- Ultrasound measurements are easy to perform (equipment can be brought to patient at the point of care) and can be repeated as required but quality of image can rely on skill of operator.

- NMR produces a 3-dimensional scan, ultrasound typically produces a 2-dimensional scan.

- Detail produced by NMR is greater than by ultrasound.

- NMR particularly useful for very delicate areas of body e.g. brain.

- NMR patients have to remain very still, ultrasound images can be more dynamic.

- Ultrasound waves do not enter the body easily and multiple reflections can reduce the clarity of the image.

- Both wave energies carry energy but the energy associated with the ultrasound is greater that the energy associated with the radio frequencies used in NMR.

- At the radio frequencies used in NMR there is no danger of resonance but some ultrasound energy can cause heating.

- Ultrasound can cause cavitation – the production of small gas bubbles which will absorb energy and can damage surrounding tissue. The frequencies and intensities used for diagnostics avoid this possibility as much as possible.

- The strong magnetic fields used in NMR present problems for patients with surgical implants and / or pacemakers.

IB Questions – option C – imaging

1. For each of the following situations, locate and describe the final image formed. Solutions should found using scale diagrams and mathematically.

   a) An object is placed 7 cm in front of a concave mirror of focal length 14 cm. [4]

   b) A diverging lens of focal length 12.0 cm is placed at the focal point of a converging lens of focal length 8.0 cm. An object is placed 16.0 cm in front of the converging lens. [4]

   c) An object is placed 18.0 cm in front of a convex lens of focal length 6.0 cm. A second convex lens of focal length 3.0 cm is an additional 18 cm behind the first lens. [4]

2. A student is given two converging lenses, A and B, and a tube in order to make a telescope.

   a) Describe a simple method by which she can determine the focal length of each lens. [2]

   b) She finds the focal lengths to be as follows:

      Focal length of lens A   10 cm

      Focal length of lens B   50 cm

      Draw a diagram to show how the lenses should be arranged in the tube in order to make a telescope. Your diagram should include:

      (i)   labels for each lens;

      (ii) the focal points for each lens;

      (iii) the position of the eye when the telescope is in use. [4]

   c) On your diagram, mark the location of the intermediate image formed in the tube. [1]

   d) Is the image seen through the telescope upright or upside-down? [1]

   e) Approximately how long must the telescope tube be? [1]

3. Explain what is meant by

   a) Material dispersion   e) A Cassegrain mounting

   b) Waveguide dispersion   f) Total Internal reflection

   c) Spherical aberrations   g) Step-index fibres

   d) Chromatic aberrations   [2 each]

4. A 15 km length of optical fibre has an attenuation of 4 dB km^{-1}. A 5 mW signal is sent along the wire using two amplifiers as represented by the diagram below.

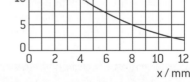

   Calculate

   a) the overall gain of the system

   b) the output power. [2]

HL _____

5. This question is about ultrasound scanning.

   a) State a typical value for the frequency of ultrasound used in medical scanning. [1]

   The diagram below shows an ultrasound transmitter and receiver placed in contact with the skin.

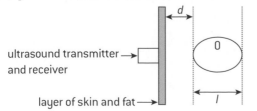

The purpose of this particular scan is to find the depth d of the organ labelled O below the skin and also to find its length, I.

   b) (i) Suggest why a layer of gel is applied between the ultrasound transmitter/receiver and the skin. [2]

On the graph below the pulse strength of the reflected pulses is plotted against the time lapsed between the pulse being transmitted and the time that the pulse is received, t.

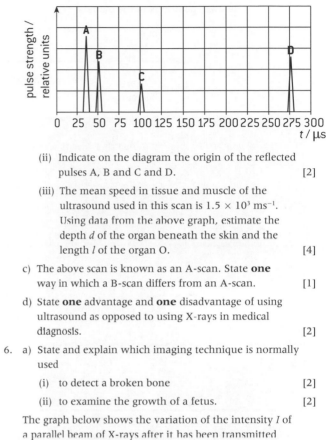

   (ii) Indicate on the diagram the origin of the reflected pulses A, B and C and D. [2]

   (iii) The mean speed in tissue and muscle of the ultrasound used in this scan is 1.5×10^3 ms^{-1}. Using data from the above graph, estimate the depth d of the organ beneath the skin and the length l of the organ O. [4]

   c) The above scan is known as an A-scan. State **one** way in which a B-scan differs from an A-scan. [1]

   d) State **one** advantage and **one** disadvantage of using ultrasound as opposed to using X-rays in medical diagnosis. [2]

6. a) State and explain which imaging technique is normally used

   (i) to detect a broken bone [2]

   (ii) to examine the growth of a fetus. [2]

The graph below shows the variation of the intensity I of a parallel beam of X-rays after it has been transmitted through a thickness x of lead.

   b) (i)   Define *half-value thickness, $x_{\frac{1}{2}}$*. [2]

   (ii) Use the graph to estimate $x_{\frac{1}{2}}$ for this beam in lead. [2]

   (iii) Determine the thickness of lead required to reduce the intensity transmitted to 20% of its initial value. [2]

   (iv) A second metal has a half-value thickness $x_{\frac{1}{2}}$ for this radiation of 8 mm. Calculate what thickness of this metal is required to reduce the intensity of the transmitted beam by 80%. [3]

Objects in the universe (1)

SOLAR SYSTEM

We live on the Earth. This is one of eight planets that orbit the Sun – collectively this system is known as the Solar System. Each planet is kept in its elliptical orbit by the gravitational attraction between the Sun and the planet. Other smaller masses such as **dwarf planets** like Pluto or planetoids also exist.

	Mercury	Venus	Earth	Mars	Jupiter	Saturn	Uranus	Neptune
diameter / km	4,880	12,104	12,756	6,787	142,800	120,000	51,800	49,500
distance to Sun / $\times 10^8$ m	58	107.5	149.6	228	778	1,427	2,870	4,497

Relative positions of the planets

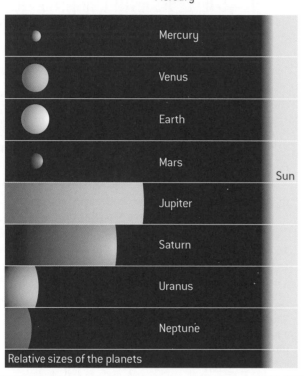

Relative sizes of the planets

Some of these planets (including the Earth) have other small objects orbiting around them called moons. Our Moon is 3.8×10^8 m away and its diameter is about 1/4 of the Earth's.

An **asteroid** is a small rocky body that drifts around the Solar System. There are many orbiting the Sun between Mars and Jupiter – the asteroid belt. An asteroid on a collision course with another planet is known as a meteoroid.

Small meteors can be vaporized due to the friction with the atmosphere ('shooting stars') whereas larger ones can land on Earth. The bits that arrive are called **meteorites**.

Comets are mixtures of rock and ice (a 'dirty snowball') in very elliptical orbits around the Sun. Their 'tails' always point away from the Sun.

VIEW FROM ONE PLACE ON EARTH

If we look up at the night sky we see the stars – many of these 'stars' are, in fact, other galaxies but they are very far away. The stars in our own galaxy appear as a band across the sky – the Milky Way.

Patterns of stars have been identified and 88 different regions of the sky have been labelled as the different **constellations**. Stars in a constellation are not necessarily close to one another.

Over the period of a night, the constellations seem to rotate around one star. This apparent rotation is a result of the rotation of the Earth about its own axis.

On top of this nightly rotation, there is a slow change in the stars and constellations that are visible from one night to the next. This variation over the period of one year is due to the rotation of the Earth about the Sun.

Planetary systems have been discovered around many stars.

NEBULAE

In many constellations there are diffuse but relatively large structures which are called nebulae. These are interstellar clouds of dust, hydrogen, helium and other ionized gases. An example is M42 otherwise known as the Orion Nebula.

VIEW FROM PLACE TO PLACE ON EARTH

If you move from place to place around the Earth, the section of the night sky that is visible over a year changes with latitude. The total pattern of the constellations is always the same, but you will see different sections of the pattern.

Objects in the universe (2)

DURING ONE DAY

The most important observation is that the pattern of the stars remains the same from one night to the next. Patterns of stars have been identified and 88 different regions of the sky have been labelled as the different **constellations**. A particular pattern is not always in the same place, however. The constellations appear to move over the period of one night. They appear to rotate around one direction. In the Northern Hemisphere everything seems to rotate about the pole star.

It is common to refer measurements to the 'fixed stars' the patterns of the constellations. The fixed background of stars always appears to rotate around the pole star. During the night, some stars rise above the horizon and some stars set beneath it.

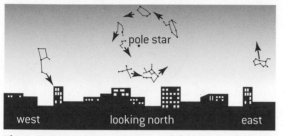

The same movement is continued during the day. The Sun rises in the east and sets in the west, reaching its maximum height at midday. At this time in the Northern Hemisphere the Sun is in a southerly direction.

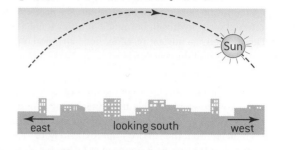

DURING THE YEAR

Every night, the constellations have the same relative positions to each other, but the location of the pole star (and thus the portion of the night sky that is visible above the horizon) changes slightly from night to night. Over the period of a year this slow change returns back to exactly the same position.

The Sun continues to rise in the east and set in the west, but as the year goes from winter into summer, the arc gets bigger and the Sun climbs higher in the sky.

UNITS

When comparing distances on the astronomical scale, it can be quite unhelpful to remain in SI units. Possible other units include the **astronomical unit (AU)**, the **parsec (pc)** or the **light year (ly)**. See page 193 for the definition of the first two of these.

The light year is the distance travelled by light in one year (9.5×10^{15} m). The next nearest star to our Sun is about 4 light years away. Our galaxy is about 100,000 light years across. The nearest galaxy is about a million light years away and the observable Universe is 13.7 billion light years in any given direction.

THE MILKY WAY GALAXY

When observing the night sky a faint band of light can be seen crossing the constellations. This 'path' (or 'way') across the night sky became known as the Milky Way. What you are actually seeing is some of the millions of stars that make up our own galaxy but they are too far away to be seen as individual stars. The reason that they appear to be in a band is that our galaxy has a spiral shape.

The centre of our galaxy lies in the direction of the constellation Sagittarius. The galaxy is rotating – all the stars are orbiting the centre of the galaxy as a result of their mutual gravitational attraction. The period of orbit is about 250 million years.

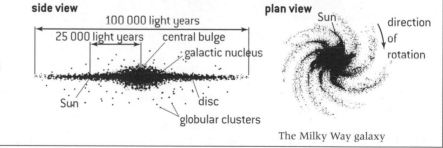

The Milky Way galaxy

THE UNIVERSE

Stars are grouped together in **stellar clusters**. These can be **open** containing 10^3 stars e.g. located in the disc of our galaxy or **globular** containing 10^5 stars. Our Sun is just one of the billions of stars in our **galaxy** (the Milky Way galaxy). The galaxy rotates with a period of about 2.5×10^8 years.

Beyond our galaxy, there are billions of other galaxies. Some of them are grouped together into **clusters** or **super clusters** of galaxies, but the vast majority of space (like the gaps between the planets or between stars) appears to be empty – essentially a vacuum. Everything together is known as the **Universe**.

1.5×10^{26} m (= 15 billion light years)	the visible Universe
5×10^{22} m (= 5 million light years)	local group of galaxies
10^{21} m (= 100,000 light years)	our galaxy
10^{13} m (= 0.001 light years)	our Solar System

The nature of stars

ENERGY FLOW FOR STARS

The stars are emitting a great deal of energy. The source for all this energy is the fusion of hydrogen into helium. See page 196. Sometimes this is referred to as 'hydrogen burning' but it this is not a precise term. The reaction is a nuclear reaction, not a chemical one (such as combustion). Overall the reaction is

$$4 \, {}^1_1\text{p} \rightarrow {}^4_2\text{He} + 2 \, {}^0_1\text{e}^+ + 2\nu$$

The mass of the products is less than the mass of the reactants. Using $\Delta E = \Delta m \, c^2$ we can work out that the Sun is losing mass at a rate of $4 \times 10^9 \, \text{kg s}^{-1}$. This takes place in the core of a star. Eventually all this energy is radiated from the surface – approximately 10^{26} J every second. The structure inside a star does not need to be known in detail.

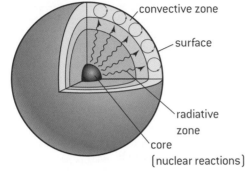

- convective zone
- surface
- radiative zone
- core (nuclear reactions)

EQUILIBRIUM

The Sun has been radiating energy for the past 4½ billion years. It might be imagined that the powerful reactions in the core should have forced away the outer layers of the Sun a long time ago. Like other stars, the Sun is stable because there is a **hydrostatic equilibrium** between this outward pressure and the inward gravitational force (see page 164).

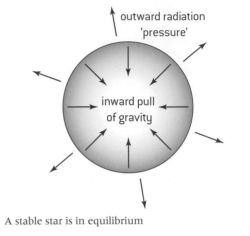

- outward radiation 'pressure'
- inward pull of gravity

A stable star is in equilibrium

BINARY STARS

Our Sun is a single star. Many 'stars' actually turn out to be two (or more) stars in orbit around each other. (To be precise they orbit around their common centre of mass.) These are called **binary stars**.

- centre of mass

binary stars – two stars in orbit around their common centre of mass

There are different categories of binary star – **visual**, **spectroscopic** and **eclipsing**.

1. A visual binary is one that can be distinguished as two separate stars using a telescope.

2. A spectroscopic binary star is identified from the analysis of the spectrum of light from the 'star'. Over time the wavelengths show a periodic shift or splitting in frequency. An example of this is shown (below).

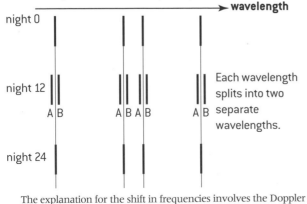

→ **wavelength**

night 0

night 12 — A|B A|B A|B A|B

Each wavelength splits into two separate wavelengths.

night 24

The explanation for the shift in frequencies involves the Doppler effect. As a result of its orbit, the stars are sometimes moving towards the Earth and sometimes they are moving away. When a star is moving towards the Earth, its spectrum will be blue shifted. When it is moving away, it will be red shifted.

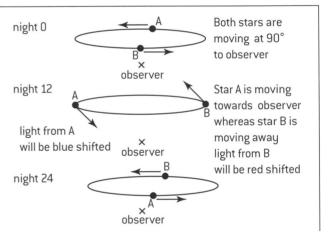

night 0 — Both stars are moving at 90° to observer

night 12 — light from A will be blue shifted — Star A is moving towards observer whereas star B is moving away light from B will be red shifted

night 24 — observer

3. An eclipsing binary star is identified from the analysis of the brightness of the light from the 'star'. Over time the brightness shows a periodic variation. An example of this is shown below.

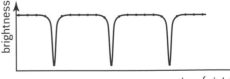

brightness / time (nights)

The explanation for the 'dip' in brightness is that as a result of its orbit, one star gets in front of the other. If the stars are of equal brightness, then this would cause the total brightness to drop to 50%.

When one star blocks the light coming from the other star, the overall brightness is reduced

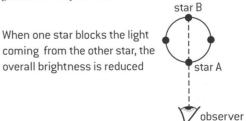

- star B
- star A
- observer

Stellar parallax

PRINCIPLES OF MEASUREMENT

As you move from one position to another objects change their relative positions. As far as you are concerned, near objects appear to move when compared with far objects. Objects that are very far away do not appear to move at all. You can demonstrate this effect by closing one eye and moving your head from side to side. An object that is near to you (for example the tip of your finger) will appear to move when compared with objects that are far away (for example a distant building).

This apparent movement is known as **parallax** and the effect can used to measure the distance to some of the stars in our galaxy. All stars appear to move over the period of a night, but some stars appear to move in relation to other stars over the period of a year.

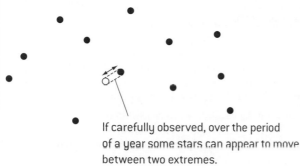

If carefully observed, over the period of a year some stars can appear to move between two extremes.

The reason for this apparent movement is that the Earth has moved over the period of a year. This change in observing position has meant that a close star will have an apparent movement when compared with a more distant set of stars. The closer a star is to the Earth, the greater will be the parallax shift.

Since all stars are very distant, this effect is a very small one and the parallax angle will be very small. It is usual to quote parallax angles not in degrees, but in seconds. An angle of 1 second of arc ('') is equal to one sixtieth of 1 minute of arc (') and 1 minute of arc is equal to one sixtieth of a degree.

In terms of angles, $3600'' = 1°$

$360° = 1$ full circle.

EXAMPLE

The star alpha Eridani (Achemar) is 1.32×10^{18} m away. Calculate its parallax angle.

$d = 1.32 \times 10^{18}$ m

$= \dfrac{1.32 \times 10^{18}}{3.08 \times 10^{16}}$ pc

$= 42.9$ pc

parallax angle $= \dfrac{1}{42.9}$

$= 0.02''$

MATHEMATICS – UNITS

The situation that gives rise to a change in apparent position for close stars is shown below.

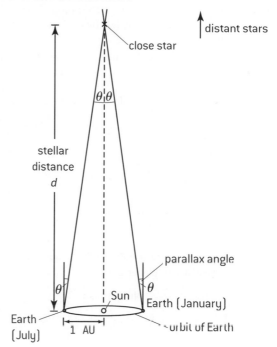

The parallax angle, θ, can be measured by observing the changes in a star's position over the period of a year. From trigonometry, if we know the distance from the Earth to the Sun, we can work out the distance from the Earth to the star, since

$$\tan \theta = \frac{\text{(distance from Earth to Sun)}}{\text{(distance from Sun to Star)}}$$

Since θ is a very small angle, $\tan \theta \approx \sin \theta \approx \theta$ (in radians)

This means that $\theta \propto \dfrac{1}{\text{(distance from Earth to star)}}$

In other words, parallax angle and distance away are inversely proportional. If we use the right units we can end up with a very simple relationship. The units are defined as follows.

The distance from the Sun to the Earth is defined to be one **astronomical unit (AU)**. It is 1.5×10^{11} m. Calculations show that a star with a parallax angle of exactly one second of arc must be 3.08×10^{16} m away (3.26 light years). This distance is defined to be one **parsec (pc)**. The name 'parsec' represents '**par**allax angle of one **sec**ond'.

If distance = 1 pc, $\theta = 1$ second

If distance = 2 pc, $\theta = 0.5$ second etc.

Or, distance in pc $= \dfrac{1}{\text{(parallax angle in seconds)}}$

$$d = \frac{1}{p}$$

The parallax method can be used to measure stellar distances that are less than **about 100 parsecs**. The parallax angle for stars that are at greater distances becomes too small to measure accurately. It is common, however, to continue to use the unit. The standard SI prefixes can also be used even though it is not strictly an SI unit.

1000 parsecs = 1 kpc

10^6 parsecs = 1 Mpc etc.

Luminosity

LUMINOSITY AND APPARENT BRIGHTNESS

The total power **radiated** by a star is called its **luminosity** (L). The SI units are watts. This is very different to the power **received** by an observer on the Earth. The power received per unit area is called the **apparent brightness** of the star. The SI units are W m^{-2}.

If two stars were at the **same distance** away from the Earth then the one with the greater luminosity would be brighter. Stars are, however, at different distances from the Earth. The brightness is inversely proportional to the (distance)^2.

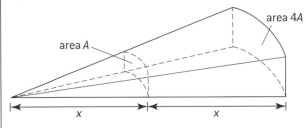

As distance increases, the brightness decreases since the light is spread over a bigger area.

distance	brightness
x	b
$2x$	$\dfrac{b}{4}$
$3x$	$\dfrac{b}{9}$
$4x$	$\dfrac{b}{16}$
$5x$	$\dfrac{b}{25}$
and so on	

inverse square

apparent brightness $b = \dfrac{L}{4\pi r^2}$

It is thus possible for two very different stars to have the same apparent brightness. It all depends on how far away the stars are.

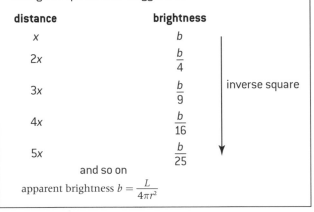

close star
(small luminosity)

distant star
(high luminosity)

Two stars can have the same apparent brightness even if they have different luminosities

ALTERNATIVE UNITS

The SI units for luminosity and brightness have already been introduced. In practice astronomers often compare the brightness of stars using the **apparent magnitude** scale. A magnitude 1 star is brighter than a magnitude 3 star. This measure of brightness is sometimes shown on star maps.

The magnitude scale can also be used to compare the luminosity of different stars, provided the distance to the star is taken into account. Astronomers quote values of **absolute magnitude** in order to compare luminosities on a familiar scale.

EXAMPLE ON LUMINOSITY

The star Betelgeuse has a parallax angle of 7.7×10^{-3} arc seconds and an apparent brightness of 2.0×10^{-7} W m^{-2}. Calculate its luminosity.

Distance to Betelgeuse d

$= \dfrac{1}{p}$

$= \dfrac{1}{7.7 \times 10^{-3}}$ pc

$= 129.9$ pc

$= 129.9 \times 3.08 \times 10^{16}$ m

$= 4.0 \times 10^{18}$ m

$L = b \times 4\pi d^2 = 4.0 \times 10^{31}$ W

BLACK-BODY RADIATION

Stars can be analysed as perfect emitters, or black bodies. The luminosity of a star is related to its brightness, surface area and temperature according to the Stefan–Boltzmann law. Wien's law can be used to relate the wavelength at which the intensity is a maximum to its temperature. See page 90 for more details.

Example:

e.g. our sun's temperature is 5,800k

So the wavelength at which the intensity of its radiation is at a maximum is $\lambda_{max} = \dfrac{2.9 \times 10^{-3}}{5800} = 500$ nm

Stellar spectra

ABSORPTION LINES

The radiation from stars is not a perfect continuous spectrum – there are particular wavelengths that are 'missing'.

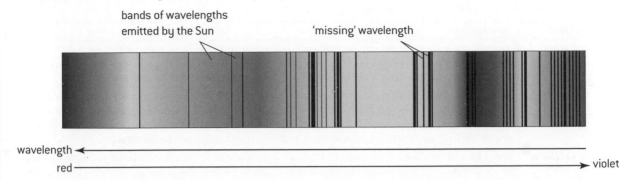

bands of wavelengths emitted by the Sun

'missing' wavelength

wavelength ← → violet
red

The missing wavelengths correspond to the absorption spectra of a number of elements. Although it seems sensible to assume that the elements concerned are in the Earth's atmosphere, this assumption is incorrect. The wavelengths would still be absent if light from the star was analysed in space.

The absorption is taking place in the outer layers of the star. This means that we have a way of telling what elements exist in the star – at least in its outer layers.

A star that is moving relative to the Earth will show a Doppler shift in its absorption spectrum. Light from stars that are receding will be **red shifted** whereas light from approaching stars will be **blue shifted**.

CLASSIFICATION OF STARS

Different stars give out different spectra of light. This allows us to classify stars by their **spectral class**. Stars that emit the same type of spectrum are allocated to the same spectral class. Historically these were just given a different letter, but we now know that these different letters also correspond to different surface temperatures.

The seven main spectral classes (in order of **decreasing** surface temperature) are O, B, A, F, G, K and M. The main spectral classes can be subdivided.

Class	Effective surface temperature/K	Colour
O	30,000–50,000	blue
B	10,000–30,000	blue-white
A	7,500–10,000	white
F	6,000–7,500	yellow-white
G	5,200–6,000	yellow
K	3,700–5,200	orange
M	2,400–3,700	red

Spectral classes do not need to be mentioned but are used in many text books.

STEFAN–BOLTZMANN LAW

The Stefan–Boltzmann law links the **total** power radiated by a black body (per unit area) to the temperature of the black body. The important relationship is that

Total power radiated $\propto T^4$

In symbols we have,

Total power radiated $= \sigma A T^4$

Where

σ is a constant called the Stefan–Boltzmann constant.

$\sigma = 5.67 \times 10^{-8}$ W m^{-2} K^{-4}

A is the surface area of the emitter (in m^2)

T is the absolute temperature of the emitter (in kelvin)

e.g. The radius of the Sun $= 6.96 \times 10^8$ m.

Surface area $= 4\pi r^2 = 6.09 \times 10^{10}$ m^2

If temperature $= 5800$ K

then total power radiated $= \sigma A T^4$

$= 5.67 \times 10^{-8} \times 6.09 \times 10^{18}$
$\times (5800^4)$

$= 3.9 \times 10^{26}$ W

The radius of the star r is linked to its surface area, A, using the equation $A = 4\pi r^2$.

SUMMARY

If we know the distance to a star we can analyse the light from the star and work out:

- the chemical composition (by analysing the absorption spectrum)
- the surface temperature (using a measurement of λ_{max} and Wien's law – see page 90)
- the luminosity (using measurements of the brightness and the distance away)
- the surface area of the star (using the luminosity, the surface temperature and the Stefan–Boltzmann law).

Nucleosynthesis

STELLAR TYPES AND BLACK HOLES

The source of energy for our Sun is the fusion of hydrogen into helium. This is also true for many other stars.
There are however, other types of object that are known to exist in the Universe.

Type of object	Description
Red giant stars	As the name suggests, these stars are large in size and red in colour. Since they are red, they are comparatively cool. They turn out to be one of the later possible stages for a star. The source of energy is the fusion of some elements other than hydrogen. **Red supergiants** are even larger.
White dwarf stars	As the name suggests, these stars are small in size and white in colour. Since they are white, they are comparatively hot. They turn out to be one of the final stages for some smaller mass stars. Fusion is no longer taking place, and a white dwarf is just a hot remnant that is cooling down. Eventually it will cease to give out light when it becomes sufficiently cold. It is then known as a **brown dwarf**.
Cepheid variables	These are stars that are a little unstable. They are observed to have a regular variation in brightness and hence luminosity. This is thought to be due to an oscillation in the size of the star. They are quite rare but are very useful as there is a link between the period of brightness variation and their average luminosity. This means that astronomers can use them to help calculate the distance to some galaxies.
Neutron stars	Neutron stars are the post-supernova remnants of some larger mass stars. The gravitational pressure has forced a total collapse and the mass of a neutron star is not composed of atoms – it is essentially composed of neutrons. The density of a neutron star is enormous. Rotating neutron stars have been identified as **pulsars**.
Black holes	Black holes are the post-supernova remnant of larger mass stars. There is no known mechanism to stop the gravitational collapse. The result is an object whose escape velocity is greater than the speed of light. See page 150.

MAIN SEQUENCE STARS

The general name for the creation of nuclei of different elements as a result of fission reactions is **nucleosynthesis**. Details of how this overall reaction takes place in the Sun do not need to be recalled by SL candidates, but HL candidates do need this information.

One process is known as the **proton–proton cycle** or **p–p cycle**.

step 1 $^1_1H + {}^1_1H \rightarrow {}^2_1H + {}^0_1e^+ + {}^0_0\nu$

step 2 $^2_1H + {}^1_1H \rightarrow {}^3_2He + {}^0_0\gamma$

step 3 $^3_2He + {}^3_2He \rightarrow {}^4_2He + 2{}^1_1p$

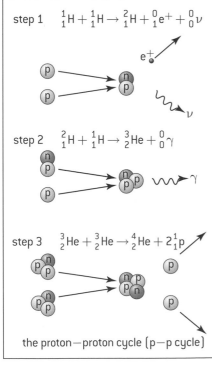

the proton—proton cycle (p—p cycle)

In order for any of these reactions to take place, two positively charged particles (hydrogen or helium nuclei) need to come close enough for interactions to take place. Obviously they will repel one another.

This means that they must be at a high temperature.

If a large cloud of hydrogen is hot enough, then these nuclear reactions can take place spontaneously. The power radiated by the star is balanced by the power released in these reactions – the temperature is effectively constant.

The star remains a stable size because the outward pressure of the radiation is balanced by the inward gravitational pull.

But how did the cloud of gas get to be at a high temperature in the first place? As the cloud comes together, the loss of gravitational potential energy must mean an increase in kinetic energy and hence temperature. In simple terms the gas molecules speed up as they fall in towards the centre to form a proto-star.

Once ignition has taken place, the star can remain stable for billions of years. See page 205 for more details.

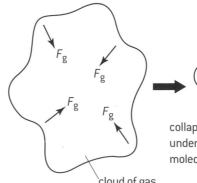

cloud of gas

collapse of cloud under gravity gives molecular KE

With sufficient KE, nuclear reactions can take place.

The Hertzsprung–Russell diagram

H–R DIAGRAM

The point of classifying the various types of stars is to see if any patterns exist. A useful way of making this comparison is the **Hertzsprung–Russell diagram**. Each dot on the diagram represents a different star. The following axes are used to position the dot.

- The vertical axis is the luminosity of the star as compared with the luminosity of the Sun. It should be noted that the scale is logarithmic.

- The horizontal axis a scale of **decreasing** temperature. Once again, the scale is not a linear one. (It is also the spectral class of the star OBAFGKM)

The result of such a plot is shown below.

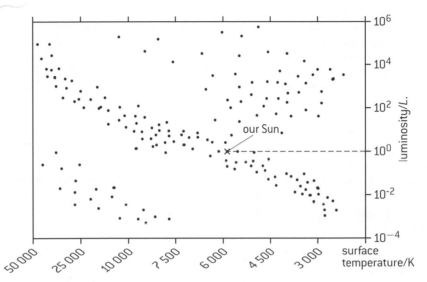

A large number of stars fall on a line that (roughly) goes from top left to bottom right. This line is known as the **main sequence** and stars that are on it are known as main sequence stars. Our Sun is a main sequence star. These stars are 'normal' stable stars – the only difference between them is their mass. They are fusing hydrogen to helium. The stars that are not on the main sequence can also be broadly put into categories.

In addition to the broad regions, lines of constant radius can be added to show the size of stars in comparison to our Sun's radius. These are lines going from top left to bottom right.

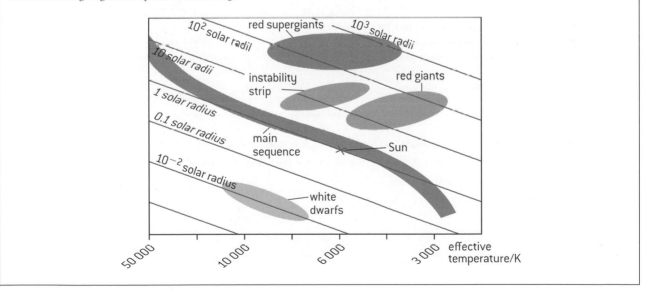

MASS-LUMINOSITY RELATION FOR MAIN SEQUENCE STARS

For stars on the main sequence, there is a correlation between the star's mass, M, and its luminosity, L. Stars that are brighter on the main sequence (i.e. higher up) are more massive and the relationship is:

$$L \propto M^{3.5}$$

Cepheid variables

PRINCIPLES

Very small parallax angles can be measured using satellite observations (e.g. Gaia mission) but even these measurement are limited to stars that are about 100 kpc away. The essential difficulty is that when we observe the light from a very distant star, we do not know the difference between a bright source that is far away and a dimmer source that is closer. This is the principal problem in the experimental determination of astronomical distances to other galaxies.

When we observe another galaxy, all of the stars in that galaxy are approximately the same distance away from the Earth. What we really need is a light source of known luminosity in the galaxy. If we had this then we could make comparisons with the other stars and judge their luminosities. In other words we need a '**standard candle**' – that is a star of known luminosity. Cepheid variable stars provide such a 'standard candle'.

A Cepheid variable star is quite a rare type of star. Its outer layers undergo a periodic compression and contraction and this produces a periodic variation in its luminosity.

A Cepheid variable star undergoes periodic compressions and contractions.

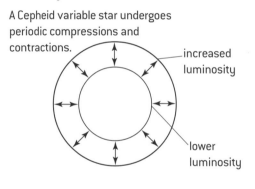

increased luminosity

lower luminosity

These stars are useful to astronomers because the period of this variation in luminosity turns out to be related to the average absolute magnitude of the Cepheid. Thus the luminosity of a Cepheid can be calculated by observing the variations in brightness.

MATHEMATICS

The process of estimating the distance to a galaxy (in which the individual stars can be imaged) might be as follows:

- Locate a Cepheid variable in the galaxy.
- Measure the variation in brightness over a given period of time.
- Use the luminosity–period relationship for Cepheids to estimate the average luminosity.
- Use the average luminosity, the average brightness and the inverse square law to estimate the distance to the star.

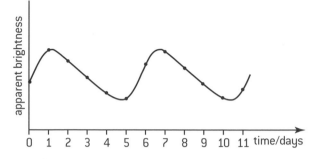

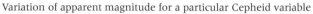

Variation of apparent magnitude for a particular Cepheid variable

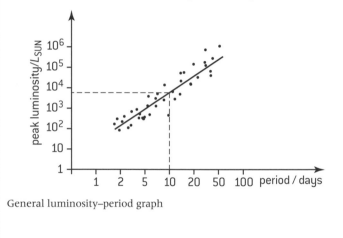

General luminosity–period graph

EXAMPLE

A Cepheid variable star has a period of 10.0 days and apparent peak brightness of 6.34×10^{-11} W m^{-2}

The luminosity of the Sun is 3.8×10^{26} W. Calculate the distance to the Cepheid variable in pc.

Using the luminosity–period graph (above)

$$\Rightarrow \text{peak luminosity} = 10^{3.7} \times L_{\text{sun}} = 5012 \times 3.8 \times 10^{26} = 1.90 \times 10^{30}\,\text{W}$$

$$L = b \times 4\pi r^2$$

$$\therefore\ r = \sqrt{\frac{L}{4\pi b}}$$

$$= \sqrt{\frac{1.90 \times 10^{30}}{4 \times \pi \times 6.34 \times 10^{-11}}}$$

$$= 4.88 \times 10^{19}\,\text{m}$$

$$= \frac{4.88 \times 10^{19}}{3.08 \times 10^{16}}\,\text{pc}$$

$$= 1590\,\text{pc}$$

Red giant stars

AFTER THE MAIN SEQUENCE

The mass–luminosity relation (page 197) can be used to compare the amount of time different mass stars take before the hydrogen fuel is used. Consider a star that is 10 times more massive than our Sun. This means that the luminosity of the larger star will be $(10)^{3.5}$ = 3,162 times more luminous that our Sun. Since the source of this luminosity is the mass of hydrogen in the star, then the larger star effectively has 10 times more 'fuel' but is using the fuel at more than 3000 times the rate. The more massive star will finish its fuel in $\frac{1}{300}$ of the time. A star that has more mass exists for a shorter amount of time.

A star cannot continue in its main sequence state forever. It is fusing hydrogen into helium and at some point hydrogen in the core will become rare. The fusion reactions will happen less often. This means that the star is no longer in equilibrium and the gravitational force will, once again, cause the core to collapse.

This collapse increases the temperature of the core still further and helium fusion is now possible. The net result is for the star to increase massively in size – this expansion means that the outer layers are cooler. It becomes a red giant star.

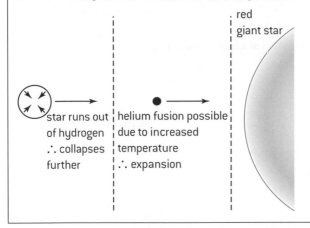

star runs out of hydrogen ∴ collapses further

helium fusion possible due to increased temperature ∴ expansion

red giant star

If it has sufficient mass, a red giant can continue to fuse higher and higher elements and the process of nucleosynthesis can continue.

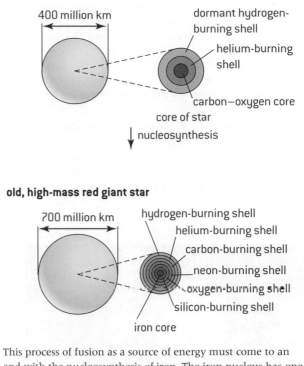

newly formed red giant star

400 million km

dormant hydrogen-burning shell

helium-burning shell

carbon–oxygen core

core of star

↓ nucleosynthesis

old, high-mass red giant star

700 million km

hydrogen-burning shell
helium-burning shell
carbon-burning shell
neon-burning shell
oxygen-burning shell
silicon-burning shell
iron core

This process of fusion as a source of energy must come to an end with the nucleosynthesis of iron. The iron nucleus has one of the greatest binding energies per nucleon of all nuclei. In other words the fusion of iron to form a higher mass nucleus would need to take in energy rather than release energy. The star cannot continue to shine. What happens next is outlined on the following page.

Stellar evolution

POSSIBLE FATES FOR A STAR (AFTER RED GIANT PHASES)

Page 199 showed that the red giant phase for a star must eventually come to an end. There are essentially two possible routes with different final states. The route that is followed depends on the initial mass of the star and thus the mass of the remnant that the red giant star leaves behind: with no further nuclear reactions taking place gravitational forces continue the collapse of the remnant. An important 'critical' mass is called the **Chandrasekhar limit** and it is equal to approximately 1.4 times the mass of our Sun. Below this limit a process called **electron degeneracy pressure** prevents the further collapse of the remnant.

If a star has a mass less than 4 Solar masses, its remnant will be less than 1.4 Solar masses and so it is below the Chandrasekhar limit. In this case the red giant forms a **planetary nebula** and becomes a **white dwarf** which ultimately becomes invisible. The name 'planetary nebula' is another term that could cause confusion. The ejected material would not be planets in the same sense as the planets in our Solar System.

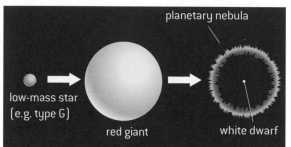

If a star is greater than 4 Solar masses, its remnant will have a mass greater than 1.4 Solar masses. It is above the Chandrasekhar limit and electron degeneracy pressure is not sufficient to prevent collapse. In this case the red supergiant experiences a **supernova**. It then becomes a **neutron star** or collapses to a **black hole**. The final state again depends on mass.

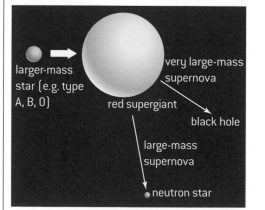

A neutron star is stable due to neutron degeneracy pressure. It should be emphasized that white dwarfs and neutron stars do not have a source of energy to fuel their radiation. They must be losing temperature all the time. The fact that these stars can still exist for many millions of years shows that the temperatures and masses involved are enormous. The largest mass a neutron star can have is called the **Oppenheimer–Volkoff limit** and is 2–3 Solar masses. Remnants above this limit will form black holes.

H – R DIAGRAM INTERPRETATION

All of the possible evolutionary paths for stars that have been described here can be represented on a H – R diagram. A common mistake in examinations is for candidates to imply that a star somehow moves along the line that represents the main sequence. It does not. Once formed it stays at a stable luminosity and spectral class – i.e. it is represented by one fixed point in the H – R diagram.

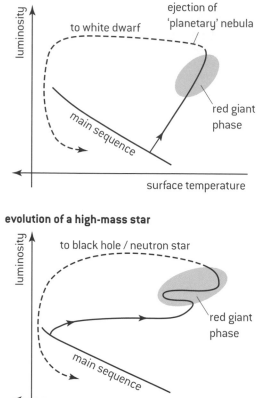

PULSARS AND QUASARS

Pulsars are cosmic sources of very weak radio wave energy that pulsate at a very rapid and precise frequency. These have now been theoretically linked to rotating neutron stars. A rotating neutron star would be expected to emit an intense beam of radio waves in a specific direction. As a result of the star's rotation, this beam moves around and causes the pulsation that we receive on Earth.

Quasi-stellar objects or quasars appear to be point-like sources of light and radio waves that are very far away. Their red shifts are very large indeed, which places them at the limits of our observations of the Universe. If they are indeed at this distance they must be emitting a great deal of power for their size (approximately 10^{40} W!). The process by which this energy is released is not well understood, but some theoretical models have been developed that rely on the existence of super-massive black holes. The energy radiated is as a result of whole stars 'falling' into the black hole.

The Big Bang model

EXPANSION OF THE UNIVERSE

If a galaxy is moving away from the Earth, the light from it will be red shifted. The surprising fact is that light from almost all galaxies shows red shifts – almost all of them are moving away from us. The Universe is expanding.

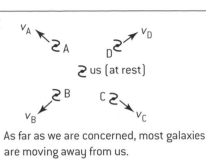

As far as we are concerned, most galaxies are moving away from us.

At first sight, this expansion seems to suggest that we are in the middle of the Universe, but this is a mistake. We only seem to be in the middle because it was we who worked out the velocities of the other galaxies. If we imagine being in a different galaxy, we would get exactly the same picture of the Universe.

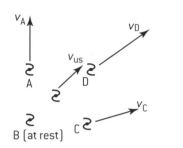

Any galaxy would see all the other galaxies moving away from it.

A good way to picture this expansion is to think of the Universe as a sheet of rubber stretching off into the distance. The galaxies are placed on this huge sheet. If the tension in the rubber is increased, everything on the sheet moves away from everything else.

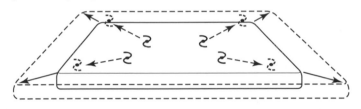

As the section of rubber sheet expands, everything moves away from everything else.

THE UNIVERSE IN THE PAST – THE BIG BANG

If the Universe is currently expanding, at some time in the past all the galaxies would have been closer together. If we examine the current expansion in detail we find that all the matter in the observable universe would have been together at the SAME point approximately 15 billion years ago.

This point, the creation of the Universe, is known as the **Big Bang**. It pictures all the matter in the Universe being crushed together (very high density) and being very hot indeed. Since the Big Bang, the Universe has been expanding – which means that, on average, the temperature and density of the Universe have been decreasing. The rate of expansion would be expected to decrease as a result of the gravitational attraction between all the masses in the Universe.

Note that this model does not attempt to explain how the Universe was created, or by Whom. All it does is analyse what happened after this creation took place. The best way to imagine the expansion is to think of the expansion of space itself rather than the galaxies expanding into a void. The Big Bang was the creation of space and time. Einstein's theory of relativity links the measurements of space and time so properly we need to imagine the Big Bang as the creation of space **and time**. It does not make sense to ask about what happened before the Big Bang, because the notion of before and after (i.e. time itself) was created in the Big Bang.

COSMIC MICROWAVE BACKGROUND RADIATION

A further piece of evidence for the Big Bang model came with the discovery of the **Cosmic microwave background (CMB) radiation** by Penzias and Wilson.

They discovered that microwave radiation was coming towards us from all directions in space. The strange thing was that the radiation was the same in all directions (**isotopic**) and did not seem to be linked to a source. Further analysis showed that this radiation was a very good match to theoretical black-body radiation produced by an extremely cold object – a temperature of just 2.73 K.

This is in perfect agreement with the predictions of Big Bang. There are two ways of understanding this.

1. All objects give out electromagnetic radiation. The frequencies can be predicted using the theoretical model of black-body radiation. The background radiation is the radiation from the Universe itself which has now cooled down to an average temperature of 2.73 K.

2. Some time after the Big Bang, radiation became able to travel through the Universe (see page 210 for details).

   It has been travelling towards us all this time. During this time the Universe has expanded – this means that the wavelength of this radiation will have increased (space has stretched). See page 210 for anisotropies in the CMB.

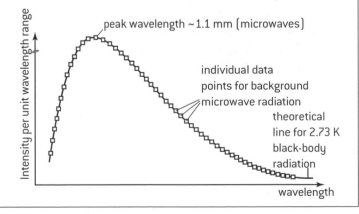

Galactic motion

DISTRIBUTIONS OF GALAXIES

Galaxies are not distributed randomly throughout space. They tend to be found clustered together. For example, in the region of the Milky Way there are twenty or so galaxies in less than 2.5 million light years.

The Virgo galactic cluster (50 million light years away from us) has over 1,000 galaxies in a region 7 million light years across. On an even larger scale, the galactic clusters are grouped into huge **superclusters** of galaxies. In general, these superclusters often involve galaxies arranged together in joined 'filaments' (or bands) that are arranged as though randomly throughout empty space.

MOTION OF GALAXIES

As has been seen on page 201 it is a surprising observational fact that the vast majority of galaxies are moving away from us. The general trend is that the more distant galaxies are moving away at a greater speed as the Universe expands. This does not, however, mean that we are at the centre of the Universe – this would be observed wherever we are located in the Universe.

As explained on page 201, a good way to imagine this expansion is to think of space itself expanding. It is the expansion of space (as opposed to the motion of the galaxies through space) that results in the galaxies' relative velocities. In this model, the red shift of light can be thought of as the expansion of the wavelength due to the 'stretching' of space.

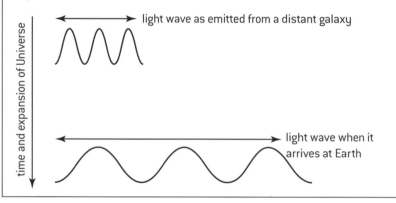

light wave as emitted from a distant galaxy

time and expansion of Universe

light wave when it arrives at Earth

MATHEMATICS

If a star or a galaxy moves away from us, then the wavelength of the light will be altered as predicted by the Doppler effect (see page 102). If a galaxy is going away from the Earth, the speed of the galaxy with respect to an observer on the Earth can be calculated from the red shift of the light from the galaxy. As long as the velocity is small when compared with the velocity of light, a simplified red shift equation can be used.

$$Z = \frac{\Delta\lambda}{\lambda_0} \approx \frac{v}{c}$$

Where

$\Delta\lambda$ = change in wavelength of observed light (positive if wavelength is increased)

λ_0 = wavelength of light emitted

v = relative velocity of source of light

c = speed of light

Z = red shift.

Example

A characteristic absorption line often seen in stars is due to ionized helium. It occurs at 468.6 nm. If the spectrum of a star has this line at a measured wavelength of 499.3 nm, what is the recession speed of the star?

$$Z = \frac{\Delta\lambda}{\lambda_0} = \frac{(499.3 - 468.6)}{468.6}$$

$$= 6.55 \times 10^{-2}$$

$$\therefore \quad v = 6.55 \times 10^{-2} \times 3 \times 10^8 \text{ m s}^{-1}$$

$$= 1.97 \times 10^7 \text{ m s}^{-1}$$

Hubble's law and cosmic scale factor

EXPERIMENTAL OBSERVATIONS

Although the uncertainties are large, the general trend for galaxies is that the recessional velocity is proportional to the distance away from Earth. This is Hubble's law.

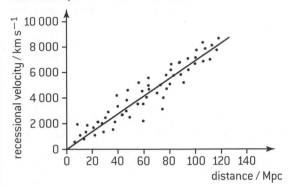

Mathematically this is expressed as

$$v \propto d$$

or

$$v = H_0 d$$

where H_0 is a constant known as the **Hubble constant**. The uncertainties in the data mean that the value of H_0 is not known to any degree of precision. The SI units of the Hubble constant are s^{-1}, but the unit of $km\ s^{-1}\ Mpc^{-1}$ is often used.

HISTORY OF THE UNIVERSE

If a galaxy is at a distance x, then Hubble's law predicts its velocity to be $H_0 x$. If it has been travelling at this constant speed since the beginning of the Universe, then the time that has elapsed can be calculated from

$$\text{Time} = \frac{\text{distance}}{\text{speed}}$$
$$= \frac{x}{H_0 x}$$
$$= \frac{1}{H_0}$$

This is an upper limit for the age of the Universe. The gravitational attraction between galaxies predicts that the speed of recession decreases all the time.

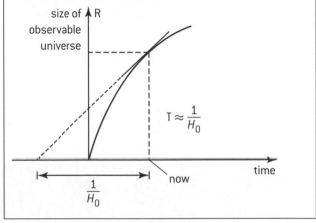

THE COSMIC SCALE FACTOR (R)

Page 202 shows how the Doppler red shift equation, $= \frac{\Delta\lambda}{\lambda_0} \approx \frac{v}{c}$, can be used to calculate the recessional velocity, v, of certain galaxies. This equation can only be used when $v \ll c$ or in other words, the recessional velocity, v, has to be small in comparison to the speed of light, c. There are however plenty of objects in the night sky (e.g. quasars) for which the observed red shift, z, is greater than 1.0. This implies that their speed of recession is greater than the speed of light. In these situations it is helpful to consider a quantity called the cosmic scale factor (R).

As introduced on page 201, the expansion of the Universe is best pictured as the expansion of space itself. The expansion of the Universe means that a measurement undertaken at some time in the distant past, for example the wavelength of light emitted by an object 10 million years ago, will be stretched and will be recorded as a larger value when measured now. All measurement will be stretched over time and this can be considered as a rescaling of the Universe (the Universe getting bigger).

The cosmic scale factor, R, is a way of quantifying the expansion that has taken place. In the above example, if the wavelength was emitted 10 million years ago with wavelength λ_0 when the scale factor was R_0, the wavelength measured today would have increased by $\Delta\lambda$ to a larger value λ ($\lambda = \lambda_0 + \Delta\lambda$). This is because the cosmic scale factor has increased by ΔR (to the larger value $R = R_0 + \Delta R$). All measurements will have increased by the ratio, $\frac{R}{R_0}$. The ratio of the measured wavelengths, $\frac{\lambda}{\lambda_0}$, is equal to the ratio of the cosmic scale factors, $\frac{R}{R_0}$, so the red shift ratio, z is given by:

$$z = \frac{\Delta\lambda}{\lambda_0} = \frac{\lambda - \lambda_0}{\lambda_0} = \frac{\lambda}{\lambda_0} - 1 = \frac{R}{R_0} - 1$$

or $z = \frac{R}{R_0} - 1$

So a measured red shift of 4 means that $\frac{R}{R_0} = 5$. If we consider R to be the present 'size' of the observable Universe, then the light must have been emitted when the Universe was one fifth of its current size.

The accelerating universe

SUPERNOVAE AND THE ACCELERATING UNIVERSE

Supernovae are catastrophic explosions that can occur in the development of some stars (see page 200). Supernovae are rare events (the last one to occur in our galaxy took place in 1604) but the large number of stars in the Universe means that many have been observed. An observer on the Earth sees a rapid increase in brightness (hence the word 'nova' = new star) which then diminishes over a period of some weeks or months. Huge amounts of radiated energy are emitted in a short period of time and, at its peak, the apparent brightness of a single supernova often exceeds many local stars or individual galaxies.

Supernovae have been categorized into two different main types (see page 207 for more details) according to a spectral analysis of the light that they emit. The light from a type II supernova indicates the presence of hydrogen (from the absorption spectra) whereas there is no hydrogen in a type I supernova. There are further subdivisions of these types (Ia, Ib, etc.) based on different aspects of the light spectrum.

Type Ia supernovae are explosions involving white dwarf stars. When these events take place, the amount of energy released can be predicted accurately and these supernovae can be used as 'standard candles'. By comparing the known luminosity of a type Ia supernova and its apparent brightness as observed in a given galaxy, a distance measurement to that galaxy can be calcuated. This technique can be used with galaxies up to approximately 1,000 Mpc away.

The expanding Universe (which is consistent with the Big Bang model) means that that the cosmic scale factor, R, is increasing. As a result of gravitational attraction, we might expect the rate at which R increases to be slowing down. Analysis of a large number of type Ia supernovae has, however, provided strong evidence that not only is the cosmic scale factor, R, increasing but the rate at which it increases is getting larger as time passes. In other words the expansion of the Universe is accelerating. The evidence from type Ia supernovae identifies this effect from a time when the universe was approximately $\frac{2}{3}$ of its current size. Note that this acceleration is different to the very rapid period of expansion of the early Universe which is called inflation.

The mechanisms that cause an accelerating Universe are not fully understood but must involve an outward accelerating force to counteract the inward gravitational pull. There must also be a source of energy which has been given the name **dark energy** (see page 212).

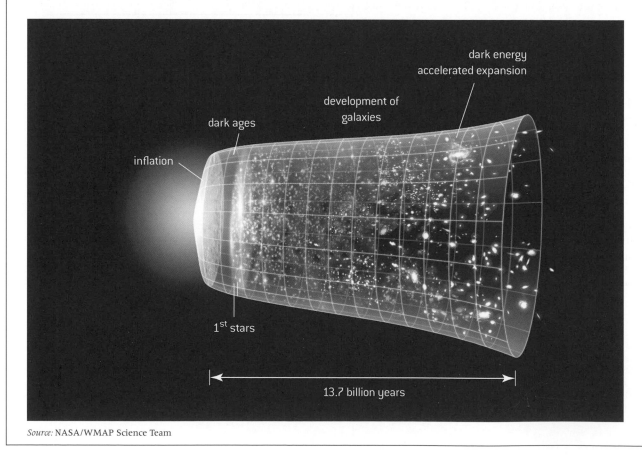

Source: NASA/WMAP Science Team

HL Nuclear fusion – the Jeans criterion

THE JEANS CRITERION

As seen on page 196, stars form out of interstellar clouds of hydrogen, helium and other materials. Such clouds can exist in stable equilibrium for many years until an external event (e.g. a collision with another cloud or the influence of another incident such as a supernova) starts the collapse. At any given point in time, the total energy associated with the gas cloud can be thought of as a combination of:

- The negative gravitational potential energy, E_P, which the cloud possesses as a result of its mass and how it is distributed in space. Important factors are thus the mass and the density of the cloud.

- The positive random kinetic energy, E_K, that the particles in the cloud possess. An important factor is thus the temperature of the cloud.

The cloud will remain gravitationally bound together if $E_P + E_K <$ zero. Using this information allows us to predict that the collapse of an interstellar cloud may begin if its mass is greater than a certain critical mass, M_J. This is the **Jeans criterion**. For a given cloud of gas, M_J is dependent on the cloud's density and temperature and the cloud is more likely to collapse if it has:

- large mass
- small size
- low temperature.

In symbols, the Jeans criterion is that collapse can start if $M > M_J$

NUCLEAR FUSION

A star on the main sequence is fusing hydrogen nuclei to produce helium nuclei. One process by which this is achieved is the proton–proton chain as outlined on page 196. This is the predominant method for nuclear fusion to take place in small mass stars (up to just above the mass of our Sun). An alternative process, called the CNO (carbon–nitrogen–oxygen) process takes place at higher temperatures in larger mass stars. In this reaction, carbon, nitrogen and oxygen are used as catalysts to aid the fusion of protons into helium nuclei. One possible cycle is shown below:

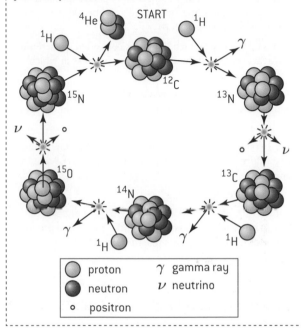

○ proton	γ gamma ray
● neutron	ν neutrino
○ positron	

TIME SPENT ON THE MAIN SEQUENCE

For so long as a star remains on the main sequence, hydrogen 'burning' is the source of energy that allows the star to remain in hydrostatic equilibrium (see page 192) and have a constant luminosity L. A star that exists on the main sequence for a time T_{MS} must in total radiate an energy E given by:

$$E = L \times T_{MS}$$

This energy release comes from the nuclear synthesis that has taken place over its lifetime. A certain fraction f of the mass of the star M has been converted into energy according to Einstein's famous relationship:

$$E = f \times Mc^2$$

$$\therefore \quad L \times T_{MS} = f \times Mc^2$$

$$T_{MS} = \frac{f \times Mc^2}{L}$$

But the mass–luminosity relationship applies, $L \propto M^{3.5}$

$$\therefore \quad T_{MS} \propto \frac{M}{M^{3.5}}$$

$$\therefore \quad T_{MS} \propto M^{-2.5}$$

Thus the higher the mass of a star, the shorter the lifetime that it spends on the main sequence

$$\frac{Time\ on\ main\ sequence\ for\ star\ A}{Time\ on\ main\ sequence\ for\ star\ B} = \left(\frac{Mass\ of\ star\ A}{Mass\ of\ star\ B}\right)^{-2.5} = \left(\frac{Mass\ of\ star\ B}{Mass\ of\ star\ A}\right)^{2.5}$$

For example our Sun is expected to have a main sequence lifetime of approximate 10^{10} years. How long would a star with 100 times its mass be expected to last?

$$Time\ on\ MS\ for\ 100\ solar\ mass\ star = 10^{10} \times \left(\frac{1}{100}\right)^{2.5} = 10^5\ years$$

Nucleosynthesis off the main sequence

NUCLEOSYNTHESIS OFF THE MAIN SEQUENCE

For so long as a star remains on the main sequence, hydrogen 'burning' is the source of energy that allows the star to continue emitting energy whilst remaining in a stable state. More and more helium exists in the core. A nuclear synthesis involving helium (helium 'burning') does release energy (since the binding energy per nucleon of the products is greater than that of the reactants) but can only take place at high temperatures.

For high mass stars, the helium burning process can begin gradually and spread throughout the core whereas in small mass stars this process starts suddenly. Whatever the mass of the star, a new equilibrium state is created: the red giant or red supergiant phase (see page 200).

A common process by which helium is converted is a series of nuclear reactions called the **triple alpha process** in which carbon is produced.

1. Two helium nuclei fuse into a beryllium nucleus (and a gamma ray), releasing energy.

    $$^{4}_{2}\text{He} + {}^{4}_{2}\text{He} \rightarrow {}^{8}_{4}\text{Be} + \gamma$$

2. The beryllium nucleus fuses with another helium nucleus to produce a carbon nucleus (and a gamma ray), releasing energy.

    $$^{8}_{4}\text{Be} + {}^{4}_{2}\text{He} \rightarrow {}^{12}_{6}\text{C} + \gamma$$

3. Some of the carbon produced in the triple alpha process can go on to fuse with another helium nucleus to produce oxygen. Again this process releases energy:

    $$^{12}_{6}\text{C} + {}^{4}_{2}\text{He} \rightarrow {}^{16}_{8}\text{O} + \gamma$$

In high and very high mass stars, gravitational contraction means that the temperature of the core can continue to rise and more massive nuclei can continue to be produced. These reactions all involve the release of energy. Typical reactions include:

Production of neon: $^{12}_{6}\text{C} + {}^{12}_{6}\text{C} \rightarrow {}^{20}_{10}\text{Ne} + {}^{4}_{2}\text{He}$

Production of magnesium: $^{12}_{6}\text{C} + {}^{12}_{6}\text{C} \rightarrow {}^{24}_{12}\text{Mg} + \gamma$

Production of oxygen: $^{12}_{6}\text{C} + {}^{12}_{6}\text{C} \rightarrow {}^{16}_{8}\text{O} + 2\,{}^{4}_{2}\text{He}$

In addition if the temperatures are high enough, neon and oxygen burning can occur:

$$^{20}_{10}\text{Ne} + \gamma \rightarrow {}^{16}_{8}\text{O} + {}^{4}_{2}\text{He}$$

$$^{20}_{10}\text{Ne} + {}^{4}_{2}\text{He} \rightarrow {}^{24}_{12}\text{Mg} + \gamma$$

Production of sulfur: $^{16}_{8}\text{O} + {}^{16}_{8}\text{O} \rightarrow {}^{32}_{16}\text{S} + \gamma$

Many reactions are possible and other heavy nuclei such as silicon and phosphorus are also produced. Some of these alternative nuclear reactions also produce neutrons, which can easily be captured by other nuclei to form new isotopes. This process of **neutron capture** is explored further below.

In very high mass stars, silicon burning can also take place which results in the formation of iron, $^{56}_{26}\text{Fe}$. As explained on page 199, iron has one of the highest binding energies per nucleon and represents the largest nucleus that can be created in a fusion process that releases energy. Heavier nuclei can be acquired, but the reactions require an energy input.

NUCLEAR SYNTHESIS OF HEAVY ELEMENTS – NEUTRON CAPTURE

Many of the reactions that take place in the core of stars also involve the release of neutrons. Since neutrons are without any charge, it is easy for them to interact with other nuclei that are present in the star. When a nucleus captures a neutron, the resulting nucleus is said to be **neutron rich**. Given enough time, most of these neutron-rich nuclei would undergo beta decay. In this process, the neutron changes into a proton, emitting an electron and an antineutrino:

$$^{1}_{0}\text{n} \rightarrow {}^{1}_{1}\text{p} + {}^{0}_{-1}\beta + \bar{\text{v}}$$

$$^{A}_{Z}\text{X} + {}^{1}_{0}\text{n} \rightarrow {}^{A+1}_{Z}\text{X} \rightarrow {}^{A+1}_{Z+1}\text{Y} + {}^{0}_{-1}\beta + \bar{\text{v}}$$

This is known as **slow neutron capture** or the **s-process**. The overall result of the s-process is a new element. Typically the

s-process takes place during the helium burning stage of a red giant star. Typically this means that elements that are heavier than helium but lighter than iron are able to be created.

The alternative process, **rapid neutron capture** or the **r-process**, takes place when the neutrons are present in such vast numbers that there is not sufficient time for the neutron-rich nuclei to undergo beta decay before several more neutrons are captured. The result is for very heavy nuclei to be created. Typically the r-process takes place during the catastrophic explosion that is a supernova. Elements that are heavier than iron, such as uranium and thorium, can only be created in this way at very high temperatures and densities.

Ⓗ Types of supernovae

SUPERNOVAE

Supernovae are among the most gigantic explosions in the Universe (see page 200). The two categories of supernova are based on their **light curves** – a plot of how their brightness varies with time and a spectral analysis of the light that they emit. Type I supernovae quickly reach a maximum brightness (and an equivalent luminosity of 10^{10} Suns) which then gradually decreases over time. Type II supernovae often have lower peak luminosities (equivalent to, say, 10^9 Suns).

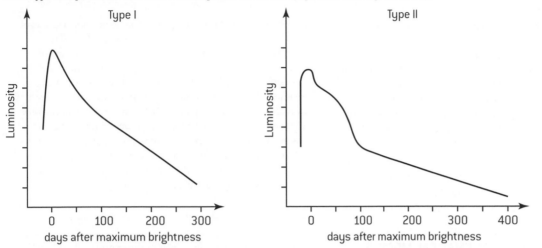

Supernovae types are distinguish by analysis of their light spectra. All type I supernovae do not include the hydrogen spectrum in the elements identified and the different subdivisions (Ia, Ib and Ic) are based on a more detailed spectral analysis:

• Type Ia shows the presence of singly ionized silicon.

• Type Ib shows the presence of non-ionized helium.

• Type Ic does not show the presence of helium.

All type II supernovae show the presence of hydrogen. The different subdivisions (IIP, IIL, IIn and IIb) again depend on the presence, or not, of different elements.

The reasons for these differences are the different mechanisms that are taking place:

	Supernova Type Ia	**Supernova Type II**
Spectra	Does not show hydrogen but does show singly ionized silicon.	Shows hydrogen.
Cause	White dwarf exploding.	Large mass red giant star collapsing.
Context	Binary star system with white dwarf and red giant orbiting each other.	Large star (greater than 8 Solar masses) at the end of its lifetime, fusing lighter elements up to the production of iron.
Process	The gravity field of the white dwarf star attracts material from the red giant star, thus increasing the mass of the white dwarf.	When the star runs out of fuel, the iron centre core cannot release any further energy by nuclear fusion. The star collapses under its own gravity forming a neutron star.
Explosion	The extra mass gained by the white dwarf takes the total mass of the star beyond the Chandrasekhar limit (1.4 Solar masses) for a white dwarf. Electron degeneracy pressure is no longer sufficient to halt the gravitational collapse. Nuclear fusion of heavier elements (up to iron) starts and the resulting sudden release of energy causes the star to explode with the matter being distributed throughout space.	Electron degeneracy pressure is not sufficient to halt the gravitational collapse of the core, but neutron degeneracy pressure is and the core becomes a stable and rigid neutron star. The rest of the infalling material bounces off the core creating a shock wave moving outwards. This causes all of the outer layers to be ejected.

HL The cosmological principle and mathematical models

THE COSMOLOGICAL PRINCIPLE

The **cosmological principle** is a pair of assumptions about the structure of the Universe upon which current models are based. The two assumptions are that the Universe, providing one only considers the large scale structures in the Universe, is **isotropic** and **homogeneous**.

An isotropic universe is one that looks the same in every direction – no particular direction is different to any other. From the perspective of an observer on Earth, this appears to be a true statement about the large scale structure of the universe, but the assumption does not only apply to observers on the Earth. In an isotropic universe **all** observers, wherever they are in the universe, are expected to see the same basic random distribution of galaxies and galaxy clusters as we do on Earth and this is true in whatever direction they observe.

A homogeneous universe is one where the local distribution of galaxies and galaxy clusters that exists in one region of the universe turns out to be the same distribution in all regions of the universe. Provided one is considering a reasonably large

section of space (e.g. a sphere of radius equal to several hundreds of Mpc), then the number of galaxies in that volume of space will be effectively the same wherever we choose to look in the universe. Recent discoveries of apparently very large scale structures in the Universe cause some astrophysicists to question the validity of the cosmological principle.

Einstein used the cosmological principle to develop a model of the Universe in which the Universe was static. He did this by proposing that the gravitational attraction between galaxies would be balanced by a yet-to-be-discovered cosmological repulsion. Subsequent analysis of the equations of general relativity showed that, if the cosmological principle is correct, the Universe must be non-static. Hubble's observational discovery of the expansion of the Universe and the existence of CMB has meant that many physicists now agree that the Universe is non-static based around the Big Bang model of an expanding universe. The cosmological principle is also linked to three possible models for the future of the Universe (see page 211).

ROTATION CURVES – MATHEMATICAL MODELS

The stars in a galaxy rotate around their common centre of mass. Different models can be used to predict how the speed varies with distance from the galactic centre.

1. Near the galactic centre

A simple model to explain the different speeds of rotation of stars near the galactic centre assumes that density of the galaxy near its centre, ρ, is constant. A star of mass m feels a resultant force of gravitational attraction in towards the centre. The value of this resultant force is the same as if the total mass M of all the stars that are closer to the galactic centre were concentrated in the centre. An important point to note is that the net effect of all the stars that are orbiting at radius that is greater than r sums to zero.

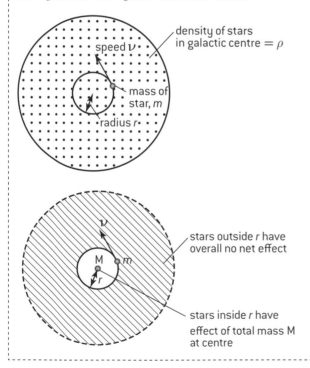

The star at a given distance r from the centre will orbit in circular motion because its centripetal force is provided by the gravitational attraction:

$$\frac{GMm}{r^2} = \frac{mv^2}{r}$$

$$\therefore v^2 = \frac{GM}{r}$$

The total mass of stars that orbit closer than of this star, M, is given by

$$M = \text{volume} \times \text{density} = \tfrac{4}{3}\pi r^3 \times \rho$$

$$v^2 = \frac{G\tfrac{4}{3}\pi r^3 \rho}{r} = \frac{4\pi G \rho}{3} r^2$$

$$\therefore v = \sqrt{\frac{4\pi G \rho}{3}} \cdot r$$

i.e. $v \propto r$

2. Far away from the galactic centre

Far away from the galactic centre, observations of the number of visible stars show that the effective density of the galaxy has reduced so much that individual stars at these distances can be considered to be freely orbiting the central mass and to be unaffected by their neighbouring stars. In this situation,

$$v^2 = \frac{GM}{r} \quad \text{where } M \text{ is the mass of the galaxy}$$

i.e. $v \propto \sqrt{\frac{1}{r}}$

Comparisons with observations of real galaxies show good agreement with mathematical model (1) but no agreement with mathematical model (2). The proposed solution is discussed on page 209.

ⓗ Rotation curves and dark matter

ROTATION CURVES

Galaxies rotate around their centre of mass and the speeds of this rotation can be calculated for individual stars from an analysis of the star's spectra. A **rotation curve** for a galaxy show how this orbital speed varies with distance from the galactic centre. Most galaxies show:

- an initial linear increase in orbital velocity with distance within the galactic centre

- a flat or slightly increasing curve showing a roughly constant speed of rotation away from the galactic centre.

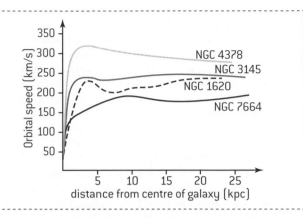

EVIDENCE FOR DARK MATTER

As shown above, observed rotation curves for real galaxies agree with theoretical models within the galactic centre ($v \propto r$) **but** the orbital velocity of stars is not observed to decrease with distance away from the centre as would be expected. Instead, the orbital velocity is roughly constant whatever the radius. If the orbital velocity v of a star is constant at different values of radius, then

since $v^2 = \dfrac{GM}{r}$

$\dfrac{M}{r} = constant$ or $M \propto r$

Thus the total mass that is keeping the star orbiting in its galaxy must be increasing with distance from the galactic centre. This is certainly not true of the visible mass (the stars emitting light)

that we can see so the suggestion is that there must be **dark matter**. In this situation it would have to be concentrated outside the galactic centre forming a halo around the galaxy. Further evidence suggests that only a very small amount of this matter could be imagined to be made up of the protons and neutrons that constitute ordinary, or **baryonic**, matter.

Dark matter:

- gravitationally attracts ordinary matter

- does not emit radiation and cannot be inferred from its interactions

- is unknown in structure

- makes up the majority of the Universe with less than 5% of the Universe made up of ordinary baryonic matter.

MACHOS, WIMPS AND OTHER THEORIES

Astrophysicists are attempting to come up with theories to explain why there is so much dark matter and what it consists of. There are a number of possible theories:

- The matter could be found in **M**assive **A**stronomical **C**ompact **H**alo **O**bjects or **MACHOs** for short. There is some evidence that lots of ordinary matter does exist in these groupings. These can be thought of as low-mass 'failed' stars or high-mass planets. They could even be black holes. These would produce little or no light. Evidence suggests that these could only account for a small proportion.

- There could be new particles that we do not know about. These are the **W**eakly **I**nteracting **M**assive **P**articles. Many experimenters around the world are searching for these so-called **WIMPs**.

- Perhaps our current theories of gravity are not completely correct. Some theories try to explain the missing matter as simply a failure of our current theories to take everything into account.

ⓗⓛ The history of the Universe

FLUCTUATIONS IN CMB

The cosmic microwave background radiation (CMB) is essentially **isotropic** (the same in all directions). This implies that the matter in the early Universe was uniformly distributed throughout space with no random temperature variations at all. If this was precisely the case then the development of the Universe would be expected to be absolutely identical everywhere and matter would be uniformly distributed throughout the Universe – it would be without any structure. We know, however, that matter is not uniformly distributed as it is concentrated into stars and galaxies.

Further analysis of the CMB reveals tiny fluctuations (**anisotropies**) in the temperature distribution of the early Universe in different directions. These temperature variations are typically a few μK compared with the background effective temperature of 2.73 K. The diagram right is an enhanced projection which highlights the minor observed variations in the CMB (with the effects of our own galaxy removed). Just like a map includes all the countries of the world, this projection shows the variation in received CMB from the whole Universe.

Variation in CMB as observed by the Wilkinson Microwave Anisotropy Probe (WMAP)

The minute differences in temperature imply minor differences in densities, which allow structures to be developed as the Universe expands.

THE HISTORY OF THE UNIVERSE

We can 'work backwards' and imagine the process that took place soon after the Big Bang.

- Very soon after the Big Bang, the Universe must have been very hot.
- As the Universe expanded it cooled. It had to cool to a certain temperature before atoms and molecules could be formed.
- The Universe underwent a short period of huge expansion (Inflation) that would have taken place from about 10^{-36} s after the Big Bang to 10^{-32} s.

Time	What is happening	Comments
10^{-45} s → 10^{-36} s	Unification of forces	This is the starting point.
10^{-36} s → 10^{-32} s	Inflation	A rapid period of expansion – the so-called inflationary epoch. The reasons for this rapid expansion are not fully understood.
10^{-32} s → 10^{-5} s	Quark–lepton era	Matter and antimatter (quarks and leptons) are interacting all the time. There is slightly more matter than antimatter.
10^{-5} s → 10^{-2} s	Hadron era	At the beginning of this short period it is just cool enough for hadrons (e.g. protons and neutrons) to be stable.
10^{-2} s → 10^3 s	Nucleosynthesis	During this period some of the protons and neutrons have combined to form helium nuclei. The matter that now exists is the 'small amount' that is left over when matter and antimatter have interacted.
10^3 s → 3×10^5 years	Plasma era (radiation era)	The formation of light nuclei has now finished and the Universe is in the form of a plasma with electrons, protons, neutrons, helium nuclei and photons all interacting.
3×10^5 years → 10^9 years	Formation of atoms	At the beginning of this period, the Universe has become cool enough for the first atoms to exist. Under these conditions, the photons that exist stop having to interact with the matter. It is these photons that are now being received as part of the background microwave radiation. The Universe is essentially 75% hydrogen and 25% helium.
10^9 years → now	Formation of stars, galaxies and galactic clusters	Some of the matter can be brought together by gravitational interactions. If this matter is dense enough and hot enough, nuclear reactions can take place and stars are formed.

COSMIC SCALE FACTOR AND TEMPERATURE

The expansion of the Universe means that the wavelength of any radiation that has been emitted in the past will be 'stretched' over time (see page 202). Thus the radiation that was emitted approximately 12 billion years ago (shortly after the Big Bang) at very short wavelengths is now being received as much longer microwaves – the CMB radiation.

The spectrum of CMB radiation received corresponds to black-body radiation at a temperature of 2.73 K. The calculation uses Wien's law to link the peak wavelength, λ_{max}, of the radiation to the temperature, T, of the black body in kelvins:

$$\lambda_{max} = \frac{2.9 \times 10^{-3}}{T}$$

$$\lambda_{max} \propto \frac{1}{T}$$

When the radiation was emitted the temperature of the universe was much hotter, the cosmic scale factor, R, was much smaller and λ_{max} was also proportionally much smaller.

Since the stretching of the Universe is the cause of the change in wavelength, then the ratio of cosmic scale factors at two different times must be the same as the ratio of peak wavelengths so

$$\lambda_{max} \propto R$$

$$\therefore \frac{1}{T} \propto R \text{ or } T \propto \frac{1}{R}$$

Ⓗ The future of the Universe

FUTURE OF THE UNIVERSE (WITHOUT DARK ENERGY)

If the Universe is expanding at the moment, what is it going to do in the future? As a result of the Big Bang, other galaxies are moving away from us. If there were no forces between the galaxies, then this expansion could be thought of as being constant.

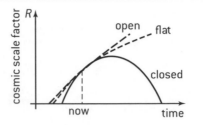

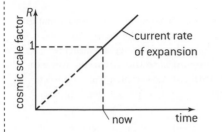

The expansion of the Universe cannot, however, have been uniform. The force of gravity acts between all masses. This means that if two masses are moving apart from one another there is a force of attraction pulling them back together. This force must have slowed the expansion down in the past. What it is going to do in the future depends on the current rate of expansion and the density of matter in the Universe.

An **open Universe** is one that continues to expand forever. The force of gravity slows the rate of recession of the galaxies down a little bit but it is not strong enough to bring the expansion to a halt. This would happen if the density in the Universe were low.

A **closed Universe** is one that is brought to a stop and then collapses back on itself. The force of gravity is enough to bring the expansion to an end. This would happen if the density in the Universe were high.

A **flat Universe** is the mathematical possibility between open and closed. The force of gravity keeps on slowing the expansion down but it takes an infinite time to get to rest. This would only happen if the Universe were exactly the right density. One electron-positron pair more, and the gravitational force would be a little bit bigger. Just enough to start the contraction and make the Universe closed.

CRITICAL DENSITY, ρ_c

The theoretical value of density that would create a flat Universe is called the **critical density**, ρ_c. Its value is not certain because the current rate of expansion is not easy to measure. Its order of magnitude is 10^{-26} kg m^{-3} or a few proton masses every cubic metre. If this sounds very small remember that enormous amounts of space exist that contain little or no mass at all.

The density of the Universe is not an easy quantity to measure. It is reasonably easy to estimate the mass in a galaxy by estimating the number of stars and their average mass but the majority of the mass in the Universe is dark matter.

The value of ρ_c can be estimated using Newtonian gravitation. We consider a galaxy at a distance r away from an observer with a recessional velocity of v with respect to the observer.

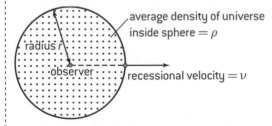

The net effect of all the masses in the Universe outside the sphere on the galaxy is zero (see page 208 for an analogous situation). The galaxy is thus gravitationally attracted in by a total mass M which acts as though it was located at the observer as shown (above).

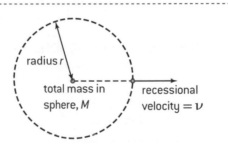

The total energy E_T of the galaxy is the addition of its kinetic energy E_K and gravitational potential energy, E_P given by:

$$E_T = E_K + E_P$$
$$E_K = \frac{1}{2}mv^2 \text{ but Hubble's law gives } v = H_0 r$$
$$\therefore E_K = \frac{1}{2}m(H_0 r)^2$$
$$E_P = -\frac{GMm}{r} \text{ but } M = \text{volume} \times \text{density} = \frac{4}{3}\pi r^3 \rho$$
$$E_P = -\frac{G4\pi r^3 \rho m}{3r} = -\frac{4G\pi r^2 \rho m}{3}$$

If E_T is positive, the galaxy will escape the inward attraction – the universe is open.

If E_T is negative, the galaxy will eventually fall back in – the universe is closed.

If E_T is exactly zero, the galaxy will take an infinite time to be brought to rest – the universe is flat. This will occur when the density of the universe ρ is equal to the critical density ρ_c.

$$\therefore \frac{1}{2}m(H_0 r)^2 = \frac{4G\pi r^2 \rho_c m}{3}$$
$$\therefore mH_0^2 r^2 = \frac{8G\pi r^2 \rho_c m}{3}$$
$$\therefore \rho_c = \frac{3H_0^2}{8\pi G}$$

Ⓗ Dark energy

COSMIC DENSITY PARAMETER

The cosmic density parameter, Ω_0 is the ratio of the average density of matter and energy in the Universe, ρ, to the critical density, ρ_c

$$\Omega_0 = \frac{\rho}{\rho_c}$$

If $\Omega_0 > 1$, the universe is closed.

If $\Omega_0 < 1$, the universe is open.

If $\Omega_0 = 1$, the universe is flat.

DARK ENERGY

Gravitational attraction between masses means that the rate of expansion of the Universe would be expected to decrease with time. Measurements using type Ia supernovae as standard candles have provided strong evidence that the expansion has not, in fact, been slowing down over time (see page 204). Observations currently indicate that the Universe's rate of expansion has been increasing.

Currently there is no single accepted explanation for this observation and, of course, it is possible that our theories of gravity and general relativity need to be modified. Perhaps we are on the brink of discovery of new physics. Whatever the cause, the reason for the Universe's accelerating expansion has been given the general name 'dark energy'.

Dark energy and dark matter are two different concepts. In both cases experimental evidence implies their existence but physicists have yet to agree a theoretical basis that explains the existence of either concept.

- Dark matter is hypothesized to explain the 'missing matter' that must exist within galaxies for the known laws of gravitational attraction to be able to explain a galaxy's rate of rotation. Dark matter **adds to the attractive force of gravity acting within galaxies** implying more unseen mass than had been previously expected, hence the name *dark mass*.

- The observation that expansion of the Universe is accelerating means that then there must be a force that is counteracting the attractive force of gravity. Dark energy **opposes the attractive force of gravity between galaxies**. The resulting increase in energy implies an unseen source of energy, hence the name *dark energy*.

EFFECT OF DARK ENERGY ON THE COSMIC SCALE FACTOR

The existence of dark energy counteracts the attractive force of gravity. This will cause the cosmic scale factor to increase over time. The graph below compares how a flat Universe is predicted to develop with and without dark energy.

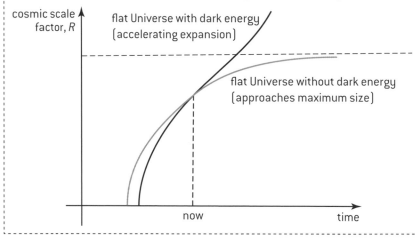

ⓗ Astrophysics research

ASTROPHYSICS RESEARCH

Much of the current fundamental research that is being undertaken in astrophysics involves close international collaboration and the sharing of resources. Scientists can be proud of their record of international collaboration. For example, at the time that the previous edition of this book was being published, the Cassini spacecraft had been in orbit around Saturn for several years sending information about the planet back to Earth and is currently (2014) continuing to produce data.

The Cassini–Huygens spacecraft was funded jointly by ESA (the European Space Agency), NASA (the National Aeronautics and Space Administration of the United States of America) and ASI (Agenzia Spaziale Italiana – the Italian Space Agency). As well as general information about Saturn, an important focus of the mission was a moon of Saturn called Titan. The Huygens probe was released and sent back information as it descended towards the surface. The information discovered is shared among the entire scientific community. Many current projects, for example the Dark Energy Survey (involving more than 120 scientists for 23 institutions worldwide), continue this process.

All countries have a limited budget available for the scientific research that they can undertake. There are arguments both for, and against, investing significant resources into researching the nature of the Universe.

Future research, such as the Euclid mission to map the geometry of the dark Universe continues to be planned.

Arguments for:

- Understanding the nature of the Universe sheds light on fundamental philosophical questions like:
  - Why are we here?
  - Is there (intelligent) life elsewhere in the Universe?

- It is one of the most fundamental, interesting and important areas for humankind as a whole and it therefore deserves to be properly researched.

- All fundamental research will give rise to technology that may eventually improve the quality of life for many people.

- Life on Earth will, at some time in the distant future, become an impossibility. If humankind's descendents are to exist in this future, we must be able to travel to distant stars and colonize new planets.

Arguments against:

- The money could be more usefully spent providing food, shelter and medical care to the many millions of people who are suffering from hunger, homelessness and disease around the world.

- If money is to be allocated on research, it is much more worthwhile to invest limited resources into medical research. This offers the immediate possibility of saving lives and improving the quality of life for some sufferers.

- It is better to fund a great deal of small diverse research rather than concentrating all funding into one expensive area. Sending a rocket into space is expensive, thus funding space research should not be a priority.

- Is the information gained really worth the cost?

CURRENT OBSERVATIONS

Three recent scientific experiments that have studied the CMB in detail have together added a great deal to our understanding of the Universe. Particular experiments of note include:

- NASA's Cosmic Background Explorer (COBE)
- NASA's Wilkinson Microwave Anisotropy Probe (WMAP)
- ESA's Planck space observatory.

Together these experiments have:

- mapped the anisotropies of the CMB in great detail and with precision
- discovered that the first generation of stars to shine did so 200 million years after the Big Bang, much earlier than many scientists had previously expected
- calculated the age of the Universe as 13.75 ± 0.14 billion years old

- calculated the Hubble constant to be 67.15 km s^{-1} Mpc^{-1}
- showed that their results were consistent with the Big Bang and specific inflation theories
- showed the Universe to be flat, $\Omega_0 = 1$
- calculated the Universe to be composed of 4.6% atoms, 23% dark matter and 71.4% dark energy.

In summary, current scientific evidence suggests that, when dark matter and dark energy are taken into consideration, the Universe:

- is flat
- has a density that is, within experimental error, very close to the critical density
- has an accelerating expansion
- is composed mainly of dark matter and dark energy.

IB Questions – astrophysics

1. This question is about determining some properties of the star Wolf 359.

   a) The star Wolf 359 has a parallax angle of 0.419 seconds.

      (i) Describe how this parallax angle is measured. [4]

      (ii) Calculate the distance in light-years from Earth to Wolf 359. [2]

      (iii) State why the method of parallax can only be used for stars at a distance of less than a few hundred parsecs from Earth. [1]

   b) The ratio [4]

      $\dfrac{\text{apparent brightness of Wolf 359}}{\text{apparent brightness of the Sun}}$ is 3.7×10^{-15}.

      Show that the ratio

      $\dfrac{\text{luminosity of Wolf 359}}{\text{luminosity of the Sun}}$ is 8.9×10^{-4}. ($1\,\text{ly} = 6.3 \times 10^{4}\,\text{AU}$)

   c) The surface temperature of Wolf 359 is 2800 K and its luminosity is 3.5×10^{23} W. Calculate the radius of Wolf 359. [2]

   d) By reference to the data in (c), suggest why Wolf 359 is neither a white dwarf nor a red giant. [2]

2. The diagram below shows the grid of an HR diagram, on which the positions of selected stars are shown. (L_S = luminosity of the Sun.)

   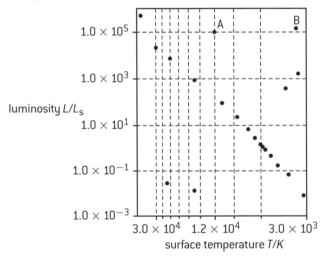

   a) (i) Draw a circle around the stars that are red giants. Label this circle R. [1]

      (ii) Draw a circle around the stars that are white dwarfs. Label this circle W. [1]

      (iii) Draw a line through the stars that are main sequence stars. [1]

   b) Explain, without doing any calculation, how astronomers can deduce that star B has a larger diameter than star A. [3]

   c) Using the following data and information from the HR diagram, show that star A is at a distance of about 800 pc from Earth.

Apparent brightness of the Sun	$= 1.4 \times 10^{3}$ W m^{-2}
Apparent brightness of star A	$= 4.9 \times 10^{-9}$ W m^{-2}
Mean distance of Sun from Earth	$= 1.0$ AU
1 pc	$= 2.1 \times 10^{5}$ AU [4]

   d) Explain why the distance of star A from Earth cannot be determined by the method of stellar parallax. [1]

3. a) The spectrum of light from the Sun is shown below.

   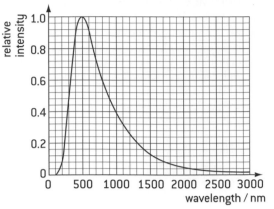

   Use this spectrum to estimate the surface temperature of the Sun. [2]

   b) Outline how the following quantities can, in principle, be determined from the spectrum of a star.

      (i) The elements present in its outer layers. [2]

      (ii) Its speed relative to the Earth. [2]

4. a) Explain how Hubble's law supports the Big Bang model of the Universe. [2]

   b) Outline **one** other piece of evidence for the model, saying how it supports the Big Bang. [3]

   c) The Andromeda galaxy is a relatively close galaxy, about 700 kpc from the Milky Way, whereas the Virgo nebula is 2.3 Mpc away. If Virgo is moving away at 1200 km s^{-1}, show that Hubble's law predicts that Andromeda should be moving away at roughly 400 km s^{-1}. [1]

   d) Andromeda is in fact moving **towards** the Milky Way, with a speed of about 100 km s^{-1}. How can this discrepancy from the prediction, in both magnitude and direction, be explained? [3]

   e) If light of wavelength 500 nm is emitted from Andromeda, what would be the wavelength observed from Earth? [3]

5. A quasar has a redshift of 6.4. Calculate the ratio of the current size of the universe to its size when the quasar emitted the light that is being detected. [3]

6. Explain the following:

   a) Why more massive stars have shorter lifetimes [2]

   b) The jeans criterion [2]

   c) How elements heavier than iron are produced by stars [2]

   d) How type 1a supernovae can be used as standard candles [2]

   e) The significance of observed anisotropies in the Cosmic Microwave background [2]

   f) The significance of the critical density of universe [2]

   g) The evidence for dark matter [2]

   h) What is meant by dark energy [2]

7. Calculate the critical density for of the universe using the Hubble constant of 71 km s^{-1} Mpc^{-1} [3]

Graphs

PLOTTING GRAPHS – AXES AND BEST FIT

The reason for plotting a graph in the first place is that it allows us to identify trends. To be precise, it allows us a visual way of representing the variation of one quantity with respect to another. When plotting graphs, you need to make sure that all of the following points have been remembered:

- The graph should have a title. Sometimes they also need a key.
- The scales of the axes should be suitable – there should not, of course, be any sudden or uneven 'jumps' in the numbers.
- The inclusion of the origin has been thought about. Most graphs should have the origin included – it is rare for a graph to be improved by this being missed out. If in doubt include it. You can always draw a second graph without it if necessary.
- The final graph should, if possible, cover more than half the paper in either direction.
- The axes are labelled with both the quantity (e.g. current) AND the units (e.g. amps).
- The points are clear. Vertical and horizontal lines to make crosses are better than 45 degree crosses or dots.

- All the points have been plotted correctly.
- Error bars are included if appropriate.
- A best-fit trend line is added. This line NEVER just 'joins the dots' – it is there to show the overall trend.
- If the best-fit line is a curve, this has been drawn as a single smooth line.
- If the best-fit line is a straight line, this has been added WITH A RULER.
- As a general rule, there should be roughly the same number of points above the line as below the line.
- Check that the points are randomly above and below the line. Sometimes people try to fit a best-fit straight line to points that should be represented by a gentle curve. If this was done then points below the line would be at the beginning of the curve and all the points above the line would be at the end, or vice versa.
- Any points that do not agree with the best-fit line have been identified.

MEASURING INTERCEPT, GRADIENT AND AREA UNDER THE GRAPH

Graphs can be used to analyse the data. This is particularly easy for straight-line graphs, though many of the same principles can be used for curves as well. Three things are particularly useful: the **intercept**, the **gradient** and the **area under the graph**.

1. Intercept

In general, a graph can intercept (cut) either axis any number of times. A straight-line graph can only cut each axis once and often it is the **y-intercept** that has particular importance. (Sometimes the y-intercept is referred to as simply 'the intercept'.) If a graph has an intercept of zero it goes through the origin.
Proportional – note that two quantities are proportional if the graph is a straight line THAT PASSES THROUGH THE ORIGIN.

Sometimes a graph has to be 'continued on' (outside the range of the readings) in order for the intercept to be found. This process is known as **extrapolation**. The process of assuming that the trend line applies between two points is known as **interpolation**.

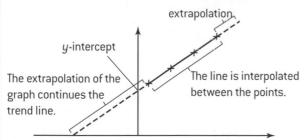

The extrapolation of the graph continues the trend line.

The line is interpolated between the points.

2. Gradient

The gradient of a straight-line graph is the increase in the y-axis value divided by the increase in the x-axis value.

The following points should be remembered:

- A straight-line graph has a constant gradient.
- The triangle used to calculate the gradient should be as large as possible.
- The gradient has units. They are the units on the y-axis divided by the units on the x-axis.
- Only if the x-axis is a measurement of time does the gradient represent the RATE at which the quantity on the y-axis increases.

The gradient of a curve at any particular point is the gradient of the tangent to the curve at that point.

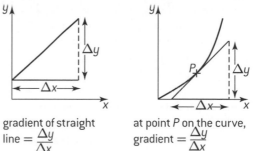

gradient of straight line $= \dfrac{\Delta y}{\Delta x}$

at point P on the curve, gradient $= \dfrac{\Delta y}{\Delta x}$

3. Area under a graph

The area under a straight-line graph is the product of multiplying the average quantity on the y axis by the quantity on the x-axis. This does not always represent a useful physical quantity. When working out the area under the graph:

- If the graph consists of straight-line sections, the area can be worked out by dividing the shape up into simple shapes.
- If the graph is a curve, the area can be calculated by 'counting the squares' and working out what one square represents.
- The units for the area under the graph are the units on the y-axis multiplied by the units on the x-axis.
- If the mathematical equation of the line is known, the area of the graph can be calculated using a process called **integration**.

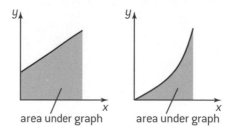

area under graph   area under graph

Graphical analysis and determination of relationships

EQUATION OF A STRAIGHT-LINE GRAPH

All straight-line graphs can be described using one general equation

$$y = mx + c$$

y and x are the two variables (to match with the y-axis and the x-axis).

m and c are both constants – they have one fixed value.

- c represents the intercept on the y-axis (the value y takes when $x = 0$)
- m is the gradient of the graph.

In some situations, a direct plot of the measured variable will give a straight line. In some other situations we have to choose carefully what to plot in order to get a straight line. In either case, once we have a straight line, we then use the gradient and the intercept to calculate other values.

For example, a simple experiment might measure the velocity of a trolley as it rolls down a slope. The equation that describes the motion is $v = u + at$ where u is the initial velocity of the object. In this situation v and t are our variables, a and u are the constants.

CHOOSING WHAT TO PLOT TO GET A STRAIGHT LINE

With a little rearrangement we can often end up with the physics equation in the same form as the mathematical equation of a straight line. Important points include

- Identify which symbols represent variables and which symbols represent constants.
- The symbols that correspond to x and y must be variables and the symbols that correspond to m and c must be constants.
- If you take a variable reading and square it (or cube, square root, reciprocal etc.) – the result is still a variable and you could choose to plot this on one of the axes.
- You can plot any mathematical combination of your original readings on one axis – this is still a variable.
- Sometimes the physical quantities involved use the symbols m (e.g. mass) or c (e.g. speed of light). Be careful not to confuse these with the symbols for gradient or intercept.

Example 1

The gravitational force F that acts on an object at a distance r away from the centre of a planet is given by the equation

$$F = \frac{GMm}{r^2}$$ where M is the mass of the planet and the m is mass of the object.

If we plot force against distance we get a curve (graph 1).

We can restate the equation as $F = \frac{GMm}{r^2} + 0$ and if we plot F on the y-axis and $\frac{1}{r^2}$ on the x-axis we will get a straight-line (graph 2).

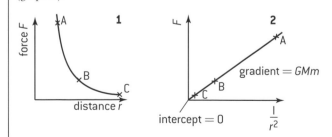

You should be able to see that the physics equation has exactly the same form as the mathematical equation. The order has been changed below so as to emphasize the link.

$$v = u + at$$
$$y = c + mx$$

By comparing these two equations, you should be able to see that if we plot the velocity on the y-axis and the time on the x-axis we are going to get a straight-line graph.

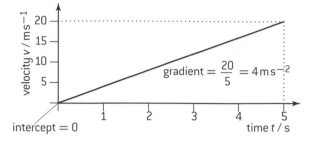

The comparison also works for the constants.

- c (the y-intercept) must be equal to the initial velocity u
- m (the gradient) must be equal to the acceleration a

In this example the graph tells us that the trolley must have started from rest (intercept zero) and it had a constant acceleration of 4.0 m s^{-2}.

Example 2

If an object is placed in front of a lens we get an image. The image distance v is related to the object distance u and the focal length of the lens f by the following equation.

$$\frac{1}{u} + \frac{1}{v} = \frac{1}{f}$$

There are many possible ways to rearrange this in order to get it into straight–line form. You should check that all these are algebraically the same.

$$v + u = \frac{uv}{f} \quad \text{or} \quad \frac{v}{u} = \frac{v}{f} - 1 \quad \text{or} \quad \frac{1}{u} = \frac{1}{f} - \frac{1}{v}$$

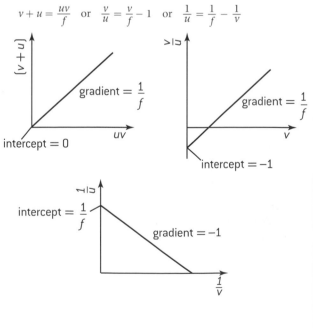

Ⓗ Graphical analysis – logarithmic functions

LOGS – BASE TEN AND BASE e

Mathematically,

If $a = 10^b$

Then $\log(a) = b$

[to be absolutely precise $\log_{10}(a) = b$]

Most calculators have a 'log' button on them. But we don't **have** to use 10 as the base. We can use any number that we like. For example we could use 2.0, 563.2, 17.5, 42 or even 2.7182818284590452353602874714. For complex reasons this last number IS the most useful number to use! It is given the symbol e and logarithms to this base are called **natural logarithms**. The symbol for natural logarithms is ln (x). This is also on most calculators.

If $p = e^q$

Then $\ln(p) = q$

The powerful nature of logarithms means that we have the following rules

$\ln(c \times d) = \ln(c) + \ln(d)$

$\ln(c \div d) = \ln(c) - \ln(d)$

$\ln(c^n) = n \ln(c)$

$\ln\left(\dfrac{1}{c}\right) = -\ln(c)$

These rules have been expressed for natural logarithms, but they work for all logarithms whatever the base.

The point of logarithms is that they can be used to express some relationships (particularly power laws and exponentials) in straight-line form. This means that we will be plotting graphs with logarithmic scales.

A normal scale increases by the same amount each time.

1 2 3 4 5 6 7 8 9 10 11

A logarithmic scale increases by the same ratio all the time.

10^0 10^1 10^2 10^3

1 10 100 1000

EXPONENTIALS AND LOGS (LOG – LINEAR)

Natural logarithms are very important because many natural processes are exponential. Radioactive decay is an important example. In this case, once again the taking of logarithms allows the equation to be compared with the equation for a straight line.

POWER LAWS AND LOGS (LOG – LOG)

When an experimental situation involves a power law it is often only possible to transform it into straight-line form by taking logs. For example, the time period of a simple pendulum, T, is related to its length, l, by the following equation.

$T = k \, l^p$

k and p are constants.

A plot of the variables will give a curve, but it is not clear from this curve what the values of k and p work out to be. On top of this, if we do not know what the value of p is, we can not calculate the values to plot a straight-line graph.

Time period versus length for a simple pendulum

The 'trick' is to take logs of both sides of the equation. The equations below have used natural logarithms, but would work for all logarithms whatever the base.

$\ln(T) = \ln(k l^p)$

$\ln(T) = \ln(k) + \ln(l^p)$

$\ln(T) = \ln(k) + p \ln(l)$

This is now in the same form as the equation for a straight line

$y = c + mx$

For example, the count rate R at any given time t is given by the equation

$R = R_0 \, e^{-\lambda t}$

R_0 and λ are constants.

If we take logs, we get

$\ln(R) = \ln(R_0 \, e^{-\lambda t})$

$\ln(R) = \ln(R_0) + \ln(e^{-\lambda t})$

$\ln(R) = \ln(R_0) - \lambda t \ln(e)$

$\ln(R) = \ln(R_0) - \lambda t$ [$\ln(e) = 1$]

This can be compared with the equation for a straight-line graph

$y = c + mx$

Thus if we plot ln (R) on the y-axis and t on the x-axis, we will get a straight line.

Gradient $= -\lambda$

Intercept $= \ln(R_0)$

Thus if we plot ln (T) on the y-axis and ln (l) on the x-axis we will get a straight-line graph.

The gradient will be equal to p
The intercept will be equal to ln (k) [so $k = e^{(\text{intercept})}$]

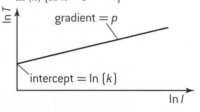

Plot of ln (time period) versus ln (length) gives a straight-line graph

Both the gravity force and the electrostatic force are inverse-square relationships. This means that the force $\propto$ (distance apart)$^{-2}$. The same technique can be used to generate a straight-line graph.

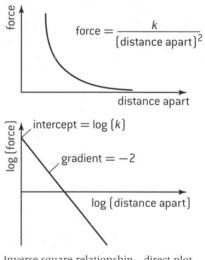

Inverse square relationship – direct plot and log-log plot

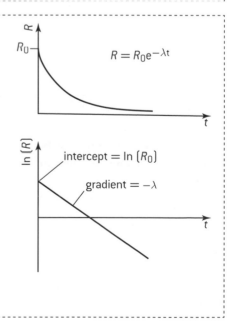

Answers

Topic 1 (Page 8): Measurements and uncertainties
1. (a)(i) $0.5 \times$ acceleration down the slope (a)(iv) 0.36 ms^{-2}
2. C 3. D 4. D 5. (b) 2.4 ± 0.1 s (c) 2.6 ± 0.2 ms^{-2}
6. (b)(i) -3; (b)(ii) 2.6×10^{-4} Nm^{-3}

Topic 2 (Page 24): Mechanics
1. C 2. D 3. B 4. B 5. (a) 520 N; (b)(i) 1.2 MJ; (b)(ii) 270 W
6. (a) equal; (b) left; (c) 20 km hr^{-1}; (e) car driver; (f) No
7. (c) 3.50 N

Topic 3 (Page 32): Thermal Physics
1. B 2. D 3. D 4. D 5. (a) (i) length = 20 m, depth = 2 m, width = 5m, temp = 25 °C; (a)(ii) $464; (b)(i) 84 days
6. (a)(i) 7.8 J K mol^{-1}

Topic 4 (Page 50): Waves
1. C 2. C 3. (a) longitudinal (b)(i) 0.5 m; (ii) 0.5 mm; (iii) 330 m s^{-1} 4. (c) (i) 2.0 Hz; (ii) 1.25 (1.3) cm; (f) (i) 4.73×10^{-7} m; (ii) 0.510 mm 5. 45°

Topic 5 (Page 64): Electricity and magnetism
1. C 2. A 3. (c) (ii) 7.2×10^{15} m s^{-2} (c) (iv) 100 v 4. (d) B; (e) (i) Equal; (ii) approx. 0.4A; (iii) lamp A will have greater power dissipation;

Topic 6 (Page 68): Circular motion and gravitation
1. A 2. A 3. C 4. (a)(ii) No; (b) 1.4 m s^{-1} 5. (b) (i) $g = G\dfrac{M}{R^2}$; (b)(ii) 1.9×10^{27} kg 6. (a) $g = G\dfrac{M}{R^2}$; (b) 6.0×10^{24} kg;

Topic 7 (Page 81): Atomic nuclear and particle physics
1. B 2. D 3. A 4. D 5. B 6. (a)(i) uud; (ii) electron is fundamental; (iii) 3 quarks or 3 antiquarks; (iv) a quark and an antiquark; 7. $u\overline{u}[\pi^0]$ 8. (b)(i) ${}^{2}_{1}H + {}^{26}_{12}Mg \rightarrow {}^{24}_{11}Na + {}^{4}_{2}He$
9. (a) ${}^{12}_{6}C \rightarrow {}^{12}_{7}N + {}^{0}_{-1}\beta + \overline{\nu}$; (b)(ii) 11600 years; 10. (a)(i) 3; (b)(i) 1.72×10^{19} 11. A

Topic 8 (Page 94): Energy production
1. (c) 15 MW (d)(i) 20% 4. (a) 1000 MW; (b) 1200 MW; (c) 17%; (d) 43 kg s^{-1} 5. (c) 1.8 MW

Topic 9 (Page 104): Wave phenomena
1. B 2. (b) 27.5 m s^{-1} 3. (a) 0.2°; 4. (a) (i) zero; (ii) π or $\dfrac{\lambda}{2}$; (iii) zero; (b) 110 nm; 5. (b) (i) 1.5×10^{-10} m; (d) (ii) 5.0×10^{19} m s^{-2}

Topic 10 (Page 111): Fields
1. A 2. C 3. C 4. (a)(i) 1.9×10^{11} J; (a)(ii) 7.7 km hr^{-1} (a)(iii) 2.2×10^{12} J; (c) 2.6 hr 5. (b)(i) 2.5 m s^{-2}

Topic 11 (Page 120): Electromagnetic induction
1. D 3. B 4. B 5. D 6. (b) 0.7 v 7. (a) 7.2×10^{-4} C; (b) 2.9×10^{-3} s; (c)(ii) 5CR = 3.5 s; (c)(iii) No

Topic 12 (Page 130): Quantum and nuclear
1. C 2. B 3. (b) ln R & t; (c) Yes; (e) 0.375 hr^{-1}; (h) 1.85 hr
5. (b)(i) 6.9×10^{-34} Js; (b)(ii) 3.3×10^{-19} J 6. 4.5×10^4 Bq

Option A (Page 151): relativity
2. (a)(i) 1.40c; (a)(ii) 0.95c; (c) 6.0×10^{19} J 3. a) 2 yrs; b) 4 yrs ; c) $x = 5$ ly; d) 0.5 c 4. (c) front; (d) T:100 m, S:87 m; (e) T:75m, S:87 m; 5. (a)(i) zero; (a)(ii) 2.7 m$_0$c^2; (b)(i) 0.923 c; (b)(ii) 2.4 m$_0$c^2; (b)(iii) 3.6 m$_0$c^2; (c) agree.

Option B (Page 170): Engineering physics
2. a) 0.95 N m; b) 25.2 J; c) 13.4 N 3. (a) No; (b) Equal; (c) 300 J; (d) -500 J; (e) 500 J; (f) 150 J; (g) 16% 4. (b) 990K; (c) (i) 1; (c) (ii) 2 & 3; (c) (iii) 3; 6. Laminar (R=1200) 7. (a) 2Hz; (b) 21 mW

Option C (Page 189): Imaging
1. (a) 14 cm behind mirror, virtual, upright, magnified ($\times 2$); (b) 24 cm behind diverging lens, real, inverted, magnified ($\times 3$); (c) 4.5 cm behind second lens, real, upright & diminished (0.25)
2. (d) upside down; (e) 60 cm; 4. (a) – 10 dB; (b) 0.5 mW
5. (a) $1 \rightarrow 20$ MHz; (b)(iii) d = 38mm, l = 130mm
6. (b)(ii) 4 mm; (b)(iii) 9.3 mm; (b)(iv) 18.6 mm

Option D (Page 214): Astrophysics
1. (ii) $d = 7.78$ ly, (c) $r = 8.9 \times 10^7$ m; 3. (a) 5800 K
4. (e) 499.83 nm 5. 14% of current size 7. 9.5×10^{-27} kg m^{-3}

Origin of individual questions

The questions detailed below are all taken from past IB examination papers and are all © IB.

Topic 1: Measurement and uncertainties
1 N99S2(S2) 2 M98H1(5) 3 N98H1(5) 4 M99H1(3)
5 M98SpH2(A2) 6 N98H2(A1)

Topic 2: Mechanics
1 M98S1(2) 2 M98S1(4) 3 M98S1(8) 5 M101S2(A2)
6 N00H2(B2) 7 M091S2(A2)

Topic 3: Thermal Physics
1 N99H1(15) 2 N99H1 (16) 3 N99H1(17) 5 M98 Sp2(B2)
6 M112H2(A5) 7 M091S2(A2)

Topic 4: Waves
1 M01H1(14) 2 N10H1(15) 3 N03S2(A3) 4 5 N04H2(B4.1)

Topic 5: Electricity and magnetism
4 N03 HL2 Q2.2

Topic 6: Circular motion and gravitation
1 N10S1(7) 2 M111H1(4) 3 M101S1(8) 4 M1112(A5)
5 M08 SpS3(A3) 6 N05H2(B2.1) – part question – sections (d) to (g)

Topic 7: Atomic, nuclear and particle physics
1 N98S1(29) 2 M99S1(29) 3 M99S1(30) 4 M98SpS1(29)
5 M98SpS1(30) 8 M98S2(Λ3) 9 M99S2(Λ3) 10 M99H2(B4)
11 M122H1(32)

Topic 8: Energy production
1 NO1S3(C1) 2 M99S3(C1) 3 M98SpS3(C3) 4 M98SpS3(C2)
5 M98S3(C2) 6 M122H2(B2.1)

Topic 9: Wave phenomena
1 N01H1(24) 2 N98H2(A5) 3 M111H3(G3) 5 N09 HL3 G4

Topic 10: Fields
3 N10H1(24) 4 N98H2(B4) 5 N01H2(A3)

Topic 11: Electromagnetic induction
1 N00H1(31) 3 M98H1(33) 4 M112H1(24) 5 N99H1(34)
6 N98H2(A4)

Topic 12: Quantum and nuclear physics
1 N10H1(34) 2 M01H1(35) 3 N00H2(A1)

Option A relativity
2 M111H3(2) 4 M00H3(G1) 5 N01H3(G2) 6 M092H3(3)

Option B Engineering physics
3 N01H2(B2) 4 N98H2(A2)

Option C imaging
2 N00H3(H1) 5 M03H3(D2)

Option D Astrophysics
1 M101H3(E1) 2 M111H3(E2) 3 N01H3(F2) 4 N98H3(F2)

Index

Page numbers in *italics* refer to question sections.